AF615560

CHEMICAL ADDITIVES FOR FUELS

CHEMICAL ADDITIVES FOR FUELS

Developments Since 1978

Edited by M.T. Gillies

NOYES DATA CORPORATION
Park Ridge, New Jersey, U.S.A.
1982

Library of Congress Catalog Card Number: 81-18939
ISBN: 0-8155-0886-7
ISSN: 0198-6888; 0270-9155
Printed in the United States

Published in the United States of America by
Noyes Data Corporation
Mill Road, Park Ridge, New Jersey 07656

Library of Congress Cataloging in Publication Data
Main entry under title:

Chemical additives for fuels.

(Chemical technology review, ISSN 0198-6888 ; 203) (Energy technology review, ISSN 0270-9155 ; 76)

Includes index.

1. Motor fuels--Additives--Patents. I. Gillies, M. T. II. Series. III. Series: Energy technology review ; 76.

TP343.C53 662'.6 81-18939
ISBN 0-8155-0886-7 AACR2

FOREWORD

The detailed descriptive information in this book is based on U.S. patents, issued from January 1979 to June 1981, that deal with fuel additives. This title contains new developments since our previous title, *Fuel Additives for Internal Combustion Engines,* published in 1978.

This book is a data-based publication, providing information retrieved and made available from the U.S. patent literature. It thus serves a double purpose in that it supplies detailed technical information and can be used as a guide to the patent literature in this field. By indicating all the information that is significant, and eliminating legal jargon and juristic phraseology, this book presents an advanced commercially oriented review of recent developments in the field of fuel additives.

The U.S. patent literature is the largest and most comprehensive collection of technical information in the world. There is more practical, commercial, timely process information assembled here than is available from any other source. The technical information obtained from a patent is extremely reliable and comprehensive; sufficient information must be included to avoid rejection for "insufficient disclosure." These patents include practically all of those issued on the subject in the United States during the period under review; there has been no bias in the selection of patents for inclusion.

The patent literature covers a substantial amount of information not available in the journal literature. The patent literature is a prime source of basic commercially useful information. This information is overlooked by those who rely primarily on the periodical journal literature. It is realized that there is a lag between a patent application on a new process development and the granting of a patent, but it is felt that this may roughly parallel or even anticipate the lag in putting that development into commercial practice.

Many of these patents are being utilized commercially. Whether used or not, they offer opportunities for technological transfer. Also, a major purpose of this book is to describe the number of technical possibilities available, which may open up profitable areas of research and development. The information contained in this book will allow you to establish a sound background before launching into research in this field.

Advanced composition and production methods developed by Noyes Data are employed to bring these durably bound books to you in a minimum of time. Special techniques are used to close the gap between "manuscript" and "completed book." Industrial technology is progressing so rapidly that time-honored, conventional typesetting, binding and shipping methods are no longer suitable. We have bypassed the delays in the conventional book publishing cycle and provide the user with an effective and convenient means of reviewing up-to-date information in depth.

The table of contents is organized in such a way as to serve as a subject index. Other indexes by company, inventor and patent number help in providing easy access to the information contained in this book.

16 Reasons Why the U.S. Patent Office Literature Is Important to You

1. The U.S. patent literature is the largest and most comprehensive collection of technical information in the world. There is more practical commercial process information assembled here than is available from any other source. Most important technological advances are described in the patent literature.

2. The technical information obtained from the patent literature is extremely comprehensive; sufficient information must be included to avoid rejection for "insufficient disclosure."

3. The patent literature is a prime source of basic commercially utilizable information. This information is overlooked by those who rely primarily on the periodical journal literature.

4. An important feature of the patent literature is that it can serve to avoid duplication of research and development.

5. Patents, unlike periodical literature, are bound by definition to contain new information, data and ideas.

6. It can serve as a source of new ideas in a different but related field, and may be outside the patent protection offered the original invention.

7. Since claims are narrowly defined, much valuable information is included that may be outside the legal protection afforded by the claims.

8. Patents discuss the difficulties associated with previous research, development or production techniques, and offer a specific method of overcoming problems. This gives clues to current process information that has not been published in periodicals or books.

9. Can aid in process design by providing a selection of alternate techniques. A powerful research and engineering tool.

10. Obtain licenses–many U.S. chemical patents have not been developed commercially.

11. Patents provide an excellent starting point for the next investigator.

12. Frequently, innovations derived from research are first disclosed in the patent literature, prior to coverage in the periodical literature.

13. Patents offer a most valuable method of keeping abreast of latest technologies, serving an individual's own "current awareness" program.

14. Identifying potential new competitors.

15. It is a creative source of ideas for those with imagination.

16. Scrutiny of the patent literature has important profit-making potential.

CONTENTS AND SUBJECT INDEX

INTRODUCTION 1

1. CARBURETOR DETERGENTS 2
Multipurpose Detergents 2
Reaction Product of a Substituted Piperazine and a Substituted Acid Lactone 2
Hydrocarbon-Polyether-Substituted Succinamic Acid Compound 4
Polyether-Maleic Anhydride Reaction Product 5
Polyether-Asparagine Reaction Product 7
Primary Aliphatic Hydrocarbon Aminoalkylene Substituted Asparagine . . 8
Substituted Asparagine plus a Carbonic Acid Ester 9
ω-N-Disubstituted Aminoalkanoic Acid N'-Amides 10
Polymers from Methacrylic Esters 12
Quaternary Ammonium Salt of a Succinimide 14
Amine Oxide Polymers 16
Quaternary Ammonium Salt of a Copolymer 18
Inner Quaternary Ammonium Salts of Copolymers 19
Alkyl- or Hydroxy-Substituted Benzyl Polyamines 20
Two-Part Acylated Polyalkyleneamine/Polyolefin Additive 22
Reaction Products of Isatoic Anhydride and N-Alkyl Trimethylene Diamines 24
Polyamine Derivatives of Oxidized Olefinic Substituted Dicarboxylic Acid Compounds 24
Glycol Polyether-Acrylic Acid-Amine Reaction Product 25
High Molecular Weight Mannich Bases 28
Nitrogen-Containing Copolymers 29
Alkenylsuccinimide 31
Imides or Amide-Imides of Nitrilotriacetic or Ethylenediamine-tetraacetic Acid 33
Polyoxyalkylene Polyamines 34
Additives to Prevent Octane Requirement Increase 36
Solid Acid Catalysts of Refractory or Halogenated Metal Oxides 36
Dialkyl Formamide 40

Polyoxyalkylene Aminocarbamates. 42
Hydrocarbylpolyoxyalkylene Aminoesters 48
Hydrocarbylpolyoxyalkylene Polyamines 50
Reaction Product of a Hydrocarbylsuccinic Anhydride and an Aminotriazole 52
Cerium(III or IV) 2-Ethylhexanoate 53

2. DETERGENTS FOR FUELS AND LUBRICATING OILS. 56
Multipurpose Fuel and Lubricant Detergents 56
Reaction Product of a Phenol plus Aldehyde plus Active Hydrogen Compound 56
Polyolefinic Copolymers 58
N,N′-Substituted Diamines 62
Alkenyl-Substituted Oxa-Amines 63
Mannich Condensation Products 64
Haloalkyl Phenol/Unsaturated Compound Condensation Product. 66
Hydroxyalkyl/Hydroxy Aromatic Condensation Product 68
Aminoalkylalkanolamine Reaction Product with Acids. 69
Imidazoline and Its Reaction Product with a Sulfonic Acid 72
Nitrophenol-Amine Condensates. 73
Overbased Manganese Salts of Organic Acids. 74
Substituted Phenol/Epichlorohydrin/Amine Adducts 76
Lubricant and Diesel Fuel Additives 81
Chelate Derived from Hydroxyalkylated Benzotriazoles 81
Aroyl Derivative of an Alkenylsuccinic Anhydride 83
Process for Preparing Highly Basic Magnesium Sulfonates. 85
Lubricants for Two-Cycle Engines. 87
Aminophenols plus Certain Other Detergent/Dispersants 87
Tertiary Carbinamine-Modified Mannich Compositions. 89
Mannich Additives Modified by Ditertiaryalkyl Phenols 91
Additives for Use in Cat-Cracked Gasoline 92

3. SLUDGE DISPERSANTS AND SEDIMENT INHIBITORS 95
Sludge Dispersants for Lubricants and Fuels 95
Haze-Free Grafted Ethylene Copolymer 95
Polyolefin Graft Polymers. 96
Acylated Polyamine-Substituted Cycloaliphatic Compounds 98
Alkyl-Guanidino Heterocyclic Compound. 100
Oxazoline-Linked Dicarboxylic Acids and Certain Amino Acids. 102
Oxazoline-Containing Dispersants Stabilized Against Oxidation with Sulfur 104
Polyurethanes from Reaction of Diisocyanate and Diols. 105
Glyoxal-Polyamine-Polybutenyl Succinic Anhydride Reaction Products 107
Oxidatively-Coupled Hydroxyaromatic Compounds. 108
Sediment Inhibitors. 110
Polymeric Reaction Products of Poly(Alkoxyalkylene)Amines and Epichlorohydrin 111
Polymeric Reaction Products of Alkoxyalkylamines and Epihalohydrins 112
Alkylaryl Sulfonic Acids. 113

4. FLOW IMPROVERS AND POUR POINT DEPRESSANTS 115
For Middle Distillate Fuel Oils 115
Copolymer of 1-Hexene and 1-Octadecene 115
Borated Mannich Bases plus a Coadditive Hydrocarbon 117
Amine Salts of Thiobis Lactone Acids plus Coadditive Hydrocarbons. . 119
Aliphatic Copolymer plus Derivative of a Succinic Acid plus a Nitrogen Compound 122
Borated Derivatives of Substituted Succinic Acids or Acid Salts 125
Combination of Polyethylenes and Polyesters. 125
Combination of Ethylene-Vinyl Ester Copolymers and Polyesters 129
Combination of Polyethylene and a Second Polymer 129
Iminodiimides of 3,3′,4,4′-Benzophenonetetracarboxylic Dianhydride 130
Three-Component Additive with Ethylene Polymer plus a Nitrogen-Containing Compound 132
Poly(Isomerized C_{12-50} Monoolefins) with Pour Point Depressant 136
Three-Component Additive Including an Antiagglomerant 137
For Heavy Petroleum Oils 140
1,2-Epoxy Alkanes/Cyclic Carboxylate Copolymers 140
Production of Low Sulfur Fuel Oil with Lowered Pour Point 142
Ethylene-Vinyl Acetate Copolymer or A-B-A-Type Block Polymer. . . . 143
Alkylated Polybenzyl Polymer 144
Vinyl Acetate-Ethylene-Acrylic Acid Terpolymer. 146
Terpolymer or Graft Polymer 147
Alkenyl Succinate Diesters 149
Copolymer of Conjugated Diene and Alkyl Acrylonitrile as Viscosity Stabilizer. 150

5. ANTIKNOCK COMPOUNDS AND OCTANE IMPROVERS. 153
Antiknock Compounds 153
Organocerium(IV) Chelate plus p-Cresol or an Alkanediol 154
Rare Earth Chelates from 2,2,7-Trimethyl-3,5-Octanedione 155
Process for Preparing Highly Overbased Metallo-Organic Complexes. . . 157
Coordination Compounds of Divalent Mn, Fe, Co and Ni 158
Divalent Tetracoordinated Cobalt Compounds 160
Hexacoordinated Metal Compounds 161
Aliphatic β-Diketone plus β-Diketone Chelate of Cerium(IV) 163
Halogenated Substituted Fulvenes. 164
Aryl o-Aminoazides. 166
Octane Improvers 168
Organic Reactive Intermediates. 169
Preparation of Methyl tert-Butyl Ether 171
Process for Making a Hydrocarbon Fraction Containing tert-Amyl Methyl Ether. 173
Preparation of Phenyl tert-Butyl Ether 176
Preparation of Methyl tert-Butyl and Methyl tert-Amyl Ethers. 176

6. CORROSION AND OXIDATION INHIBITORS. 179
Anticorrosion Additives for Fuels and Lubricating Compounds 179
Certain Alkenylsuccinic Ester Acids and Their Amine Salts 179
Oxazoline Reaction Products 180
Magnesium-Containing Dispersions from Magnesium Carboxylates 181
Magnesium-Containing Dispersions from Magnesium Carbonate 184

Dispersions Containing Calcium Carbonate 185
Zinc-Containing Dispersions from Zinc Carbonate 186
Metal Salts of Organosulfonic Acids 188
Overbased Manganese Salts of Organic Acids 189
Certain Amide and Polyamine Derivatives of Lactam Carboxylic Acids 192
Ether Diamine Salts of N-Acylsarcosines 193
Methylol Polyester Derivatives of C_{12} to C_{22}-Substituted Succinic Anhydrides 194
Bridged Phenol Metal Salt/Halocarboxylic Acid Condensate Additives 196
Trisubstituted, Hydrocarbon-Soluble Chromium Compounds 199
Tartarimides 200
Oxazolonium Hydroxides 202
Containing a Hydrocarbon-Substituted Lactono-Imidazoline Reaction Product 204
Thixotropic Magnesium-Containing Complexes 205
Containing Acyl Glycine Oxazolines 207
Hydrocarbon-Substituted Methylol Phenols 208

Pipeline Corrosion Inhibitors 210
Polymerized Fatty Acids plus Mannich Polymer 210
Reaction Product of a Hydrocarbylsuccinic Anhydride and an Aminotriazole 211
Containing a Polymerized Unsaturated Aliphatic Monocarboxylic Acid 213

Cold-End Additives—Ethylene Polyamines plus Alkanolamines 214

Antioxidants 217
Alkenylsuccinic Acid or Anhydride/Aniline-Aldehyde Resin 217
Polyalkylene Amine plus Mannich Base 218
Metal Complexes of Thiobis(Alkylphenols) 219
End-Capped Phenols with 1 to 10 Methylene-Bridged o-Hydrocarbyl Phenol Units 223

7. COMBUSTION IMPROVERS AND FRICTION REDUCERS 225

Friction Reducers 225
Molybdenum Complexes of Hydroxy Amines 226
Molybdenum Complex of Thiobisphenols 228
Alkanol or Phenol Solutions of Molybdenum-Thiobisphenol Complexes 230
Molybdenum Complexes of Lactone Oxazoline Dispersants 231
Alkyl Phosphate Esters 233
Oil-Soluble Alkylene Glycol Ester Derivatives 234
Sulfurized Fatty Acid Amide or Ester of Diethanolamine 236
N-Hydroxymethyl Aliphatic Hydrocarbylamides 237
Aliphatic Hydrocarbyl Sulfinyl or Sulfonyl Methane 239
Aliphatic Hydrocarbyl Sulfinyl or Sulfonyl Alkanol 240
N-Aliphatic Hydrocarbyl Hydroxyalkyl Sulfinyl or Sulfonyl Succinimide 242
Reaction Product of Molybdenum Compound, Phenol plus an Amine 243

Combustion Improvers 245
Picric Acid with Ferrous Sulfate 245
Reaction Product of a Polyamine and an Acrylate Ester 246

Tertiary Diamines plus Ethanol. 248
Iron(II) Chelate and Picric Acid/Organic Base Complex 250
β-Carboxyethyl Polysiloxane. 251

8. ANTISTATS AND OTHER ADDITIVES 253
Antistatic Agents. 253
Reaction Product of α-Olefin/Maleic Anhydride Copolymer with Amines 253
Porphyrin Compound plus a Metal Alkyl Sulfosuccinate. 255
Mixtures Containing an Aminomethylene Sulfonic Acid. 255
Polymeric Amines plus α-Olefin/Acrylonitrile Copolymers 257
Exhaust Emission Reducers. 258
Compounds of Si and/or Ca, Mg, Mn, etc. plus Ash-Containing Resin. .258
Organic Magnesium and Manganese Compounds to Reduce Sulfur Emissions 260
Irradiation with Low Radioactivity. 261
Addition of Wax Oxidates to Diesel Fuel 261
Tracer Compounds 262
Chlorohydrocarbons and Chlorocarbons 262
Morpholino-Substituted Naphthylaminopropanes. 264
Miscellaneous Additives. 265
Boron Compound plus Amine as Biocide 265
Starter Fluid for Very Cold Weather 266
Multicomponent Demulsifier for Hydrocarbons 267
Dyes Prepared by Diazotization and Coupling. 269
Metal Deactivator for Aviation Fuel 270
Fuel Containing Insecticide Dispersed in Engine Exhaust Fumes 272

9. FUEL COMPOSITIONS AND SUPPLEMENTS 274
Gasoline-Alcohol Combinations 274
Gasoline plus Methyl Alcohol plus Ammonia 274
Mixture Solubilized with Ethyl or Methyl tert-Butyl Ether. 274
Fuel Supplement of Methyl and Ethyl Alcohols, Xylene plus Alkali Metal Hydroxide 276
Fuel Supplement Containing a Lower Alkanol plus Alkali Metal Hydroxide. 277
Methanol, Alkyl-Substituted Benzene, Heptane and Chlorinated Benzene. 278
Dry Single-Phase Gasohol 280
Alcohol Fuels for Diesel Engines. 282
Containing a Fatty Acid Amine or Ester of Diethanolamine 282
Dimerized Fatty Acid plus an Ester of a Phosphorus Acid 284
Dimerized Fatty Acid plus up to 25% Fuel Oil Fraction 285
Straight Chain Aliphatic Primary Amine 285
N-Hydroxy Oleamide 286
C_{12-30} Hydrocarbyl Succinic Acid or Anhydride 287
C_{8-20} Aliphatic Monocarboxylic Acids. 287
Gasoline-Water Combinations 288
Nonionic Ethoxylated Alkyl Phenol Surfactant 288
Coal Slurries in Fuel Oil with a Mixture of Surfactants 289
Oleic Acid Derivatives as Emulsifiers. 291
Surface Active Agent to Control Nitrogen Oxide Emissions 292
Five-Ingredient Surfactant. 293

Other Compositions 295
Sulfonated Aromatic Compound Condensed with Formalin as Coal-Fuel Oil Stabilizer 295
Light Fuel from Cellulose 296
Addition of Hydrogen Carriers to Ammonia and Amine Fuels 297
Alcohol as Petroleum Fuel plus Carbon Black. .300

COMPANY INDEX. 302
INVENTOR INDEX 303
U.S. PATENT NUMBER INDEX 306

INTRODUCTION

This book brings together descriptions of the preparation and testing of over 200 additives for fuel and lubricating oils and processes for formulating hundreds of fuel compositions patented between January 1979 and July 1981. It is clear that the primary thrust of research and development in this field during that time period has been toward attempting to provide fuel compositions which contain deposit control additives to improve fuel efficiency by effectively controlling or inhibiting deposit formation in the intake systems (carburetor, valves, etc.) of internal combustion engines. This must be accomplished without contributing to combustion chamber deposits which cause increased octane requirements for the engines and without producing exhaust emissions which are harmful to catalytic converters or environmental conditions. Chapters 1, 2, 3, 5 and 7 of this book relate to such additives.

Another problem which is always of importance is that of corrosion. Additives which can impart effective corrosion-inhibiting properties to fuel and oil compositions while at the same time imparting other useful characteristics such as detergency are particularly desirable. These rust and oxidation inhibitors are covered in Chapter 6.

Chapter 4 deals with additives to improve flow and pour properties of middle distillate and heavy petroleum oils, particularly during storage and/or transport. Chapter 8 is concerned with various specialized additives such as antistatic compounds, tracers, dyes, etc.

The last chapter describes compositions containing other fuels in addition to or in place of petroleum compounds. Such fuels as gasohol fall into this category.

The book, therefore, provides a comprehensive picture of technological developments made in the field of fuel additives over the past two and one-half years.

-1-

CARBURETOR DETERGENTS

MULTIPURPOSE DETERGENTS

Internal combustion gasoline engine design is undergoing important changes to meet stricter emission standards set for engine and exhaust gas emissions. A significant change in engine design is the feeding of blow-by gases from the crankcase zone of the engine into the intake air-fuel mixture near the throttle plate of the carburetor rather than venting these gases to the atmosphere as was practiced formally. This blow-by gas contains substantial amounts of deposit-forming substances and it is known to form deposits in and around the throttle plate area of the carburetor.

Another significant change in engine design and operation is the recirculation of a part of the exhaust gases to the fuel air intake of the engine. These exhaust gases also have a pronounced deposit-forming tendency. The carburetor deposits produced by the blow-by gases and the recycled exhaust gases restrict the flow of air through the carburetor at idle and at low speeds so that an over-rich fuel mixture results. This condition produced rough engine idling and/or stalling and leads to the release of excessive hydrocarbon emissions to the atmosphere.

Reaction Product of a Substituted Piperazine and a Substituted Acid Lactone

The additive for a motor fuel composition suggested by *W.M. Cummings and J.C. Powell; U.S. Patent 4,132,531; January 2, 1979; assigned to Texaco Inc.* which is effective as a carburetor detergent in gasoline, comprises the reaction product of a substituted piperazine and a hydrocarbon substituted lactone reaction product which, in turn, is produced by reacting an alkenylsuccinic acid with an acid catalyst or protonating agent under substantially anhydrous esterification conditions.

More specifically, the detergent additive is the reaction product of a substituted piperazine represented by the formula shown on the following page and a hydrocarbon-substituted acid lactone reaction product.

$$\begin{array}{c} | \quad\; | \\ CH-CH \\ NH \qquad\qquad N-R-NH_2 \\ CH-CH \\ | \quad | \quad | \\ R' \end{array}$$

In the above structural formula R is a hydrocarbyl radical having from 2 to 4 carbon atoms and R' is hydrogen or an alkyl radical having from 1 to 3 carbon atoms. The hydrocarbon-substituted acid lactone reaction product which is reacted with the piperazine is the reaction product of an alkenylsuccinic acid, in which the alkenyl radical has a molecular weight from about 300 to 3,000; which has been reacted with a concentrated mineral acid or protonating agent under substantially anhydrous reaction conditions at a temperature ranging from about 50° to 100°C.

The starting reactant from which the first reaction product is obtained is an alkenylsuccinic acid represented by the formula:

$$\begin{array}{l} \quad\;\; H \\ \quad\;\; | \\ R-C-COOH \\ \quad\;\; | \\ H_2C-COOH \end{array}$$

in which R is an alkenyl radical having a molecular weight ranging from about 300 to 3,000. The alkenyl radical itself is derived from the polymerization of propylene or isobutylene or mixtures thereof until a polymer of from about 300 to 3,000 average molecular weight preferably from about 700 to 2,000 average molecular weight is produced. This is reacted with maleic anhydride to produce an alkenylsuccinic anhydride which is thereafter hydrolyzed to an alkenyl succinic acid. The polymerization of olefins, the reaction of the olefin polymer with maleic anhydride and subsequent hydrolysis to alkenylsuccinic acid are well-known conventional processes and require no detailed description.

The prescribed alkenylsuccinic acid is mixed with a catalyst to form a reaction mixture which is heated to an elevated temperature to effect lactone formation. The catalyst which is employed may be any protonating agent or electron pair acceptor, i.e., any material which can provide a hydrogen ion or accept a pair of electrons to catalyze the reaction.

A variety of protonating agents or electron pair acceptors can be employed in the process. Included among these are mineral acids such as sulfuric acid and perchloric acid. Organic acids, including p-toluene sulfonic acid hydrate, boron trifluoride etherate and solid sulfonic acid ion exchange resins are also suitable.

A preferred temperature range for this process is from about 70° to 98°C. An important feature in this step is that it be conducted under substantially anhydrous conditions. The reactant solvent and the catalyst or the protonating agent must all be selected so as to insure substantially anhydrous and preferably anhydrous reaction conditions.

In the second step for preparing the additive composition of the process the acid lactone reaction product from the first step is reacted with the prescribed piperazine to produce a reaction product characterized as an amide.

The particular substituted piperazines which are employed for preparing the additive are generically described as aminoalkylpiperazines. Typical reactants include aminoethylpiperazine, aminopropylpiperazine, aminobutylpiperazine, etc.

It is convenient to conduct this reaction in an inert diluent or solvent which will facilitate refluxing of the reactant within the indicated temperature range. In general, an inert hydrocarbon or mixture of hydrocarbons which is an effective solvent for the reactants and of a suitable boiling range is the preferred medium for effecting this reaction.

The substituted piperazine and the substituted lactone reaction product are reacted employing approximately equimolar amounts of each reactant. These proportions can be varied somewhat but without advantage since the reactant in excess generally remains unreacted. Thus, the proportion of reactants employed is conveniently expressed as from about 1 mol of the substituted piperazine per mol of the hydrocarbyl or hydrocarbon-substituted lactone reaction product. It will be appreciated that the hydrocarbon radical on the substituted lactone reaction product can be saturated or unsaturated and that in either case the hydrocarbon or alkenyl radical will have approximately the same average molecular weight, i.e., about 300 to 2,000.

Carburetor detergency tests on a fuel composition containing 20 PTB (pounds of additive per 1,000 barrels of fuel) showed that the additive provided a high level of detergency and is suitable for use as a premium fuel additive.

Hydrocarbon-Polyether-Substituted Succinamic Acid Compound

A class of aliphatic hydrocarbon-polyether-substituted succinamic acid compounds are provided by *P.H. Moss and E.L. Yeakey; U.S. Patent 4,144,035; March 13, 1979; assigned to Texaco Development Corporation* as carburetor detergents and corrosion inhibitors to be employed in a liquid hydrocarbon fuel for an internal combustion engine.

A preferred aliphatic hydrocarbon-polyether-substituted succinamic acid compound of this process is represented by the formula:

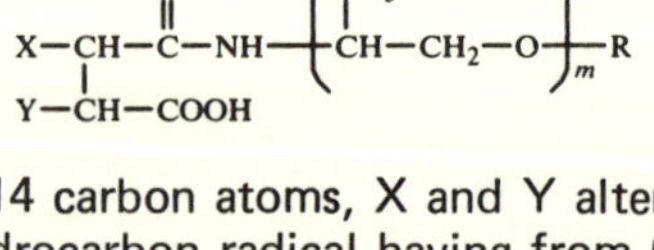

$$\begin{array}{l} X-CH-\overset{\displaystyle O}{\overset{\|}{C}}-NH-\left(\overset{\displaystyle CH_3}{\overset{|}{C}H}-CH_2-O\right)_m-R \\ \quad\;\; | \\ Y-CH-COOH \end{array}$$

in which R has from 8 to 14 carbon atoms, X and Y alternatively represent hydrogen and an aliphatic hydrocarbon radical having from 6 to 16 carbon atoms, and m is an integer from 2 to 3.

Particularly preferred compounds are those in which the aliphatic hydrocarbon radicals are branched-chain, saturated hydrocarbon radicals.

Methods for preparing the additive are well-known. In a preferred method, an olefinically unsaturated hydrocarbon is reacted with maleic anhydride to produce an alkenylsuccinic anhydride. This is then reacted with an aliphatic polyetheramine of suitable chain length in 1 to 1 mol proportions at a temperature below about 95°C to produce an N-alkylpolyether alkenylsuccinamic acid.

Examples of the effective carburetor detergents and corrosion inhibitors of this type include the following:

N-(octyldipropoxy)-tetrapropenylsuccinamic acid
N-(octyltripropoxy)-tetrapropenylsuccinamic acid
N-(octyltetrapropoxy)-tetrapropenylsuccinamic acid
N-(decyldibutoxy)-hexadecenylsuccinamic acid
N-(decyltributoxy)-hexadecenylsuccinamic acid
N-(decyltetrabutoxy)-hexadecenylsuccinamic acid
N-(dodecyldipropoxy)-dipropenylsuccinamic acid
N-(dodecyldipropoxy)-tripropenylsuccinamic acid
N-(octadecyldipropoxy)-tripropenylsuccinamic acid
N-(octadecyltripropoxy)-tripropenylsuccinamic acid
N-(hexadecyldipropoxy)-octadecenylsuccinamic acid

The additive is effective in an amount ranging from about 0.0002 to 0.2 weight percent based on the total fuel composition. An amount ranging from about 0.001 to 0.01 weight percent is preferred, the latter amounts corresponding to about 3 and 30 PTB respectively.

Polyether-Maleic Anhydride Reaction Product

W.M. Cummings; U.S. Patent 4,144,034; March 13, 1979; assigned to Texaco Inc. provides a class of polyetheramine substituted maleic anhydride reaction products as carburetor detergents and corrosion inhibitors when employed in a liquid fuel for an internal combustion engine. The reaction products are characterized by having a plurality of propylene oxide radicals and exhibit surprisingly effective carburetor detergency and corrosion inhibiting properties.

The fuel composition containing the additive prevents or mitigates the problem of corrosion and deposits lay-down in the carburetor of an internal combustion engine. When a gasoline containing the additive is employed in a carburetor which already has a substantial build-up of deposits from prior operations, a rather severe test of the detergency property of a fuel composition, this gasoline is effective for removing substantial amounts of the preformed deposits.

The aliphatic hydrocarbon etheramine and maleic anhydride reaction product is represented by the formula:

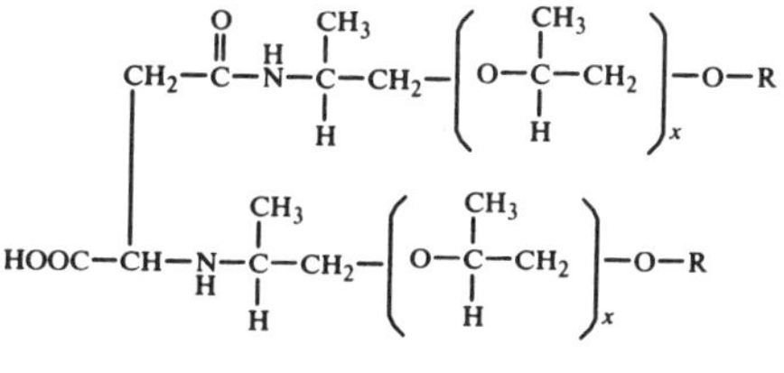

$$\begin{array}{l} CH_2-\overset{\overset{O}{\|}}{C}-\overset{H}{N}-\underset{H}{\overset{CH_3}{\overset{|}{\underset{|}{C}}}}-CH_2-\left(O-\underset{H}{\overset{CH_3}{\overset{|}{\underset{|}{C}}}}-CH_2\right)_x-O-R \\ \quad | \\ HOOC-CH-\underset{H}{N}-\underset{H}{\overset{CH_3}{\overset{|}{\underset{|}{C}}}}-CH_2-\left(O-\underset{H}{\overset{CH_3}{\overset{|}{\underset{|}{C}}}}-CH_2\right)_x-O-R \end{array}$$

in which R represents an aliphatic hydrocarbon radical having from 6 to 20 carbon atoms and x has a value from 1 to 3. A preferred reaction product for the fuel composition of the process is one in which R is a saturated aliphatic hydrocarbon radical having from 10 to 18 carbon atoms.

Methods for preparing the additive of the process are well-known. In a preferred method, a polyetheramine is reacted with maleic anhydride to produce the reaction product. Approximately 2 mols of the polyetheramine are reacted with a mol of maleic anhydride at a temperature ranging from about room temperature up to 95°C to produce the reaction product.

Example: 24.5 g (0.25 mol) of maleic anhydride were added to 160 g of a mineral oil having a viscosity in centistokes at 210°F of about 4. 132 g (0.50 mol) of an aminated bispropoxylated C_{8-14} alcohol was added to the oil solution of the maleic anhydride forming a reaction mixture. The mixture was heated to about 104°F, and maintained at this temperature for about 3½ hr. The mixture was then cooled.

The polyether-maleic anhydride reaction product was an N,N'-(1,4-dimethyl-3-oxa-5-C_{8-14} alkyl oxypentyl)asparagine and is represented by the following formula:

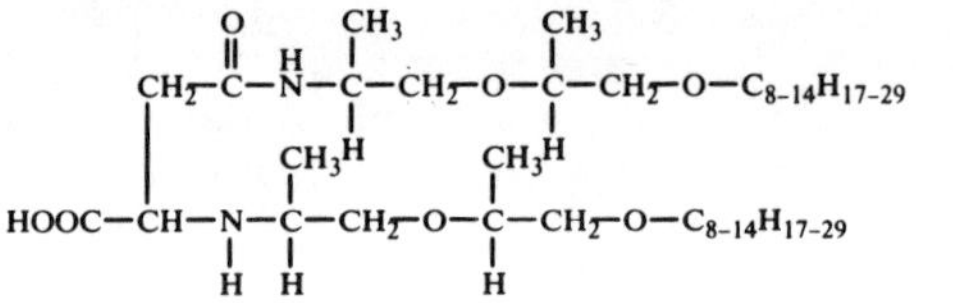

```
                O      CH3          CH3
                ‖  H    |            |
              CH2-C-N--C--CH2-O--C--CH2-O-C8-14H17-29
               |        |            |
               |       CH3H         CH3H
               |        |            |
HOOC-CH-N--C--CH2-O--C--CH2-O-C8-14H17-29
           |  |            |
           H  H            H
```

The additive was tested for its effectiveness as a carburetor detergent in the Carburetor Detergency Test. This test is run on a Chevrolet V-8 engine mounted on a test stand using a modified four barrel carburetor. The two secondary barrels of the carburetor are sealed and the feed to each of the primary barrels arranged so that an additive fuel can be run in one barrel and the base fuel run in the other. The primary carburetor barrels were also modified so that they had removable aluminum inserts in the throttle plate area in order that deposits formed on the inserts in this area would be conveniently weighed.

In the procedure designed to determine the effectiveness of an additive fuel to remove preformed deposits in the carburetor, the engine is run for a period of time, usually 24 to 48 hr, using the base fuel as the feed to both barrels with engine blowby circulated to an inlet in the carburetor body. The weight of the deposits on both sleeves is determined and recorded. The engine is then cycled for 24 additional hours with a suitable reference fuel being fed to one barrel, additive fuel to the other and blowby to the inlet in the carburetor body. The inserts are then removed from the carburetor and weighed to determine the difference between the performance of the additive and reference fuels in removing the preformed deposits.

After the aluminum inserts are cleaned, they are replaced in the carburetor and the process repeated with the fuels reversed in the carburetor to minimize differences in fuel distribution and barrel construction. The deposit weights in the two runs are averaged and the effectiveness of the fuel composition is compared to the reference fuel which contains an effective detergent additive. The difference in effectiveness is expressed in percent, a positive difference indicating that the fuel composition of the process was more effective than the commercial fuel composition.

The base fuel employed with the described detergent additive in the detergency tests was a premium grade gasoline having a Research Octane Number of about 95% and contained 4.0 cc of tetraethyl lead per gallon. This gasoline consisted of about 28% aromatic hydrocarbons, 10.5% olefinic hydrocarbons and 61.5% paraffinic hydrocarbons and boiled in the range from 90°F to 379°F.

The carburetor detergency test results obtained with the fuel composition of the process in comparison to two commercial detergent fuel compositions referred to as Reference Fuel A and Reference Fuel B, are set forth in the table below.

Carburetor Detergency Test

Run No.	Additive Fuel Composition	Percent Effective
1	Base fuel + 25 PTB of Ex. vs Ref. Fuel A*	+25
2	Base fuel + 10 PTB of Ex. vs Ref. Fuel A	+3
3	Base fuel + 40 PTB of Ex. vs Ref. Fuel B**	-14

Note: PTB = pounds of additive per 1,000 barrels of fuel.
*Contains 15 PTB of commercial detergent.
**Contains 173 PTB of commercial detergent.

The foregoing tests show that the fuel composition containing the described additive was highly effective in its carburetor detergency property and that its performance is comparable to or superior to commercial detergent fuel compositions.

Polyether-Asparagine Reaction Product

In further work, *W.M. Cummings; U.S. Patent 4,144,036; March 13, 1979; assigned to Texaco Inc.* provides polyether substituted reaction products of an asparagine as carburetor detergents and corrosion inhibitors when employed in a liquid hydrocarbon fuel for an internal combustion engine. The reaction products are characterized by having a plurality of alkyloxytrimethylene radicals attached to the asparagine base and exhibit surprisingly effective carburetor detergency and corrosion inhibiting properties.

The polyether reaction product of the asparagine is represented by the formula:

$$\begin{array}{l} \quad\quad\quad\quad\;\; O \\ \quad\quad\quad\quad\;\; \| \\ \quad\quad\; CH_2-C-NH-CH_2CH_2CH_2-O-R \\ \quad\quad\; | \\ HOOC-CH-NH-CH_2CH_2CH_2-O-R \end{array}$$

in which R represents an aliphatic hydrocarbon radical having from 10 to 20 carbon atoms. A preferred reaction product for the fuel composition of the process is one in which R is a saturated aliphatic hydrocarbon radical having from 13 to 18 carbon atoms.

Methods for preparing the additive of the process are well-known. In a preferred method, an etheramine is reacted with maleic anhydride to produce the reaction product. Approximately 2 mols of the etheramine are reacted with a mol of maleic anhydride at a temperature ranging from about room temperature up to 95°C to produce the reaction product.

Example: 49 g (0.5 mol) of maleic anhydride were added to 240 g of a mineral oil having a viscosity in centistokes at 210°F of about 4. 265 g (1.0 mol) of tridecoxypropylamine (Armour etheramine-13) was added to the oil solution of the maleic anhydride forming a reaction mixture. This mixture was heated to about 104°F, and maintained at this temperature for about 1.5 hr. The mixture was then cooled.

The etheramine anhydride reaction product obtained was N,N'-di(tridecoxypropyl)asparagine and is represented by the formula:

$$\begin{array}{l} \qquad\quad\; O \\ \qquad\quad\; \| \\ \qquad CH_2-C-NH-CH_2CH_2CH_2-O-C_{13}H_{27} \\ HOOC-CH-NH-CH_2CH_2CH_2-O-C_{13}H_{27} \end{array}$$

The additive was tested for its effectiveness as a carburetor detergent in the Carburetor Detergency Test described in the previous patent.

The carburetor detergency test results obtained with the fuel composition containing the described additive in comparison to two commercial detergent fuel compositions referred to as Reference Fuel A and Reference Fuel B, are set forth in the table below.

Carburetor Detergency Test

Run No.	Additive Fuel Composition	Percent Effective
1	Base fuel + 10 PTB of Ex. vs Ref. Fuel A*	-7
2	Base fuel + 40 PTB of Ex. vs Ref. Fuel B**	-14

*Contains 15 PTB of a commercial detergent.
**Contains 173 PTB of a commercial detergent.

Primary Aliphatic Hydrocarbon Aminoalkylene Substituted Asparagine

J.B. Biasotti, P. Dorn, S. Herbstman and K.L. Dille; U.S. Patent 4,204,841; May 27, 1980 and S. Herbstman and P. Dorn; U.S. Patent 4,207,079; June 10, 1980; both assigned to Texaco Inc. describe a detergent motor fuel composition comprising a mixture of hydrocarbons in the gasoline boiling range containing a primary aliphatic hydrocarbon aminoalkylene-substituted asparagine, which is produced by reacting about 2 mols of an N-primary alkylalkylenediamine with 1 mol of maleic anhydride. These additives possess good corrosion inhibiting properties and exhibit outstanding carburetor detergency properties.

The preferred asparagine for this detergent component is represented by the formula:

$$\begin{array}{l} \qquad CH_2-C(=O)-N(H)-CH_2CH_2CH_2-N(H)-R \\ \qquad\;\; | \\ {}^{-}OOC-CH-N^{+}H_2-CH_2CH_2CH_2-N(H)-R \end{array}$$

in which R is a primary aliphatic hydrocarbon radical having from 16 to 20 carbon atoms.

Examples of specific asparagine additives for this fuel composition include the following: N,N'-di(3-n-oleylamino-1-propyl)asparagine; N,N'-di(3-n-dodecylamino-1-propyl)asparagine; N,N'-di(3-octylamino-1-propyl)asparagine; N,N'-di(3-stearylamino-1-propyl)asparagine; N,N'-di(3-decylamino-1-propyl)asparagine; N,N'-di(3-

laurylamino-1-propyl)asparagine; and N,N'-di(3-behenylamino-1-propyl)asparagine.

The most preferred N-alkylalkylenediamine additive is represented by the formula: $R-NH-CH_2CH_2CH_2-NH_2$ in which R is a straight chain primary alkyl aliphatic hydrocarbon radical having from 16 to 20 carbon atoms. Examples of suitable N-alkylalkylenediamine additives which can be beneficially employed in combination with the prescribed substituted asparagine include N-oleyl-1,3-propanediamine, N-lauryl-1,3-propanediamine, N-stearyl-1,3-propanediamine and N-dodecyl-1,3-propanediamine.

The additive composition is a mixture of the two components described above. In general, the additive composition comprises from 30 to 70 weight percent based on the total weight of the additive composition of the primary aliphatic hydrocarbon aminoalkylene substituted asparagine, component A, and the balance of the prescribed N-alkylalkylenediamine compound, or component B. It is preferred to employ the additive components in approximately 50 to 50 weight percent amounts based on diluent free materials.

Example 1: An additive is prepared by admixing N,N'-di(3-n-oleylamino-1-propyl)asparagine with N-oleyl-1,3-propanediamine in 50 to 50 weight percent amounts based on diluent free materials.

Example 2: An additive is prepared by admixing N,N'-di(3-laurylamino-1-propyl)asparagine with N-oleyl-1,3-propanediamine in 50 to 50 weight percent amounts based on diluent free materials.

The additive composition is employed in the motor fuel composition of the process in a concentration ranging from about 0.001 to 0.003 weight percent based on the weight of the motor fuel composition. The most preferred concentration is about 0.002 weight percent, or a dosage equivalent to about 6 PTB or 6 lb of additive per 1,000 bbl of gasoline.

The effect on carburetor detergency of the fuel composition of this formulation was determined in the Buick Carburetor Detergency Test. This test is run on a Buick 350 CID V-8 engine equipped with a 2 barrel carburetor. The engine is mounted on a test stand and has operating EGR and PCV systems. Approximately 300 gal of fuel and 3 qt of oil are required for each run.

Prior to each run the carburetor is completely reconditioned. Upon completion of the run the throttle plate deposits are visually rated according to a CRC Varnish rating scale (Throttle Plate Merit Rating) where 1 describes heavy deposits on the throttle plate and 10 a completely clean plate.

Results of testing demonstrated that the fuel composition containing the additive composition described was surprisingly effective for achieving carburetor throttle plate cleanliness as measured by the CRC Varnish rating scale in the Buick Carburetor Detergency Test, the rating of the base fuel containing 6 PTB of the product of Example 1 being 9.3 as compared to a rating of 3.6 given the base fuel containing no additive.

Substituted Asparagine plus a Carbonic Acid Ester

A substituted asparagine is also used as an additive to promote gasoline detergency by *M.S. Kablaoui; U.S. Patent 4,231,758; November 4, 1980; assigned to Texaco Inc.,* this time combined with a carbonic acid ester.

The substituted asparagine is represented by the formula:

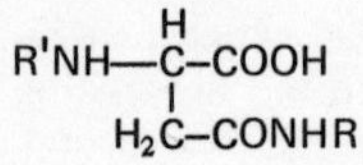

$$\begin{array}{l} \quad\quad\;\; H \\ R'NH\!-\!\overset{|}{C}\!-\!COOH \\ \quad\quad\;\; | \\ \quad\quad H_2C\!-\!CONHR \end{array}$$

in which R and R' preferably represent the same or different secondary alkyl or alkylene radicals having from 12 to 18 carbon atoms.

Carbonic acid ester compounds which exhibit no carburetor detergency properties when employed in a motor fuel composition yet which surprisingly cooperate with the substituted asparagine to enhance its carburetor detergency include diethyleneglycol-bis-2-ethoxyethyl carbonate and diethyleneglycol-bishexyl carbonate.

In general, these additive components are added to a fuel composition in minor amounts. The most effective concentration of this additive component ranges from about 0.002 to 0.10 weight percent. The carbonic acid ester compound which was discovered to cooperate with the substituted asparagine compound is preferably employed in an amount ranging from about 0.05 to 0.25 volume percent of the gasoline composition.

The additive combination was tested for its effectiveness as a carburetor detergent in the Buick Carburetor Detergency Test.

The tests show that a carbonic acid ester has absolutely no effect on the carburetor detergency of a motor fuel composition. In contrast, when a carbonic acid ester was used in combination with a substituted asparagine, a substantial improvement in carburetor detergency was provided over those in which the substituted asparagine was used without the carbonic acid ester.

ω-N-Disubstituted Aminoalkanoic Acid N'-Amides

Although some of the prior art detergent products show a good cleaning effect in the carburetor, they suffer from the disadvantage that they deposit the dirt which they have removed from the carburetor onto the hot inlet valves, where it decomposes or carbonizes to cause the inlet valves to stick. It is thus very important to find additives for gasoline which clean the carburetor and inlet channels and also show no tendency to decompose or become deposited on the very hot inlet valves.

H. Graefje, G. Nottes, H. Mueller, G. Daumiller and H. Hoffmann; U.S. Patent 4,153,425; May 8, 1979; assigned to BASF AG, Germany have found that ω-N-disubstituted aminoalkanoic acid N'-amides of the formula:

$$(1) \quad \begin{matrix} R^1 \\ R^2 \end{matrix}\!\!>\!N\!-\!(CH_2)_n\!-\!CO\!-\!N\!<\!\!\begin{matrix} R^3 \\ R^4 \end{matrix}$$

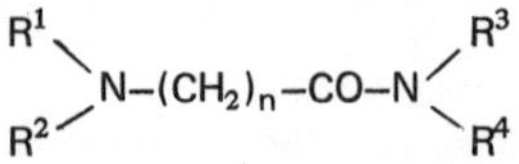

in which R^1, R^2, R^3 and R^4 are the same or different and denote branched-chain or straight-chain alkyl of from 7 to 20 carbon atoms, R^2 and R^4 may also denote aryl or aralkyl, preferably phenyl, alkylphenyl or benzyl, and n is a number from

3 to 5, the total number of carbon atoms in R^1 and R^2 or in R^3 and R^4 being at least 16, show very good valve-cleaning properties when dissolved in small amounts in fuels for internal combustion engines.

In the compounds of the formula, the radicals R^1, R^2 and R^3 and R^4 are preferably identical and contain, preferably from 12 to 14 carbon atoms and have a molecular weight of at least 500. The upper limit of the chain length of the radicals is determined only by practical considerations relating to the reaction rate at which such high molecular weight secondary amines may be synthesized. This limit usually lies at C_{20}.

The amines R–NH–R to be used in the preparation of ω-dialkylaminoalkanoic acid dialkylamines may have radicals having linear or branched chains, and are readily obtained from commercially available materials. R is the same as R^1 through R^4 above. Preparation of the ω-N-disubstituted aminoalkanoic acid-N'-amides of formula (1) may be effected in known manner by reaction of substantially stoichiometric amounts of amines of the formulas R^1–NH–R^2 and R^3–NH–R^4 with caprolactone, valerolactone and, preferably, butyrolactone at 105° to 300°C and at pressures up to 50 bars.

Specific examples of the compounds are: ditridecylaminobutyroditridecylamide, didodecylaminobutyrodidodecylamide, dioleylaminobutyrodioleylamide, distearylaminobutyrodistearylamide, and dipalmitylbutyrodipalmitylamide. These gasoline additives are added to gasoline in amounts preferably from 50 to 200 ppm.

An important technical advantage of these additives is that they do not attack the light metal alloys of which the automobile carburetors are usually made and, furthermore, even provide a certain degree of protection against corrosion by moisture, alcohols and polyglycol-containing additives for the prevention of carburetor icing.

Example: A BASF single-cylinder test engine having a cubic capacity of 332 cm^3 (cylinder diameter = 65 mm, stroke = 100 mm) was run over periods of 50 hr at a constant speed of 2,000 rpm and a fuel rate of 1.6 ℓ/hr. The engine was modified so as to return 10% of the exhaust gases to the crank case below the level of the oil, from which it was passed to the air filter of the suction line of the carburetor.

The automobile gasoline used was one consisting of a mixture of catalytically cracked gasoline, platformate, straight-run gasoline, and 5% of pyrolysis gasoline having the usual lead content. The engine was opened up on completion of the test run. The inlet valve was found to be covered with asphaltlike or cokelike residues. This deposit was carefully removed and the difference in weight between the polluted and cleaned inlet valve was found to be 164 g.

However, when 1,000 mg/ℓ of ditridecylaminobutyroditridecylamide were added to the gasoline prior to the test run, the valve was found to be covered by only a small amount of asphaltlike residue on completion of the run, the valve shaft being clean. The residue, determined from the difference in weight of the inlet valve before and after cleaning, was found to weigh 26 g.

The corresponding weight found on using 1,000 mg/ℓ of ditetradecylaminobutyroditetradecylamide was 32 g. The valve shaft and the appearance of the inlet valve were found to be very good.

Polymers from Methacrylic Esters

Numerous detergents for inclusion in fuels have heretofore been suggested. For instance, it has long been known to include in gasoline a detergent based upon an amine such as a condensation product of a secondary amine with an anhydride such as maleic anhydride. Other known detergents include materials such as polyisobutylamines. Unfortunately, while effective, these known detergents are quite expensive, thus raising the overall price of the gasoline. It became desirable to provide a detergent which can be synthesized at substantially less cost and is similarly effective in providing detergency in the lines through which the fuels pass. It became particularly desirable to provide a detergent for gasoline which would remove deposits in the carburetor of an automobile and, at the same time, would prevent deposit buildup on such carburetor.

Stabilized fuels have been proposed containing copolymers made from esters of unsaturated carboxylic acids. Thus, Catlin et al showed in U.S. Patent 2,737,452, a stabilized fuel containing an oil soluble basic amino nitrogen-containing addition type polymer made from a plurality of polymerizable olefinically unsaturated compounds. The copolymer of Catlin et al has units of at least 8 carbon atoms in the ester chain, for Catlin et al discloses that polymers of aminoethyl methacrylate (where the chain length is far less than 8 carbon atoms) are without effect in the formation of sediment in fuel oils even when such homopolymers contain long hydrocarbon chains.

Undoubtedly, the Catlin et al stabilizer is effective for stabilizing catalytically cracked fuel oils. It has become desirable, however, to provide a stabilizer derived from an ester of a carboxylic acid and a nitrogen containing comonomer thereof which is especially effective to remove the buildup of deposits on an engine carburetor.

In accordance with work reported by *C.M. Cusano, I.D. Rubin, R.E. Jones and P.F. Vartanian; U.S. Patent 4,161,392; July 17, 1979; assigned to Texaco Inc.* an inexpensive stabilizer which is effective as a detergent to remove deposit buildup from a carburetor, is provided by a copolymer comprising the olefin polymerization product of:

(A) a C_{1-6} saturated or unsaturated, substituted or unsubstituted, aliphatic or aromatic ester of an unsaturated aliphatic mono-, di- or polycarboxylic acid of chain length C_{1-6} in an amount of between 5 and 30 weight percent;

(B) a C_{8-20} saturated or unsaturated, substituted or unsubstituted, aliphatic or aromatic ester of an unsaturated mono-, di- or polyaliphatic C_{1-6} carboxylic acid in the amount of 50 to 81 weight percent; and

(C) an ethylenically unsaturated compound containing a basic nitrogen in a side chain.

It has been found that an improved gasoline detergent is provided if the copolymer contains units derived from an ester whose ester group contains only up to 6 carbon atoms. It has been surprisingly found that if a portion of the amine-free components of the Catlin et al stabilizer are replaced with units of shorter ester group length, that improved detergency is obtained. This detergent is effective in removing up to 75% of the deposits on an automotive engine carburetor.

Component A is, therefore, a necessary and important component of the copolymer and is present in the copolymer in an amount preferably from 15 to 25 weight percent. Component A is free of amino nitrogen.

Component B is present in an amount preferably of 60 to 75 weight percent.

Component C usually makes up the balance of the polymer and can be present in an amount of preferably 7 to 20 weight percent.

Component A and Component B can be made from the same acids or different acids, the components differing in the length of the ester group. They can be made from mono-, di- or polyaliphatic carboxylic acids of C_{1-6} chain length. Particularly preferred acids include methacrylic acid, acrylic acid, fumaric acid, maleic acid and butenic acid. Where substituted esters are employed, the substituents on the ester group can be halogen, cyano, hydroxy, thiohydroxy and acetyl. The amount of any halo substituent and of sulfur is, of course, limited by recognized upper limits for these materials.

With respect to the nature of the ester group of the carboxylic acid esters, it is preferred that the ester group be an alkyl group. The copolymer has a molecular weight preferably between 1,000 and 3,500.

The nitrogen-containing component, Component C, can be any of those heretofore known for use in stabilizers for motor fuels. These include, in particular, those ethylenically unsaturated compounds which contain a basic amino nitrogen in a side chain which will supply nitrogen in an amount of 0.3 to 3.5% by weight to the overall polymer. The preferred amines include the polymerizable, ethylenically unsaturated compounds having a basic tertiary amino group. Particularly outstanding amines are 4-vinyl-pyridine and dimethylaminoethyl methacrylate.

This copolymer is useful in fuel oils, especially gasoline. When employed, it is included in the fuel in an amount between 0.005 and 1.0 weight percent, preferably 0.05 to 0.75 weight percent.

Example: Into a 1 ℓ resin kettle fitted with external heating, thermometer, nitrogen inlet, stirrer and condenser, there was charged 272 g of a commercially available long chain ester of methacrylic acid where the ester chain was between 12 and 15 carbon atoms in length with an average chain length of 13.6. A material identified as Neodol 25L methacrylate supplied such esters. The resin kettle was additionally charged with 82 g of commercially available longer chained ester of methacrylic acid which supplied esters of 16 to 20 carbon atom length with average chain length of 17.7. Alfol 1620 methacrylate supplied such esters. The Neodol 25L and Alfol 1620 together made up the Component B. The resin kettle was supplied with 100 g of n-butylmethacrylate which supplied Component A. Component C was supplied by 45 g of dimethylaminoethyl methacrylate (DMAEMA).

Into the reaction mixture was also introduced 5 g of n-dodecylmercaptan. The reaction mixture was heated to 95°C under a nitrogen atmosphere and 0.4 g of azobisisobutyronitrile initiator was added. The progress of the polymerization was followed by determining the refractive index of samples. Additional azobisisobutyronitrile in the amount of 0.2 g and 0.1 g was added after 2 hr and 3½ hr of heating respectively. Heating was continued for a total of 6 hr at 95° to 100°C.

A product was obtained having only 0.6% residual monomer. The same was tested to determine its effect in removing deposit of buildup according to the Chevrolet Carburetor Detergency Test described in a previous patent.

The efficacy of the various stabilizers is shown in the table below.

Ex.	C_{12}-C_{16} Methacrylic Acid Esters (wt %)	C_{16}-C_{20} Methacrylic Acid Esters (wt %)	Butyl Methacrylate (wt %)	DMAEMA (wt %)	4-Vinyl Pyridine (wt %)	Deposit Buildup (mg)	Deposit Removal (mg)	Efficacy (%)
1	57.5	17.5	21	–	4	20.4	6.3	31
2	54.5	16.5	20	–	9	18.8	13.4	71
3	57.5	17.5	21	4	–	19.1	10.5	55
4	54.5	16.5	20	9	–	19.0	12.5	66
5	50	15	18	18	–	21.7	16.2	75
A*	63.7	21.3	–	15	–	23.8	7.2	30

*Comparative Example

From the data set forth in the table, it is evident that the presence of shorter chain esters remarkably improves the ability of the detergent to remove deposits from the engine carburetor. Note that where the polymer contains 9% by weight of components of dimethylaminoethyl methacrylic acid with 20 weight percent of units of butylmethacrylate, a 66% efficacy is achieved. Where, however, the polymer is free of such shorter chain esters of methacrylic acid and contains 15 weight percent dimethylaminoethyl methacrylate, the removal efficacy is only 30%. Thus, while the non-butylmethacrylate-containing polymer has almost twice the amount of nitrogen-containing component, it is less than one half as effective as a polymer containing the shorter chained ester.

Quaternary Ammonium Salt of a Succinimide

The motor fuel composition described by *P.F. Vartanian; U.S. Patent 4,171,959; October 23, 1979; assigned to Texaco Inc.* comprises a mixture of hydrocarbons in the gasoline boiling range containing a minor amount of a quaternary ammonium salt of a succinimide represented by the formula:

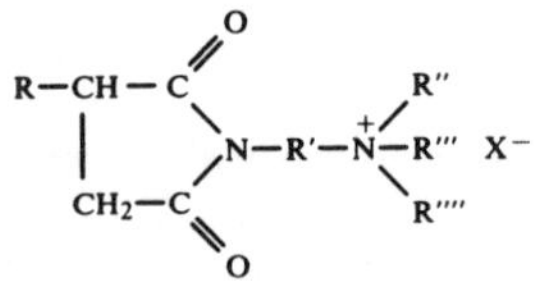

```
               O
              //
R—CH—C              R''
      |      \        +/
      |       N—R'—N—R'''  X−
      |      /        \
  CH2—C               R''''
              \\
               O
```

in which R, R', R''' and R'''' are hydrocarbon radicals and X is an anion. More specifically, R is a hydrocarbon radical having a molecular weight ranging from about 280 to 1,800, R' is a divalent hydrocarbon radical having from 2 to 10 carbon atoms, R'' and R''' are hydrocarbon radicals having from 1 to 6 carbon atoms or are interconnected to form a heterocyclic ring consisting of from 3 to 6 atoms selected from the group consisting of carbon, nitrogen and oxygen atoms, R'''' is a hydrocarbon radical having from 1 to 6 carbon atoms, and X is the anion of an acid, i.e., a halide or an organic acid such as sulfonate or carboxylate.

The quaternary ammonium salt of a succinimideamine compound is prepared by reacting a succinimideamine compound having a residual amine function with a compound which will quaternize an amine. The resulting quaternary ammonium compound exhibits outstanding carburetor detergency when employed in a gasoline motor fuel composition.

The aliphatic hydrocarbon radical represented by R can be a saturated or unsaturated hydrocarbon radical. It will generally be an alkenyl or polyalkylene radical prepared by polymerizing an olefin, preferably a C_3 or C_4 olefin such as propylene or isobutylene, to form a polymer of the prescribed molecular weight.

Examples of quaternary ammonium salts of succinimides which perform well as detergent additives for gasoline, besides the two which are prepared as examples, are the following compounds (the numbers in the parentheses following the name of the type of polymer employed refer to the molecular weight of this radical, preferred being in a range from 325 to 425 or from 750 to 1,300): polyisobutenyl(1200)-N,N,N-trimethylpropa-1,3-diaminosuccinimide quaternary ammonium iodide; polyisobutenyl(850)-N,N,N-trimethyletha-1,2-diaminosuccinimide quaternary ammonium iodide; polypropenyl(700)-N,N,N-triethylpropa-1,3-diaminosuccinimide quaternary ammonium bromide; polyisobutenyl(335)-N-methyl-N-(3-aminopropyl)piperazinosuccinimide quaternary ammonium iodide; and polypropenyl(800)-N-methyl-N-(2-aminoethyl)piperazinosuccinimide quaternary ammonium chloride.

Example 1: Polyisobutenyl(335)-N,N,N-trimethylpropa-1,3-diaminosuccinimide quaternary ammonium iodide is prepared as follows. To 110 g of polyisobutenyl (335)-N,N-dimethylpropa-1,3-diaminosuccinimide having a total base number (TBN) of 79.4 in 100 ml of xylene solvent is added 35 g (ca 2:1 based on active succinimide concentration) of methyl iodide. The mixture was heated to 90°C for 1 hr. The solvent and any unreacted methyl iodide was then stripped by reduced pressure distillation to give a substantial yield of the quaternized product having a total base number of 21.4.

Example 2: Polyisobutenyl(335)-N-methyl-N-(3-aminopropyl)morpholinosuccinimide quaternary ammonium iodide is prepared as follows. To 100 g (0.17 mol) of polyisobutenyl(335)-N-(3-aminopropyl)morpholinosuccinimide having a total base number of 66.3 in 100 ml of xylene solvent is added 25 g (0.17 mol) of methyl iodide. The mixture was heated to 90°C for 1 hr. The solvent and any unreacted methyl iodide was stripped off by distillation under reduced pressure to yield 105 g of the quaternized product having a total base number of 13.8.

In general, the additive is added to the base fuel in a minor amount, i.e., an amount effective to provide carburetor detergency to the fuel composition. The additive is highly effective in an amount ranging from about 0.003 to 0.25 weight percent based on the total fuel composition. The most preferred concentration ranges from about 0.005 to 0.10 weight percent.

The additive was tested for its effectiveness as a carburetor detergent in the previously described (see U.S. Patent 4,144,034) Carburetor Detergency Test.

The carburetor detergency test results obtained with the base fuel, comparison fuels and the fuel composition of this process are set forth in the table below. The additive concentration in the comparison fuels, Runs 2, 3, 4, and 5, were at a concentration of 50 PTB.

Chevrolet Carburetor Detergency Test, Phase III*

Run No.	Additive (50 PTB)	. . .Deposit (mg). . . Buildup	Removal	Effectiveness (%)
1	Base fuel (no carb. deter. added)	16.8	-1.7	-10
2	Polyisobutenyl(335)-N,N-dimethyl-propa-1,3-diamine succinimide	20.8	-13.8	-66
3	Quaternary ammonium iodide salt of Ex. 1	38.4	26.8	+70
4	Polyisobutenyl(335)-N-(3-aminopropyl)-morpholine succinimide	17.1	-15.9	-93
5	Quaternary ammonium iodide salt of Ex. 2	27.0	17.6	+65
6	Premium of fuel composition A**	36.7	13.3	+36
7	Premium fuel composition B**	21.8	17.1	+78

*Clean-up type test
**Runs 6 and 7 employed premium commercial detergent fuel compositions.

The foregoing data show that the detergent fuel compositions containing the described additives, illustrated by Runs 3 and 5, exhibit a surprising improvement over comparison fuel compositions containing the precursor succinimide and over the base fuel. The performance of these fuel compositions also is essentially equivalent to or substantially surpasses that of premium commercial detergent fuel compositions.

Amine Oxide Polymers

Commercially available types of gasoline detergents are particularly costly in production, thus adding to the final cost of gasoline at the pump. It has become desirable, therefore, to provide less expensive, yet effective, gasoline detergents. More especially, it has become desirable to provide more effective detergents which are polymeric in nature and are derived from both nitrogen-containing and nitrogen-free comonomers.

P.F. Vartanian and A.L. Ippolito; U.S. Patent 4,179,271; December 18, 1979; assigned to Texaco Inc. have found that by oxidizing a polymer having an amine group there is provided an amine oxide reaction product which is especially effective as a gasoline detergent.

The polymers which are reacted in accordance with the process are of the type disclosed by Catlin et al, U.S. Patent 2,737,452. By reacting such known copolymers with an oxidizing agent, there is formed an amine oxide which is especially effective, when employed as a gasoline detergent, in removing deposits from an automotive carburetor and, perhaps more significantly, in inhibiting deposit buildup upon such an automotive carburetor. Particularly preferred amine oxides are amine oxides of primary, secondary, and tertiary amino nitrogen-containing polymers, especially amine oxides of polymers containing a tertiary amino nitrogen atom.

The process can perhaps be appreciated when reference is made to an equation showing the conversion of a particularly desirable polymer to its corresponding amine oxide:

$$\text{Methacrylate-4-vinyl pyridine polymer} + \left\{\begin{array}{l} H_2O_2 \\ CH_3COOH \\ KMnO_4 \\ C_{2-8}\text{ alkenyl peracid} \\ C_{1-8}\text{ alkyl peracid} \end{array}\right. \rightarrow \text{Amino oxide polymer}$$

In the polymer above, the amine group is supplied employing, as a comonomer of the copolymer, 4-vinylpyridine. The nitrogen atom of the pyridine component is oxidized to form the corresponding amine oxide polymer.

It is particularly preferred to react an oxidizing agent with a polymer which itself is the olefin polymerization product of:

(A) a nitrogen amine free ester of a C_{1-6} olefinically unsaturated aliphatic mono-, di- or polycarboxylic acid; and

(B) an olefinically unsaturated comonomer containing a basic nitrogen atom in a side chain, especially a tertiary nitrogen atom.

This amine oxide polymer is especially useful in a fuel as a detergent. Generally speaking, the detergent is present in the fuel in an amount of between 0.005 and 1 weight percent, preferably 0.05 to 0.75 weight percent.

Example 1: A nitrogen-containing polymer prepared from 4-vinylpyridine, butylmethacrylate and a mixture of long chained esters of methacrylic acid was prepared by the following technique. Into a 1 ℓ resin kettle fitted with external heating, thermometer, nitrogen inlet, stirrer and condenser, there was charged 54.5% by weight of Neodol 25L methacrylate (a C_{12-15} side chain ester of methacrylic acid having an average chain length of 13.6 carbon atoms), 16.5 weight percent of Alfol 1620 methacrylate (a C_{16-20} side chain ester of methacrylic acid having an average ester chain length of 17.1 carbon atoms), 20% by weight butylmethacrylate and 9% by weight 4-vinylpyridine.

The reactants were admixed with n-dodecylmercaptan and heated to 95°C, under a nitrogen blanket in the presence of azobisisobutyronitrile as polymerization initiator. Polymerization was permitted to continue until the polymerization was substantially complete. The polymerization was effected for a period of about 6 hr at 95° to 100°C.

Example 2: A mixture of 100 g of a polymer having a molecular weight of 2,600 prepared in accordance with the procedure set forth above, 40 g of acetic acid and 30 ml of 30% hydrogen peroxide were stirred together at 70°C for 20 hr. At the end of the heating time, the acetic acid and peroxide were removed by reduced pressure distillation to yield an amine oxide product.

Example 3: In a procedure analagous to Example 1, a polymer having a molecular weight of 2,900 was prepared from 20 weight percent butylmethacrylate, 54.5% Neodol 25L methacrylate, 16.5% Alfol 1620 methacrylate and 9% dimethylaminoethyl methacrylate.

Example 4: 150 g of the polymer of Example 3 were mixed with 60 g of acetic acid and 30 ml of 30 weight percent hydrogen peroxide and stirred together at 75°C for 20 hr. The acetic acid and peroxide were removed by reduced pressure distillation to yield the amine oxide product.

In order to evaluate the ability of the amine oxide polymer to remove deposits from a carburetor, a Chevrolet Carburetor Detergency Test, Phase III, was performed.

Set forth below are the values obtained using an amine oxide copolymer in accordance with this formulation vis-a-vis the basic polymer in the nonoxide form.

Chevrolet Carburetor Detergency Test, Phase III

Additive 100 PTB Conc.	.... Deposit (mg)..... Buildup	Removal	Effectiveness (%)
Example 1 (nonoxide)	18.8	13.4	71
Example 2 (amine oxide polymer)	23.3	16.4	70
Example 3 (nonoxide)	19.0	21.6	66
Example 4 (amine oxide polymer)	27.5	21.6	79

Quaternary Ammonium Salt of a Copolymer

P.F. Vartanian and J.B. Biasotti; U.S. Patent 4,183,732; January 15, 1980; assigned to Texaco Inc. have, as a result of further research, combined the features of the previous two patents to produce a fuel additive which exhibits considerably improved detergency. This improved utility is provided by a quaternary ammonium salt of a copolymer which is the olefin polymerization product of:

(A) a nitrogen amine-free ester of a C_{1-6} olefinically unsaturated aliphatic mono, di- or polycarboxylic acid having a tertiary nitrogen atom; and

(B) an olefinically unsaturated comonomer containing a basic amino nitrogen in a side chain.

The quaternary ammonium salt of this process is particularly useful in gasoline, where it not only removes deposits from an engine carburetor, but actually prevents deposit buildup on the components of such carburetor. Generally speaking, the detergent is present in the fuel in an amount of between 0.005 and 1 weight percent, preferably 0.05 to 0.75 weight percent.

A nitrogen-containing polymer is prepared from 4-vinylpyridine butylmethacrylate and a mixture of long-chain esters of methacrylic acid as shown in Example 1 of the previous patent.

100 g of the polymer so-prepared were mixed with 12 g of methyl iodide and 150 ml of benzene in a 500 ml round bottom three neck flask fitted with a reflux condenser and a motor driven stirrer. The mixture was heated at reflux for 30 min and the benzene solvent thereafter was removed by distillation under pressure. The resultant quaternary ammonium salt of the nitrogen-containing polymer was thereafter evaluated for its ability to prevent deposit buildup on the components of an automotive carburetor. To test the same, the quaternary ammonium salt was evaluated in accordance with the 1973 Buick Carburetor Detergency Test. The test cycle is representative of normal road conditions. Approximately 300 gal of fuel and 3 qt of oil were required for each run.

The quaternary ammonium salt of the polymer was evaluated against the same polymer which had not been formed into a quaternary ammonium salt form, and the data showed that the quaternary ammonium salt is at least twice as effective as the nonquaternary ammonium salt of the same polymer.

In order to evaluate the ability of the quaternary ammonium salt to remove deposits on a carburetor, a Chevrolet Carburetor Detergency Test Phase III, was performed.

The results obtained in two runs are averaged and the effectiveness of the additive fuel in removing deposits is expressed in percent. Set forth below are the values obtained for using a quaternary ammonium salt of a copolymer containing a basic amino nitrogen vis-a-vis nonquaternary ammonium salts of the same polymer.

Additive	. . . Deposit (mg). . . .		Effectiveness
	Buildup	Removal	(%)
(A) Base fuel	16.8	-1.7	-10
(B) Quaternary ammonium salt of polymer of 9% 4-vinylpyridine	17.3	12.1	71
(C) Nonquaternary ammonium salt of polymer of Additive B	18.8	13.4	71
(D) Quaternary ammonium salt of polymer of 9% dimethylaminoethyl methacrylate	18.6	13.1	70
(E) Nonquaternary ammonium salt of polymer of Additive D	19.0	12.5	66

While perhaps less dramatic than the data obtained in the 1973 Buick Carburetor Detergency Test, it is evident that these quaternary ammonium salts are generally more effective in removing deposits on a Chevrolet carburetor than the basic polymers themselves. In the test, the base fuel was a commercially available premium base fuel. The additives were supplied to the fuel in an amount of 100 PTB of fuel. The base polymers comprise 20% by weight butylmethacrylate, 54% by weight Neodol 25L methacrylate, 16.5% Alfol 1620 methacrylate and 9% of the indicated nitrogen compound.

Inner Quaternary Ammonium Salts of Copolymers

In another variation, *P.F. Vartanian and J.B. Biasotti; U.S. Patent 4,201,554; assigned to Texaco Inc.* have provided as gasoline additives to promote detergency inner quaternary ammonium salts of copolymers. These compounds are reaction products of polymers having an amine group and an α,β-unsaturated C_{3-6} aliphatic carboxylic acid.

The polymers are those already described in the two previous patents. The process may be more readily understood through reference to the following equation:

$$+\ CH_2{=}CH{-}COOH \longrightarrow$$

N

Methacrylate-4-vinyl-pyridine polymer

$N^{\oplus}$ $^{\ominus}O$

$CH_2{-}CH_2{-}C{=}O$

In the polymer above, the amine group is supplied employing 4-vinylpyridine as copolymer in the preparation of the basic polymer. It can be seen that an inner quaternary ammonium salt is formed by an unusual and unpredictable reaction.

Particularly contemplated α,β-unsaturated C_{3-6} aliphatic carboxylic acids are: maleic acid, acrylic acid, methacrylic acid, butenic acid and fumaric acid.

A nitrogen-containing polymer prepared from 4-vinylpyridine, butylmethacrylate and a mixture of long chained esters of methacrylic acid was prepared by the technique described in U.S. Patent 4,179,271, abstracted on p 16 of this volume.

Example 1: A mixture of 100 g of the polymer abovedescribed having an average molecular weight of 2,600 and 13 g of maleic acid were heated at 190°C for 2 hr. There was recovered a product which was the reaction product of the maleic acid and the nitrogen-containing polymer.

Example 2: Example 1 was repeated except that instead of reacting the basic polymer with maleic acid, 100 g thereof were reacted with 25 g of acrylic acid in the presence of 0.4 g of hydroquinone (to inhibit polymerization of the acrylic acid). The reaction was conducted at 174°C for 3 hr and excess acrylic acid was then removed by reduced pressure distillation to yield the desired product.

In order to evaluate the ability of the inner quaternary ammonium salt to remove deposits on a carburetor, a Chevrolet Carburetor Detergency Test, Phase III was performed.

Set forth below are the values obtained using an inner quaternary ammonium salt of a copolymer in accordance with the process.

	. . . . Deposit (mg). . . .		Effectiveness
Additive, 100 PTB Conc.	Buildup	Removal	(%)
Example 1	29.6	16.3	55
Example 2	29.9	22.8	76

To evaluate the ability of the inner quaternary ammonium salt to prevent deposit build-up on an automotive carburetor, a Buick Carburetor Detergency Test was conducted. The inner quaternary ammonium salt showed itself to be almost twice as effective in this test as the corresponding polymer in noninner quaternary ammonium salt form.

Alkyl- or Hydroxy-Substituted Benzyl Polyamines

Certain condensation products of an alkylphenol, an aldehyde and an amine, commonly known as Mannich Bases, are effective carburetor detergents in gasoline. In common with most other carburetor detergents, however, these Mannich Bases can cause significantly increased intake valve deposits, particularly when used in low concentrations.

B.H. Garth; U.S. Patent 4,172,707; October 30, 1979; assigned to E.I. Du Pont de Nemours & Company describes a class of Mannich Bases which provide carburetor detergency. In contrast to the Mannich Bases of the art, these compositions at low concentrations have a minimal effect upon the intake valve deposit weight increase and at high concentrations provide control or even reduction of the deposit weights. These compositions have the formula:

$$ZNHC_3H_6N\begin{matrix} \diagup C_nH_{2n} \diagdown \\ \diagdown C_nH_{2n} \diagup \end{matrix}N(C_3H_6)_x(C_2H_4)_y(NH)_cZ$$

wherein Z is an alkyl- and hydroxy-substituted benzyl group wherein the alkyl has 20 to 1,000 carbon atoms; n is 2; x, y, and c are each 0 or 1; x + y = 0 or

1; with the proviso that when x is 0 and y is 0, then c is 0; when x is 1 and y is 0, then c is 1 and when x is 0 and y is 1, then c is 1. Preferred compositions are those wherein the alkyl group in Z has 50 to 200 carbon atoms, most preferably 50 to 75 carbon atoms.

Gasoline compositions may be made which contain from about 0.005 to 0.06 weight percent (12.5 to 150 pounds per thousand barrels) of these amine compositions of this process. The amine compositions can also be used in lubricating oils to provide detergency. Normally, the lubricating oils will contain from about 0.5 weight percent to 15 weight percent of the amine.

The amine compositions are preferably prepared by the Mannich reaction wherein an alkylphenol, a suitable amine and formaldehyde are mixed and heated to a temperature in the range of about 80° to 200°C, for a time sufficient for the reaction to occur. While not required, an acidic catalyst such as hydrochloric acid or sulfuric acid can be used. The reaction mixture is kept at the reaction temperature until sufficient water of condensation has been evolved and removed.

The reaction can be carried out in the absence of a solvent but it is preferable to use one, especially one which distills with water azeotropically, Suitable solvents are hydrocarbons boiling in the gasoline boiling range of about 32° to 205°C and include among others: hexane, cyclohexane, n-octane, isooctane, benzene, toluene, xylene and mixtures thereof. The amount of solvent used is not critical but when used it is usually employed in an amount of about 10 to 90% by weight of the total reaction mixture.

The use of a solvent not only facilitates the reaction but some or all of the solvent can be retained with the reaction product to provide the additive composition as a solution. Such a solution makes handling and incorporation of the composition into gasoline easier. The solution will preferably contain 50 to 75% by weight of Mannich Bases in a single solvent or a mixture of solvents such as those listed above.

The alkylphenol useful in the preparation of the composition is preferably a monoalkylphenol where the alkyl group has 20 to 1,000 carbon atoms. Where the compositions are intended to provide control of quick-heat intake manifold deposits it is preferred, although not necessary, that the alkylphenol have an alkyl group of 50 to 1,000 carbon atoms and that at least about 60% of the alkyl group be located para to the phenolic hydroxyl group. Such an alkylphenol can be prepared by the alkylation of phenol with a monoolefin of 50 to 1,000 carbon atoms using boron trifluoride catalyst and a reaction temperature below about 65°C, preferably in the range of 40° to 50°C. The most preferred olefins are homopolymers of propylene having 50 to 75 carbon atoms.

Included among the suitable amines are alkylene polyamines wherein there are at least three amino-nitrogens and each nonterminal amino-nitrogen is a tertiary amino-nitrogen, as in diethylenetriamine, triethylenetetramine, tetraethylenepentamine, etc. where each nonterminal amino-nitrogen has an alkyl substituent of 1 to 20 carbon atoms.

Preferred polyalkylenepolyamines are di-1,3-propylenetriamines of the formula $H_2N{-}(CH_2)_3{-}N(R){-}(CH_2)_3{-}NH_2$ where R is methyl, ethyl, propyl, or butyl.

Also among suitable amines are those containing a piperazine ring where n = 2 in the ring structure.

Representative examples of such amines are: N-aminoethylpiperazine, N,N'-bis(aminoethyl)piperazine, N-aminopropylpiperazine and N,N'-bis(aminopropyl)piperazine. All of the suitable amines discussed above are available commercially or can be prepared readily by methods well-known in the art.

Example: *Preparation of* $ZNHC_3H_6NCH_3C_3H_6NHZ$ – A 75% by weight solution in toluene of polypropylenephenol (prepared from polypropylene MW 840 and phenol), 300 g, methyliminobispropylamine, 14.5 g, and 36% by weight aqueous formaldehyde solution, 17 g, were placed in a reaction flask equipped with an agitator, a reflux condenser and a Dean-Stark water separator. The reaction mass was refluxed for 11 hr during which time 14.5 ml of water was collected and removed. Infrared analysis of the reaction mixture indicated that the polypropylenephenol had reacted. An additional 21.5 g of toluene was added to the reaction mixture to provide the reaction product as a 70% by weight solution in toluene.

Tests performed on gasoline mixtures containing 14 PTB of the compound of the example showed that this compound is an effective carburetor detergent. When used at a rate of 70 PTB the composition provided reduction in the intake valve deposit over a control fuel. Used at 70 PTB, the compound of the example did not add to the ORI (octane requirement increase) of the fuel.

Two-Part Acylated Polyalkyleneamine/Polyolefin Additive

H.J. Scheule and R.E. Starn, Jr.; U.S. Patent 4,173,456; November 6, 1979; assigned to E.I. Du Pont de Nemours & Co. have developed a multifunctional gasoline additive composition comprising:

(1) a hydrocarbon-soluble, acylated, polyalkyleneamine of the formula

$$R^1-\overset{\overset{O}{\|}}{C}-\overset{\overset{H}{|}}{N}-(C_nH_{2n}\overset{\overset{R}{|}}{N})_xH$$

wherein
R is selected from at least one of H and R'–C(O)–;

R^1 is C_{9-23} hydrocarbyl;

n is 2, 3, or a mixture of 2 and 3; and

x is 2 to 6; and

(2) from about 1 to 10 parts per part of the polyalkyleneamine of a normally liquid, hydrocarbon-soluble polymer of a C_{2-6} olefin, having an average molecular weight of about 400 to 3,000.

Preferred compositions are those wherein at least two R groups are R'–C(O)–; R' is a C_{9-21} hydrocarbyl; n is 2 or 3; x is 4; and the polymer is polypropylene having an average molecular weight of about 500 to 1,000 in an amount of about 3 to 8 parts of the polymer per part of polyalkyleneamine.

The additive composition comprises from 30 to 75% by weight of the two component additive, (1) and (2), and from 25 to 70% of a hydrocarbon solvent.

Preferably, the polyalkyleneamine is used in the range of 0.0008 weight percent (2 PTB, 2 pounds per thousand barrels) to about 0.02 weight percent (50 PTB) and the hydrocarbon polymer is used in the range of about 0.004 weight percent (10 PTB) to 0.08 weight percent (200 PTB).

The effectiveness of the described composition in keeping a carburetor clean (carburetor detergency) was demonstrated in a carburetor keep-clean test (Onan) carried out in a single cylinder engine to which a controlled amount of exhaust gas from another engine was added with the air supplied to the test carburetor. The test carburetor throat consisted of a two-piece stainless steel liner fitted around the throttle plate shaft. The liner was removed for inspection and rating. The engine was operated under cycling conditions of 1 min idling and 3 min part throttle over a 2 hr test period. A visual rating of 10 for a clean carburetor and 0 for a dirty carburetor was used. A rating of 7 or greater is considered good.

Examples 1 through 5 and Comparisons A and B: The acylated polyalkyleneamine was a triamide of tetraethylenepentamine and tall oil fatty acid which was prepared as follows: tetraethylenepentamine was reacted with tall oil fatty acid in the molar ratio of 1 mol of tetraethylenepentamine to 3 mols of acid; the temperature was maintained in the range of about 65° to 95°C. Characterization of the triamide of tetraethylenepentamine prepared as described would indicate the presence of amide groups and the absence of imidazoline groups. A detailed description of one process for making a triamide of tetraethylenepentamine which is the same as that employed in this example, is found in U.S. Patent 3,894,849.

The triamide was formulated into a gasoline additive composition (Example 1) which contained 13.6% by weight of the triamide, 50% by weight of polypropylene of 800 MW, 1.0% by weight of a dehaze agent based on oxyalkylenated alkylphenol, 1.1% by weight of a corrosion inhibitor based on long-chain carboxylic acids and 34.3% by weight of xylene solvent. For comparative purposes, a gasoline additive formulation was prepared (Comparison Composition) containing 48.5% by weight of the abovedescribed triamide; 7.3% by weight of the abovedescribed dehaze agent; 8.0% by weight of the above corrosion inhibitor and, as solvents, 112.2% by weight of isopropanol and 24.0% by weight of xylene. The corrosion inhibitor, dehaze agent and solvent do not affect carburetor detergency and so are not particularly significant in these comparative evaluations.

Amounts of the composition of Example 1 and the Comparison Composition were used in the fuel to give the compositions described as Examples 2 to 5 and Comparisons A and B. Comparative carburetor detergency test results are shown.

Example	Additive	Additive	Triamide (PTB)	Polypropylene	Onan Rating
–	Base fuel only	–	–	–	4.5
–	Polypropylene MW = 800	12.5	–	12.5	5.6
A	Comparison	5	2.5	–	6.2
B	Comparison	7	3.4	–	6.8
2	Example 1	10	1.4	5	7.2
3	Example 1	15	2.0	7.5	7.2
4	Example 1	17.8	2.4	8.7	8.2
5	Example 1	25	3.4	12.5	8.6

Testing for antiicing characterictics showed that polypropylene enhances the performance of the acylated polyalkyleneamine considerably more than expected.

Reaction Products of Isatoic Anhydride and N-Alkyl Trimethylene Diamines

The process described by *R.L. Sung and R.W. Von Allmen; U.S. Patent 4,177,041; December 4, 1979; assigned to Texaco Inc.* relates to reaction products of isatoic anhydride and N-alkyl trimethylene diamines consisting of anthranilamides and ureas useful as carburetor detergents in gasoline.

The compounds essentially consist of a mixture of products obtained by refluxing in an inert solvent such as dimethylformamide, isatoic anhydride (1) and an N-alkyl trimethylene diamine (2) and then separating the resulting products. The preparation of these products is illustrated theoretically by the following general equation:

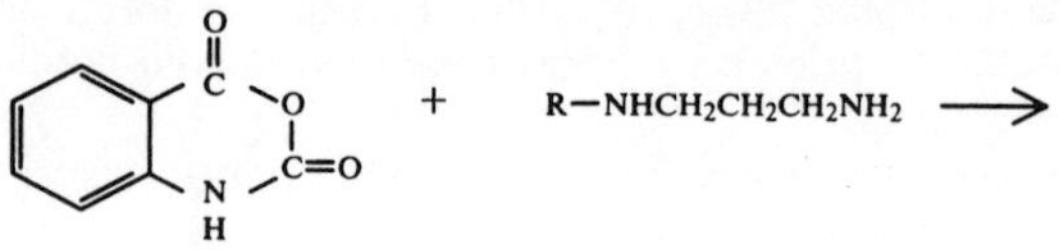

$$\text{(isatoic anhydride)} \quad + \quad R{-}NHCH_2CH_2CH_2NH_2 \longrightarrow$$

(1) (2)

$$C_6H_4(COOH)(NH{-}C(=O){-}NH(CH_2)_3NHR) \quad \text{or} \quad C_6H_4(C(=O){-}NH(CH_2)_3NHR)(NH_2)$$

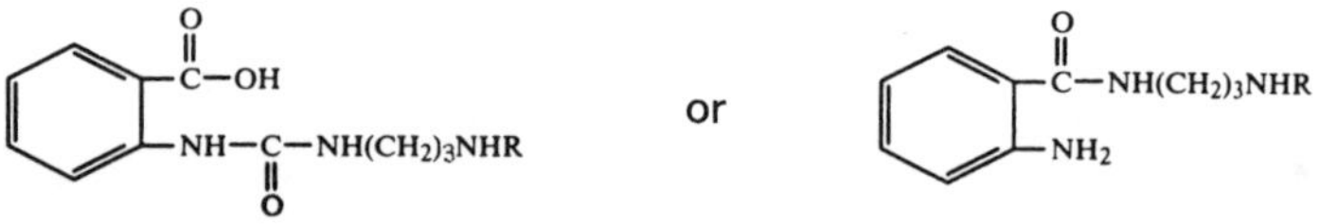

(3) (4)

R is a $C_{6\text{-}18}$ alkyl group derived from a fatty acid and the diamine preferably is one of the Duomeens ("C," "S," "T," or "O") produced by Armour Chemical Co. These compounds are N-coco-, N-soya-, N-tallow-, or N-oleyl-1,3-diaminopropanes.

As for the products, (3) is an anthranilamide of a Duomeen, and (4) is N,N'-(o-carboxyphenyl)-Duomeen urea. A mixture of compounds (3) and (4) may be incorporated by any suitable blending in the gasoline. A particularly preferred concentration of the additive is from 0.02 to 0.1 wt % [from 50 to 250 PTB (pounds of additive per 1,000 barrels of gasoline)].

Polyamine Derivatives of Oxidized Olefinic Substituted Dicarboxylic Acid Compounds

The objectives of *J.B. Hanson; U.S. Patent 4,203,730; May 20, 1980; assigned to Standard Oil Company of Indiana* are to provide a class of highly active gasoline detergents which will give excellent carburetor and engine cleanliness at low treatment levels and which during operation of the engine leave minimum deposits on carburetor and valve surfaces.

These objectives are accomplished by reacting a polyamine with an oxidized substantially undegraded olefinic substituted dicarboxylic acid compound. The oxidation of the olefinic substituted dicarboxylic acid compound is believed to introduce oxygen containing polar functional groups into the olefinic substituent by the reaction of at least one olefinic bond with an oxidizing agent without causing substantial degradation of the olefin substituent. These oxygen-containing groups apparently add polarity and appear to cause the improved detergency of the polyamine derivative of the oxidized dicarboxylic acid compound.

Oxidizing agents which can be used to oxidize the unsaturated substituted dicarboxylic acid compound are the conventional ones. Examples include oxygen, air, sulfur oxides such as sulfur dioxide, sulfur trioxide, etc., nitrogen oxides including nitrogen dioxide, nitrogen trioxide, nitrogen pentoxide, etc., peroxides such as hydrogen peroxide, sodium peroxide, percarboxylic acids and ozone. Air, air with added oxygen or diluted air with reduced oxygen concentration containing less than the naturally occurring amount of oxygen are the preferred agents for reasons of economy, availability, and safety.

The olefinic substituted dicarboxylic acid compounds can be prepared by the reaction of an olefinic compound and an unsaturated dicarboxylic acid compound. The unsaturated dicarboxylic acid compounds are compounds containing from about 4 to about 10 carbon atoms containing two carboxyl functions either free or in an anhydride form and at least one unsaturated olefinic bond. Maleic anhydride, maleic acid, and fumaric acid are preferred for reasons of high reactivity and low cost.

Olefinic compounds used to prepare the olefinic substituted dicarboxylic acid compound are substantially hydrocarbon compounds containing from about 20 to 400 carbon atoms containing 1 to 4 double bonds.

The preferred olefinic substituted dicarboxylic acid compound for reasons of economy and ease of oxidation is a polyalkenyl succinic acid or anhydride having a molecular weight from about 250 to 6,000, selected from the group consisting of polyisobutenyl succinic anhydride, polybutenyl succinic anhydride and polypropenyl succinic anhydride.

Examples of suitable polyalkylene polyamines are ethylenediamine, hexamethylenediamine, diethylenetriamine, triethylenetetramine, tetraethylenepentamine, and tripropylenetetramine. Other polyamines which are useful are the Duomeens having the formula $R-NH(CH_2)_3NH_2$ wherein R is a hydrocarbyl group having from about 2 to 25 carbon atoms. The additive can be blended in gasoline at an amount from about 0.1 to about 1,000 PTB or 0.38 to 3,800 ppm based on the gasoline.

The performance of products of this type was evaluated by a carburetor detergency test procedure and a turbine oil rust test. It was found that both the carburetor detergency and the rust inhibition of the diethylenetriamine derivatives of the oxidized polyalkenyl succinic anhydride are superior in carburetor detergency to either the unoxidized product or the base fuel containing no additives.

Glycol Polyether-Acrylic Acid-Amine Reaction Product

W.M. Cummings; U.S. Patents 4,210,425; July 1, 1980; and 4,251,670; Feb. 17, 1981; both assigned to Texaco Inc. has developed an oil-soluble ashless detergent and corrosion inhibitor for hydrocarbon fuels and lubricant compositions.

These inhibitors are the reaction products of a glycol polyether or polypropylene glycol represented by the formula:

$$\text{(1)} \qquad HO-\underset{H}{\overset{CH_3}{\overset{|}{\underset{|}{C}}}}-CH_2-O(CH_2-\underset{H}{\overset{CH_3}{\overset{|}{\underset{|}{C}}}}-O)_n-CH_2-\underset{H}{\overset{CH_3}{\overset{|}{\underset{|}{C}}}}-OH$$

in which n represents a number from about 10 to 90 with acrylic or methacrylic acid, and an amine, represented by the formula RNHR' in which R is a monovalent hydrocarbon radical having from about 4 to 30 carbon atoms or a radical selected from the group represented by the formulas: $-(CH_2CH_2NH)_x-H$ or $-(CH_2CH_2CH_2NH)_y-H$ in which x and y are integers from 1 to 6 and R' is hydrogen or a monovalent hydrocarbon radical having from 1 to 8 carbon atoms.

In formula (1) n represents a number preferably from about 13 to 80 and has a value to provide a glycol polyether having a molecular weight ranging from about 700 to 5,500, most preferably from about 1,100 to 2,500. The acrylic acid component reacts with the glycol polyether to form an acrylic ester of the glycol polyether. It is preferred to employ from about 1 to 2 mols of the acrylic acid per mol of the glycol polyether.

This reaction is conducted by admixing the reactants and maintaining them at a temperature ranging from about room temperature up to a temperature below the decomposition temperature of the reactants and the reaction product.

It is preferred to dissolve the reactants in an inert hydrocarbon and to employ a catalyst to speed up the reaction. p-Toluene sulfonic acid is particularly effective for promoting the noted reaction. A small amount of an oxidation inhibitor, such as hydroquinone, is also useful for reducing the formation of undesirable by-products. It is preferred to effect the reaction at the reflux temperature of the solvent employed which, in the case of the solvent xylene is about 140°C. On completion of the preparation of the intermediate reaction product, the reaction mixture is cooled and the solvent is removed under reduced pressure and the intermediate reaction product is reacted with the amine.

Examples of suitable amines include butylamine, hexylamine, 2-ethylhexylamine, decylamine, tetradecylamine, octadecylamine, N-ethyl-N-decylamine, N-ethyl-N-dodecylamine, dibutylamine, diamylamine, dihexylamine, di(2-ethylhexyl)amine, N-ethyl-N-2-ethylhexylamine. The preferred primary amines are those in which the hydrocarbon radical has from 6 to 18 carbon atoms.

Suitable polyamines include ethylenediamine, 1,3-propanediamine, diethylenetriamine, triethylenetetramine, tetraethylenepentamine, pentaethylene-hexamine, N-oleyl-1,3-propylenediamine, N-lauryl-1,3-propylenediamine, N-tallow-1,3-propylenediamine, N-stearyl-1,3-propylenediamine, N-tallow-1,3-ethylenediamine and N-soya-1,2-ethylenediamine.

The preferred general mol ratios for combining the glycol polyether, acrylic acid and amine can be expressed as in the range of about 1:1-2:1-2 respectively with the acrylate ester group and the amine being present in approximately equimolar amounts. The most preferred mol ratio for the components is about 1:1:1 respectively. The following examples illustrate the preparation of the intermediate reaction product from the reaction of a glycol polyether and an acrylic acid.

Example 1: 1,000.0 g of the polyether of glycol represented by the formula:

$$HO-CH_2-\underset{H}{\overset{CH_3}{\underset{|}{\overset{|}{C}}}}-CH_2-O(CH_2-\underset{H}{\overset{CH_3}{\underset{|}{\overset{|}{C}}}}-O)_n-CH_2-\underset{H}{\overset{CH_3}{\underset{|}{\overset{|}{C}}}}-OH$$

where n is 13 to 15, having a molecular weight of about 1,200 was dissolved in 100 ml of xylene. 60 g of acrylic acid, 10.0 g of p-toluenesulfonic acid and 2.0 g of hydroquinone were added to the xylene solution of the reactant. This mixture was heated for about 4 hours at a reflux temperature of about 140°C. The reaction mixture was cooled and the solvent removed under reduced pressure to yield 1,050 g of product having the following analysis: Saponification No. D-94, 42.2; Neutralization No. D-974, 3.8; and Hydroxyl No., 57.

Example 2: 200 g of a glycol polyether represented by Formula (1) in which n is 22 to 24, having a molecular weight of about 2,000, was dissolved in 100 ml of xylene. 7.5 g of acrylic acid, 2.0 g of p-toluenesulfonic acid and 0.5 g of hydroquinone were added to the xylene solution. This mixture was heated at a reflux temperature of about 140°C for 4 hours. The mixture was then cooled and the solvent removed under reduced pressure to yield 207 g of product having the following analysis: Saponification No., 26.7; Neutralization No., 3.66; and Hydroxyl No., 36.

Example 3: 1,000 g of the glycol polyether of Example 2 were reacted with 40.0 g of acrylic acid following the procedure of Example 2. 1,041 g of product were recovered having the following analysis: Saponification No., 23; Neutralization No., 4.72; and Hydroxyl No., 34.

Example 4: 127.0 g of the glycol polyether-acrylate ester from Example 1 above was reacted with 27.0 g of tallowamine heated to about 120°C and held at this temperature for 1 hour. 147.4 g of product were recovered having the following analysis: Saponification No., 28.5; Neutralization No., 2.06; nitrogen, 0.90%; and Total Base No., 33.9.

Example 5: 300 g of the glycol polyether-acrylate ester from Example 3 above, were reacted with 40.0 g of tallowamine following the procedure of Example 4 above. 339 g of product were recovered having the following analysis: Saponification No., 20.4; Neutralization No., 3.94; nitrogen, 0.65%; and Total Base No., 32.2.

The reaction product is useful as a detergent and/or a corrosion inhibitor for hydrocarbon fuels, mineral oils and lubricating oil compositions. In general, the reaction product of the process is effective as a carburetor detergent and/or inhibitor in a motor fuel composition comprising a mixture of hydrocarbons in the gasoline boiling range, i.e., from about 90° to 425°F. Broadly effective concentrations as detergent and corrosion inhibitors in fuels range from about 0.0005 to 0.20 wt % of the reaction product based on the weight of the gasoline composition. A preferred concentration of the reaction product in gasoline is an amount ranging from about 0.002 to 0.01 wt % which corresponds to about 5 and 25 PTB respectively.

In oil compositions the reaction product can be employed at concentrations ranging from about 0.0005 to 10 wt % with a preferred concentration being from 0.01 to 5 wt % based on the total weight of the oil composition.

The effectiveness of the reaction product as a carburetor detergent was tested in a Chevrolet Carburetor Detergency Test designed to remove preformed deposits from the throttle plate area of a carburetor. The base fuel employed in this test was a premium grade gasoline having a Research Octane Number of about 97 and contained about 3.0 cc of tetraethyl lead per gallon. This gasoline consisted of about 30.5% aromatic hydrocarbons, 8.5% olefinic hydrocarbons and 61.0% paraffinic hydrocarbons and boiled in the range from about 89° to 356°F.

The results obtained in the Chevrolet Carburetor Detergency Test using the above-described base fuel and the indicated additives are set forth in the table below.

Chevrolet Carburetor Detergency Test

Run	Additive Concentration	Effectiveness %
1	Base fuel + 100 PTB of 1,200 MW glycol polyether	-26
2	Base fuel + 100 PTB of Example 4	+23
3	Base fuel + 150 PTB of Example 5	+27

Run 1 consisting of the base fuel and a 1,200 molecular weight glycol polyether possesses significant carburetor detergent properties. A nonadditive fuel would be expected to be about 40% less effective than the reference fuel of Run 1. In contrast, Runs 2 and 3 demonstrated a substantial improvement in the carburetor detergency of a motor fuel containing the reaction products of this process.

High Molecular Weight Mannich Bases

This process pertains to improved gasoline hydrocarbon fuels containing a detergent additive capable of substantially removing and preventing buildup of deposits on carburetor surfaces and intake valve systems in a gasoline-powered engine system.

J.H. Udelhofen and R.W. Watson; U.S. Patent 4,231,759; November 4, 1980; assigned to Standard Oil Company (Indiana) provide a liquid hydrocarbon combustion fuel containing an amount sufficient to impart improved detergency and antirust properties thereto of an additive composition comprising the Mannich condensation product of (1) a high molecular weight alkyl-substituted hydroxyaromatic compound wherein the alkyl has a molecular weight of from about 600 to about 3,000, (2) an amine which contains an HN< group and, (3) an aldehyde, wherein the respective molar ratio of reactants is 1:0.1-10:0.1-10. The Mannich condensation product may be employed alone where carburetor cleanliness is desired or in combination with a suitable essentially nonvolatile hydrocarbon as a carrier fluid where intake valve cleanliness is also desired.

Preferably, the alkyl-substituted hydroxy aromatic compound is polypropylphenol or polybutylphenol; the amine is dimethylamine, dimethylaminopropylamine; tetraethylenepentamine, triethylenetetramine or diethylenetriamine; and the aldehyde is formaldehyde or p-formaldehyde.

Example 1: Mannich condensation products were prepared by the reaction of selected alkylphenols and alkylene polyamines with aqueous formaldehyde substantially according to the following generalized procedure. To a mixture of X mols of alkylphenol and Y mols of amine, heated to about 180° to 200°F, was charged Z mols of aqueous formaldehyde over a period of 30 minutes while maintaining the temperature of the mixture below 200°F. The mixture temperature was maintained at 180° to 200°F with stirring for an additional 30 minutes. Xylene solvent was optionally added. The reaction mixture was then heated to 300° to 350°F and held at the elevated temperature for 2 hours while blowing with an inert gas to assure removal of all water. The reaction product was then cooled, filtered and diluted with xylene to provide a concentration level of 40 to 50 wt % active Mannich product. Products were prepared substantially as set forth above to have essentially the compositions listed in the table below.

Experiment Number	. . . . . Alkyl Phenol . . . . Alkyl Group (MW)	Mols (X)	. . . . Amine . . . . Diamine	Mols (Y)	Formaldehyde Mols (Z)
1	Polybutyl (1,500)	1	Tetraethylene pentamine	1	2
2	Polybutyl (1,500)	1	Pentaethylene hexamine	2	2
3	Polybutyl (900)	1	Ethylenediamine	1	3
4	Polybutyl (900)	1	Diethylene triamine	1	3
5	Polybutyl (900)	1	Aminoethyl aminoethanol	2	2
6	Polybutyl (350)	1	Ethylenediamine	1	1
7	Polybutyl (250)	1	Aminoethyl aminoethanol	1	1
8	Polypropyl (850)	1	Diethylene triamine	1	3
9	Polypropyl (600)	1	Diethylene triamine	1	3

Example 2: Detergent performance of a series of additive formulations at 3 or 6 PTB, employing Mannich products prepared as described in Example 1, was measured by weighing and also visually rating both a thin metal specimen, fitted into the carburetor throttle plate bore area, and the throttle plate of a 1965 Ford 6-cylinder engine after 20 hours of continuous operation on MS-08 test fuel. Each performance rating included a measure of total deposit weight (in mg) on the metal insert and the throttle plate as well as a cleanliness rating (10 = clean). Data are presented in the table below.

. . . . Mannich Product . . . . Experiment No.	PTB	Deposit Weight (mg)	Cleanliness (10 = Clean)
1	3	5.7	8.4
1	6	6.0	9.4
2	6	7.8	9.2
4	6	2.2	9.7
5	6	6.9	9.2
6	6	12.0	8.8
7	6	6.9	8.5
Base fuel		19.5	6.7

The desired combination of a low deposit weight and a high cleanliness rating is especially noted when employing the Mannich product as prepared in Experiment 4 of Example 1.

Nitrogen-Containing Copolymers

W.J. Trepka and R.J. Sonnenfeld; U.S. Patent 4,238,202; December 9, 1980; assigned to Phillips Petroleum Co. have developed hydrocarbon fuel compositions that exhibit excellent detergent properties. The detergent properties are obtained by incorporating a small effective amount of an additive which is a nitrogen-containing organic compound-grafted, hydrogenated, conjugated diene/monovinylarene copolymer. These nitrogen-containing copolymers must have a number average molecular weight below about 10,000. For most usages, the preferred range is about 2 to 20 lb (0.91 to 9.1 kg) per 1,000 barrels of fuel.

In the following discussions, for simplicity and convenience, styrene is used as representative of as well as the preferred monovinylarene, and 1,3-butadiene as representative of as well as the preferred conjugated diene. Similarly, butadiene/styrene copolymers are discussed as representative of the applicable conjugated diene/monovinylarene copolymers generally.

Suitable copolymers will contain preferably about 50 to 65% by weight copolymerized styrene, the balance then being copolymerized butadiene. These copolymers preferably are substantially random copolymers, but can contain significant blocks of polystyrene and/or blocks of polybutadiene and/or blocks of random or random tapered butadiene/styrene.

The butadiene/styrene copolymers employed to prepare the additives should have a number average molecular weight in the approximate range of preferably about 5,500 to 8,500. The copolymers, as far as the polymerized butadiene portion, will have a vinyl unsaturation content prior to hydrogenation of about 20 to 95%, preferably about 20 to 70%.

The butadiene/styrene copolymer can be metalated and grafted prior to or following hydrogenation. The metalation step conveniently employs an organolithium compound in conjunction with a polar compound to introduce lithium atoms along the copolymeric structure. Representative organo lithium compounds include, for example: methyllithium, isopropyllithium, sec-butyllithium, n-butyllithium, tert-butyllithium, n-dodecyllithium, pentyllithium, α- and β-naphthyllithium, any biphenyllithium, styryllithium, benzyllithium, any indenyllithium, 1-lithio-3-butene, 1-lithiocyclohexene-3, 1,4-dilithiobenzene, 1,3,5-trilithiopentane, and the like. The polar compound promoters include a variety of tertiary amines, bridgehead amines, ethers, and metal alkoxides.

The lithiated copolymers are treated in solution, and without quenching in any manner to destroy the lithium sites, with a nitrogen-containing organic compound. The nitrogen-containing organic compounds suitable for use in this step can be described by the general formulas $X–Q–(NR^3_2)_n$ or $Y[Q–(NR^3_2)_n]_m$ wherein each R^3 is the same or different alkyl, cycloalkyl, or aryl radicals, or combination thereof; and Q is a hydrocarbon radical having a valence of n + 1 and is a saturated aliphatic, saturated cycloaliphatic, or aromatic radical, or combination thereof. X is a functional group capable of reacting on a one-to-one basis with one equivalent of polymer lithium.

An example gives very detailed instructions for preparing a 41/59 butadiene/styrene block copolymer by n-butyllithium initiated, solution polymerization, the polymer then being metalated and allowed to react with 4-dimethylamino-3-methyl-2-butanone. The nitrogen-containing copolymer was then hydrogenated until no detectable vinyl unsaturation was left. This copolymer was tested as a fuel additive in a described laboratory deposit test.

Briefly, the test gasoline (2 ml/min) was mixed with a flow of air (30 ft^3/hr) to form a gasoline-air mixture. This mixture was discharged from a nozzle as a spray against an aluminum deposit pan of known weight. The deposit pan was preheated to 375°F (190°C) and maintained at that temperature while spraying exactly 250 ml of the base gasoline stock, with and without the additive, onto the surface of the pan. After terminating the gasoline flow, the pan was allowed to cool to room temperature, and the pan then washed twice in boiling heptane and then rinsed with heptane and air dried. Weighing of the dried pan provided the weight of deposits in milligrams per 250 ml of base gasoline stock.

Results obtained using the abovedescribed test on the base gasoline stock without and with 0.125 g of the hydrogenated nitrogen-containing butadiene/styrene copolymer per 250 ml of the base gasoline stock showed a deposit of 28.4 mg for the base gasoline and only 0.3 mg for the base gasoline with the described additive.

Alkenylsuccinimide

The additive developed by *H.J. Andress, Jr.; U.S. Patent 4,240,803; December 23, 1980; assigned to Mobil Oil Corp.* to supply detergency to a liquid hydrocarbon fuel composition is an alkenylsuccinimide prepared by reacting an alkenylsuccinic acid or anhydride, wherein the alkenyl is derived from a mixture of C_{16-28} olefins, with a polyalkylene polyamine of the formula $NH_2-(RNH)_nR-NH_2$ wherein R is alkylene having from 1 to 5 carbon atoms and n is from 0 to 10.

The alkenylsuccinic anhydride can be made in accordance with a prior art process involving the thermal condensation of an olefin mixture with maleic anhydride. This is conveniently carried out at from about 150° to about 250°C, preferably about 175° to 225°C.

It has been found that, for unexpected effectiveness as a liquid hydrocarbon detergent, it is essential that the alkenyl group attached to the succinic acid or anhydride be derived from a mixture of C_{16-28} olefins. This essential olefin mixture is the bottoms from an olefin oligomerization, and such mixture will have the following composition:

Ingredient	Percent by Weight	Other
Olefin (chain length)		
C_{16}	2, max	
C_{18}	5-15	
C_{20}	42-50	
C_{22}	20-28	
C_{24}	6-12	
C_{26}	1-3	
C_{28}	2, max	
Alcohol	10, max	
Paraffin	5, max	
Iodine No.		74, min
Peroxide, ppm		10, max
Olefin types by NMR		
Vinyl	28-44	
Branched	30-50	
Internal	26-42	

Because of the source of the olefin mixture, one does not always get the same product from successive batches. However, each mixture will have a composition falling within the ranges stated and will be equally effective for use in this formulation. While the exact composition is not known, the number average molecular weight can be determined so proportions of reactant can be readily determined.

Suitable polyamines include methylenediamine, ethylenediamine, diethylenetriamine, dipropylenetriamine, triethylenetetramine, tetraethylenepentamine, pentamethylene hexamine, hexaethylene heptamine, undecaethylene dodecamine, and the like. In general, the amount of additive will be from about 0.00001% to about 15% of the additive by weight of the fuel. Expressed in another way the most preferred will range from about 1 lb to about 10 PTB.

Example 1: A mixture of 600 g (2.0 mols) of the abovedescribed olefin mixture and 198 g (2.0 mols) of maleic anhydride was stirred at about 200° to 210°C for 7 hours and at about 235° to 240°C for 3 hours to form the alkenylsuccinic anhydride. A mixture of 170 g (0.9 mol) of tetraethylenepentamine and 500 ml of toluene diluent was added to the alkenylsuccinic anhydride at 75°C. The mixture was gradually refluxed to about 175°C and held until the evolution of water ceased. The final product was obtained by topping under reduced pressure.

Example 2: A mixture of 400 g (1.0 mol) of the alkenylsuccinic anhydride (prepared according to Example 1) and 73 g (0.5 mol) of diethylenetriamine was diluted with 250 ml toluene and refluxed to about 175°C until the evolution of water ceased. The final product was obtained by topping under reduced pressure.

Example 3: A mixture of 349 g (0.5 mol) of alkenylsuccinic anhydride and 47.2 g (0.25 mol) of tetraethylenepentamine was refluxed with toluene diluent to about 200°C until evolution of water ceased. Topping under reduced pressure gave the final product. The alkenylsuccinic anhydride used in this example was made by dimerizing the olefin mixture of Example 1 by adding n-butanol to such mixture and bubbling in boron trifluoride at about 85°C for about 1 hour and then reacting the resulting dimer with maleic anhydride under conditions similar to those described in Example 1.

Example 4: A mixture of 675 g (0.5 mol) of a polyisobutenylsuccinic anhydride and 47.2 g (0.25 mol) of tetraethylenepentamine was stirred at 180°C until the evolution of water ceased. The product represents a typical fuel detergent. The inhibitors were blended in a gasoline comprising 40% catalytically cracked component, 40% catalytically reformed component, and 20% alkylate of about 90° to 410°F boiling range and tested in a Ford Carburetor Detergent test with the following results.

Compound	Concentrate (lb/1,000 bbl)	Reduction in Deposits (%)
Base fuel	0	0
Base fuel + Ex. 1	5	62
	10	79
Base fuel + Ex. 2	10	50
Base fuel + Ex. 3	5	30
	10	60
Base fuel + Ex. 4	10	15

Imides or Amide-Imides of Nitrilotriacetic or Ethylenediaminetetraacetic Acid

A fuel composition has been developed by *H.-H. Vogel, K. Oppenlaender and K. Starke; U.S. Patent 4,242,101; December 30, 1980; assigned to BASF AG, Germany* which is both a highly effective valve cleaner and a carburetor cleaner. This fuel for gasoline engines contains small amounts of both

(a) Imides or amide-imides, or mixtures of imides and amide-imides, obtained from nitrilotriacetic acid and/or from ethylenediaminetetraacetic acid and amines or amine mixtures of 7 to 18 carbon atoms, of the formula (1)

$$(1)\qquad \left[\begin{matrix} XCOCH_2 \\ XCOCH_2 \end{matrix}\!\!>\!N{-}CH_2{-}CH_2\right]_{1-m}\!\!{-}N\!\begin{matrix} CH_2CO \\ {-}CH_2CO{-}N{-}R \\ (CH_2COX)_m \end{matrix}$$

where the radicals X are identical or different radicals –HN–R, or adjacent radicals X are a radical >NR which completes a ring

$$-N\begin{matrix} CH_2CO \\ \\ CH_2CO \end{matrix}N-R,$$

m is 0 or 1 and R is an unbranched or branched aliphatic radical of 7 to 18 carbon atoms, as valve and carburetor cleaners, and

(b) Mixtures of highly hydrogenated petroleum distillates, in the boiling range of the kerosine fraction and highly refined solvent raffinates with viscosities of from 50 to 500 mm^2/s at 40°C, the weight ratio of these constituents being from 20:80 to 80:20, keep the carburetors and valves of gasoline engines clean, thereby achieving better combustion of the fuel in the engine.

The compounds according to (a), of the formula (1), are obtained by conventional methods, for example, by reacting nitrilotriacetic acid or ethylenediaminetetraacetic acid with an amine or mixture of amines of the formula $R{-}NH_2$ at from 150° to 220°C, as a rule from 160° to 200°C. Depending on the desired product, the amine or amine mixture is employed in a molar ratio of 2:1 (giving the cyclic diimide) or of 3:1 (giving the amide-imide) relative to ethylenediaminetetraacetic acid, or in a ratio of 2:1 (giving the amide-imide) relative to nitrilotriacetic acid, or in slightly larger amounts than this. In this way, predominantly amide-imides or imides are obtained in every case, in addition to minor amounts of amides, i.e., where all the carbonyl groups are substituted by one amide radical each.

Specific examples of suitable amines are the following, in which the alkyl radicals may be interrupted by nitrogen or oxygen: 2-ethylhexylamine, n-dodecylamine, n-tridecylamine, n-pentadecylamine, stearylamine, 2-amino-5-dimethylaminopentane and 1-(2-ethylhexoxy)propyl-3-amine.

Examples of conventional carburetor cleaners (c) suitable for use in combination with the additives (a) and (b) according to the method are amides of C_{12-20}-fatty acids with polyamines of 2 to 4 nitrogen atoms and 2 to 8 carbon atoms, e.g., the diamides derived from diethylenetriamine and 2 mols of oleic acid, from dipropylenetriamine and 2 mols of stearic acid, from diethylenetriamine and 2 mols of palmitic acid and from methyldipropylenetriamine and 2 mols of lauric acid. Further suitable reaction products are those of the acid with aminoethylpropylenediamine or with bisaminopropylpropylenediamine. The carburetor cleaners (c) are as a rule employed in a weight ratio, relative to the sum of components (a) and (b), of 1:0.01-0.2.

The fuel composition may also include (d) alkyl-substituted sterically hindered phenols as antioxidants and (e) residues from an Oxo alcohol synthesis, carried out with C_{3-5}-olefins as a solubilizing agent.

Testing showed that the fuel additives—particularly components (a) and (b)—when used in engine testing showed excellent valve and carburetor cleanliness after 40 hours running. Tests also showed that when the materials (a) and (b) were added to a fuel after deposition on valves and carburetor had already been made by using a fuel with no additives, considerable cleaning action took place in removing the deposit.

Polyoxyalkylene Polyamines

W.K. Langdon; U.S. Patent 4,261,704; April 14, 1981; assigned to BASF Wyandotte Corp. has developed a process in which amine-based ethers are prepared by reacting (a) an amine with a (b) halogen-containing polyoxyalkylene polyol or a polyoxyalkylene glycol monoether derived from the reaction of a hydroxyl-containing compound having 1 to 8 hydroxyl groups and a halogen-containing compound. The preferred amine reactant (a) is ethylenediamine.

The amine-based ethers are prepared, generally, by reacting the hydroxyl-containing compound and the halogen source in the presence of a suitable catalyst for about 0.5 to 5 hours at a temperature ranging from about 30° to about 100°C. Thereafter, the amine is reacted with the polyol or monoether at a temperature ranging from about 100° to about 150°C for about 1 to 5 hours at a ratio of at least about 2 mols of amine per chlorine equivalent. Generally, substantial amounts of excess amine, i.e., generally about twice the stoichiometric requirement, is employed to encourage monosubstitution of the polyoxyalkylene moiety on the polyamine. The products generally correspond to the formula:

$$R[(OCH_2\overset{\overset{R_1}{|}}{C}H)_n{-}OCH_2\overset{\overset{OH}{|}}{C}HCH_2Z]_x$$

wherein:

R is the residue, after removal of one or more hydroxyl groups, of a hydroxyl-containing compound having 1 to 8 hydroxyl groups;

R_1 is either H, CH_3 or C_2H_5;

Z is the residue, after removal of one or more amino hydrogens, of an amine selected from the group consisting of NHR''_2 in which R'' is hydrogen or a 1 to 8 carbon alkyl or hydroxylalkyl group, an alkylene diamine or a polyalkylene polyamine;

n is a number ranging from about 1 to about 100; and

x is a number from 1 to 8.

The described polyamines are useful as fuel detergent additives for preventing buildup of carburetor deposits. The preferred oxyalkylated hydroxyl-containing compound comprises a butylene oxide adduct of dodecylphenol. The polyoxyalkylene hydroxyl-containing compound is halogenated by either (a) the direct halogenation thereof or (b) the reaction of the compound with an epihalohydrin to form a corresponding ether. Where direct halogenation is utilized, suitable halogens are either chlorine or bromine.

Example: *(A) Adduct preparation* – Into a 2-ℓ flask was added 1,231 parts (0.6 mol) of a polyoxypropylene glycol of about 2,050 mol weight and 2.5 parts of boron trifluoride etherate as a catalyst. With suitable heating means, the mixture was heated to about 55°C. Thereafter, 12 parts of epichlorohydrin was added to the flask over a 70 minute period. The temperature within the reaction vessel was maintained between 55° and 65°C with a water cooling bath. The reaction mixture was stirred for 2 hours while maintaining the temperature within the range of 55° to 65°C.

Thereafter, 5 parts of sodium bicarbonate was added to the reaction vessel to destroy the catalyst. The product was stripped under reduced pressure at a temperature of 75°C and under a vacuum of <5 mm of mercury (Hg). An insignificant amount of volatiles were stripped off, indicating complete reaction of the epichlorohydrin.

(B) Polyamine synthesis – Into a 3-ℓ, 4-neck flask equipped with agitation means thermometer, addition funnel and reflux condenser was added 600 parts of ethylenediamine. The vessel was then heated to the reflux temperature of 118°C. The adduct of step (A) was then added to the vessel through the addition funnel over a 1 hour period while maintaining the reflux temperature.

During the addition period, the temperature within the vessel rose to 131°C. After the addition was completed, the reaction mixture was stirred for 1 hour at reflux temperature. Thereafter, the reaction mixture was allowed to stand for about 16 hours.

Thereafter, the amine hydrochloride formed in the flask was neutralized by the addition of 300 parts of 17.7% sodium hydroxide. Then, excess ethylenediamine and water were removed from the flask by distillation at a temperature of about 130°C at atmospheric pressure. After the distillation was completed, residual water and amine were removed by vacuum stripping at a temperature of 110° to 120°C at a vacuum of <5 mm of Hg.

After the vacuum stripping was completed, a crude product was recovered. The crude product comprised a mixture of amine-terminated polyether and sodium chloride. The sodium chloride was separated from the crude product by filtration. 1,553 parts of a final product, having a light amber color, was recovered. The liquid product was found, upon analysis, to contain 2.75% of amino hydrogen.

The following patents, covered in other sections of this book, are also noted here, since the additives described therein are useful as detergents as well as corrosion inhibitors, friction reducers, etc.: U.S. Patents 4,153,564; 4,157,243; 4,185,965; 4,216,099; 4,257,780; 4,266,944; and 4,273,891.

ADDITIVES TO PREVENT OCTANE REQUIREMENT INCREASE

In recent years, numerous fuel detergents or "deposit control" additives have been developed. These materials when added to hydrocarbon fuels employed in internal combustion engines effectively reduce deposit formation which ordinarily occurs in carburetor ports, throttle bodies, venturis, intake ports and intake valves. The reduction of these deposit levels has resulted in increased engine efficiency and a reduction in the level of hydrocarbon and carbon monoxide emissions.

A complicating factor has, however, recently arisen. With the advent of automobile engines that require the use of nonleaded gasolines (to prevent disablement of catalytic converters used to reduce emissions), it has been difficult to provide gasoline of high enough octane to prevent knocking and the concomitant damage which it causes. The difficulty is caused by octane requirement increase, herein called "ORI," which is due to deposits formed in the combustion chamber while the engine is operating on commercial gasoline.

The basis of the ORI problem is as follows: each engine, when new, requires a certain minimum octane fuel in order to operate satisfactorily without pinging and/or knocking. As the engine is operated on any gasoline, this minimum octane increases and, in most cases, if the engine is operated on the same fuel for a prolonged period will reach equilibrium. This is apparently caused by an amount of deposits in the combustion chamber. Equilibrium is typically reached after 5,000 to 15,000 miles of automobile operation.

Octane requirement increase measured in particular engines with commercial gasolines will at equilibrium vary from 5 or 6 octane units to as high as 12 or 15 units, depending upon the gasoline compositions, engine design and type of operation. The seriousness of the problem is thus apparent. A typical 1975 or 1976 automobile with a research octane requirement of 85 when new may after a few months of operation require 97 research octane gasoline for proper operation, and little unleaded gasoline of that octane is available. The ORI problem exists in some degree with engines operated on leaded fuels. U.S. Patents 3,144,311 and 3,146,203 disclose lead-containing fuel compositions having reduced ORI properties.

It is believed, however, by many experts that the ORI problem, while present with leaded gasolines, is much more serious with unleaded fuel because of the different nature of the deposits formed with the respective fuels, the size of increase, and because of the lesser availability of high-octane nonleaded fuels. This problem is compounded by the fact that the most common means of enhancing the octane of unleaded gasoline, increasing its aromatic content, also appears to increase the eventual octane requirement of the engine. Furthermore, some of the presently used nitrogen-containing deposit control additives with mineral oil or polymer carriers appear to contribute significantly to the ORI of engines operated on unleaded fuel.

Solid Acid Catalysts of Refractory or Halogenated Metal Oxides

Various studies on automobile engines operating on nonleaded fuel show an octane requirement increase (ORI) of from 5 to 11 research octane numbers (RON), from 2 to 4 RON higher than tests with leaded fuel. Furthermore, these RON values for stabilized octane requirements approached or exceeded an RON value of 100.

Implications of these studies are significant. First, it seems obvious that in view of the large ORI indicated commercially unleaded 91 RON gasolines specified by government regulations will not provide adequate antiknock performance for many automobile engines. This may force the production of unleaded gasolines having significantly higher RON ratings than those now used—an enormous economic and technical task. Further, the production of lead-free gasoline of 96 or greater RON will expend more energy than is realized in gas economy by having such a fuel available. Consequently, any gasoline composition or method which will eliminate the ORI phenomenon or provide octane requirement reduction (ORR) for both leaded and unleaded fuels in the internal combustion engine would be an exceedingly valuable advance in the art.

The enormous economic incentives for minimizing the equilibrium octane requirements of internal combustion engines have led to extensive research efforts in the field. In the gasoline industry, interest has centered mainly on the possibility of an additive that would provide ORR or prevent ORI. Although some of these materials are effective for other functions, such as the polyalkenyl succinimide and amine-containing detergent-dispersant additives, and have thereby reduced deposit formation, none has modified combustion-chamber deposit properties to effect ORR to the appreciable extent required by commercial market conditions.

A.R. Gatti, J.S. Hokanson and K. Nozaki; U.S. Patent 4,145,297; March 20, 1979; assigned to Shell Oil Co. have found that when minor amounts of certain high surface area, finely divided solid acid catalysts of critically high, nonvolatile surface acidity are used as gasoline and lubricant oil additives, a significant reduction in ORI is produced. This reduction in ORI can be embodied in the prevention or substantial inhibition of ORI in new or originally clean spark-ignition internal combustion engines in which the deposit equilibrium octane requirement (OR) level has not been reached or, alternatively, can take the form of a reduction in ORI, i.e., OR can be lowered, in a spark-ignition internal combustion engine operated or used to the extent of reaching the deposit equilibrium level inducing ORI.

These high surface area, refractory solid acid catalysts have: (a) an average particle size in the range of about 0.01 to about 10 μ; and (b) surface acid centers formed by metallic, hydrocarbon insoluble cations of sufficiently high acidity such that the average nonvolatile surface acidity is 1 or less, expressed in pKa units. The metal-containing, solid acid catalysts, which preferably comprise refractory metal oxides or halogenated metal oxides having the aforementioned high surface acidity, are essentially nonreactive with the hydrocarbon carrier, i.e., fuel or lubricating oil, prior to exposure to combustion conditions in the engine, and further, exhibit substantial thermal and chemical stability under combustion conditions in the engine.

The hydrocarbon fuels and lubricating oils containing these additives are quite effective in preventing or reversing ORI in spark-ignition, internal combustion engines which are, or have been, operated on either lead-free or leaded hydrocarbon fuels.

The specific classes of metal-containing, solid acid catalysts which possess the aforedescribed critically high, average, nonvolatile surface acidity are well known in the art, having been employed previously in a variety of acid catalyzed processes such as catalytic cracking, polymerization, alcohol dehydration and dimerization. Suitable metal-containing solid acid catalysts include refractory metal oxides having the aforementioned high surface acidity such as binary oxides of aluminum, zirconium, titanium and iron with silicon and halogenated metal oxides such as aluminum oxide. Other useful metal-containing acidic agents encompass a variety of binary metal oxides having highly acidic properties such as those listed in Tanabe K. et al, *Bull. of the Chem. Soc. of Japan* 47, 1064 (1974), acid clays and acidic ion exchange resins.

Since water is known to have a neutralizing effect on surface acid centers in many of the aforementioned solids, it is preferable to maintain the catalytic agents in a dried, substantially anhydrous state prior to use.

For reasons relating to availability and cost-effectiveness, it is preferred that the solid acid catalyst utilized be selected from the class consisting of binary oxides of silicon and aluminum, i.e., silica-aluminas, and halogenated aluminum oxides, i.e., chlorinated or fluorinated aluminas. In this regard, it is most preferred that the solid acid catalyst be silica-alumina with a silica-to-alumina weight ratio of from 0.1 to 500 (SiO_2/Al_2O_3) or a fluorinated alumina containing up to 10% by weight fluorine. For optimum effectiveness as ORI inhibiting agents, it is also essential that the solid acid catalysts possess high surface areas. Suitably, this surface area of the finely divided solid acid in the fuel or lubricant intake charge is at least 100 m^2/g with surface areas in the range of 100 to 700 m^2/g being preferred. Such high surface areas are characteristic of the aforedefined solid acid catalysts.

These solid acid catalysts are effective ORI inhibiting agents in hydrocarbon fuels and lubricating oils in total quantities which generally comprise only a very minor amount of the fuel or lubricant composition. While the specific concentration of solid acid catalyst employed to obtain the desired ORI inhibiting effect is dependent on a variety of factors such as the introducing media employed, i.e., fuel or lubricating oil, the surface area of the solid acid, its acid strength and particle size, etc., it can be generally stated that the concentration range of solid acid catalyst is established by that which produces a concentration sufficient to yield about 0.1 to 500 mg of acid catalyst in the combustion chamber.

Accordingly, considerably higher concentrations of this acid catalyst in the introducing media are required when it is introduced in the crankcase lubricating oil. Acid catalyst in the range of about 0.003 to 15 g/qt of motor oil is a useful amount sufficient to produce the abovementioned weight range of acid catalyst per gallon of gasoline in the combustion chamber. For the preferred solid acid catalysts such as silica-alumina, which possess a large number of acid centers of high acidity pKa's of –8.3 or less, it is generally preferred that the concentration of solid acid catalyst in the gasoline intake charge not exceed about 400 mg/gal, with concentrations in the range of about 1 to about 250 mg/gal being most preferred.

The hydrocarbon fuel and lubricating oil compositions of the process are heterogeneous dispersions or suspensions wherein the solid acid catalyst agent in finely divided form is dispersed or suspended in the base liquid carrier. These heterogeneous compositions can be prepared in any conventional manner.

For example, the finely-divided acid catalyst may be produced from chemical entities of larger particle size by known comminution techniques such as the use of a micronizer and the like. At the desired range of particle size, the solid acid may then be conveniently dispersed or suspended in the desired liquid media, i.e., motor fuel or lubricating oil, to give stable dispersions or suspensions by use of ultrasonic or other known mixing techniques.

To aid in the initial dispersion and to promote the ultimate stability of the final suspension, it is desirable to employ one or more conventional surface active agents including acidic, amphoteric and neutral or nonionic materials. Any preparation technique employed should preferably be carried out under substantially anhydrous conditions to avoid the deactivating effects of water on the acid centers present on the surface of the solid acid catalyst used.

Example: A test program was carried out using a single cylinder CFR engine to demonstrate both the ORI inhibiting effect in a clean engine and the reversal of ORI effect (octane requirement reduction or ORR) in an engine having accumulated combustion chamber deposits with a gasoline composition containing the additive.

The solid acid catalyst composition employed in this test was an Omega zeolite (12.5% Al_2O_3/87.5% SiO_2) having an average surface acidity of about -5.6 pKa, pore volume of 0.2 cc/g, surface area of about 500 m^2/g and an average particle diameter of about 1 μ achieved by grinding with a Sturtevant 2" laboratory micronizer. To obtain the desired gasoline composition this subdivided, solid acid catalyst was suspended in an unleaded gasoline in the amount of about 50 mg/gal utilizing an ultrasonic agitator. The base gasoline employed was a heavy catalytically cracked fraction-rich fuel containing about 38% by volume aromatics, about 1% by volume olefins and about 61% by volume of saturates. This base fuel had an initial boiling point of 34°C, a 50% boiling point of 101°C and a final boiling point of 207°C by ASTM distillation method D-86.

To determine the effect of the gasoline composition according to the process, the CFR engine (modified by the addition of a Tillotson Carburetor ahead of the CFR carburetor) was operated under accelerated deposit accumulation conditions for an approximate test period of 100 hours using both the gasoline composition and the unleaded base fuel intermittently through the test. These test fuels were charged to the engine via an electric fuel pump from nitrogen-blanketed storage tanks thereby permitting continuous operation. For engine lubricating, a 400 MVI basestock with no additives was used to limit the number of test variables.

The operating and rating procedure used in the test involved setting the motored compression pressure on a clean, warmed-up engine to 192 psig by varying the cylinder height. The micrometer reading was then set at 0.352" with appropriate corrections for ambient conditions as per ASTM Compression Ratio Method. The engine was fired on base fuel with ignition set for 13° BTDC (Before Top Dead Center). The air/fuel ratio was set for maximum knock, and the head height was adjusted to trace knock utilizing a knock meter. The head height reading was converted to KLCR (knock limited compression ratio) using ASTM tables on the method. The engine was allowed to run at an ignition setting of 5° ATDC in order to accumulate deposits and ORI. It was periodically rated by returning the ignition setting to 13° BTDC and proceeding as described above. This KLCR method allows rating of the engine to be carried out rapidly and without changing fuel. The relationship between KLCR and octane requirement increase due to deposit formation can be obtained by running the engine on base fuel and periodically comparing the KLCR as determined above with the PRF (primary reference fuel) requirement determined at constant compression ratio.

A summary of the test results, including a designation of the operating periods with each test fuel is given in Figure 1.1. In this figure, which plots operating time versus engine octane requirement, the engine performance line is segregated into line segments by points **A-E** to indicate engine operation with the gasoline containing the solid acidic additive according to the process (line segments **A-B** and **C-D**) and operation with the base gasoline (line segments **B-C** and **D-E**). The observed OR for the first 20 hours of operation (segment **A-B**) illustrates the prevention or substantial inhibition of ORI with a fuel composition according to the process.

Figure 1.1: Effect of Solid Acid Catalyst on ORI

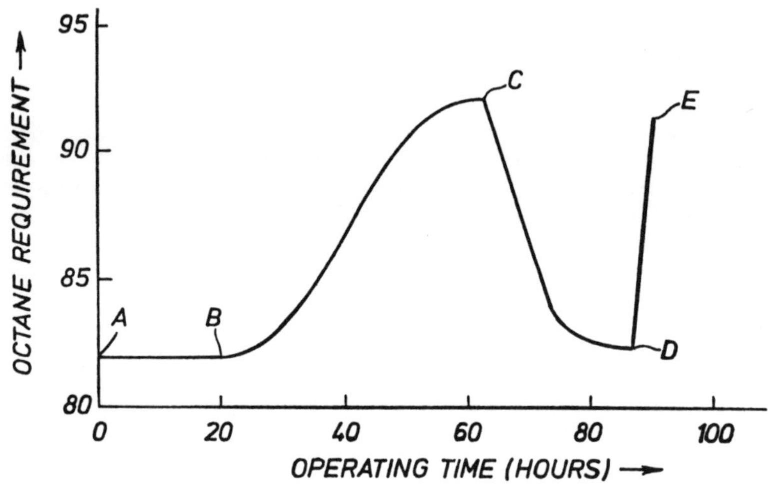

Source: U.S. Patent 4,145,297

In contrast, the recorded OR measurements for line segments **B-C** and **D-E** indicate substantial ORI where the engine was operated on base fuel. The OR results recorded beginning at point **C** and extending along line segment **C-D** show a reduction of octane requirement (ORR effect) during the 60 to 90 hour test period with the gasoline composition after OR had been allowed to increase to a rather high level (during line segment **B-C** operation) with the base fuel.

Dialkyl Formamide

K.A. Frost, Jr.; U.S. Patent 4,217,111; August 12, 1980; assigned to Chevron Research Co. has found that certain dialkyl formamide additives for gasoline contribute a minimum of increased octane requirement in engines in which they are used compared with engines operated with conventional fuels. These dialkyl formamides have the formula

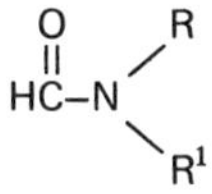

$$\overset{O}{\overset{\|}{HC}}-N\begin{matrix}\diagup R \\ \diagdown R^1\end{matrix}$$

in which R and R^1 are alkyl of 1 to 9 carbon atoms and the sum of the carbon atoms in R and R^1 is from 7 to 11. The compositions will contain not more than about 0.1 g of lead per gallon, preferably not more than about 0.005 g/gal. The quantity of formamide employed in order to moderate and stabilize the octane requirement of the engine will usually be in the range of about 800 to 10,000 ppm, preferably from 1,000 to 3,000 ppm.

When it is desired to remove previously formed deposits from the engine combustion chamber, the engine is operated with a fuel composition containing a very high concentration of the formamides, up to about 15 wt %. These clean-up fuel compositions will thus contain amounts of the formamides in the range of 3 to 15, preferably 5 to 12 wt %.

Examples of the dialkyl formamide octane requirement reducing (ORR) additives which may be employed include N-methyl-N-hexyl formamide, N-methyl-N-octyl formamide, N-methyl-N-decyl formamide, N-ethyl-N-pentyl formamide, N-ethyl-N-heptyl formamide, N-ethyl-N-nonyl formamide, N-propyl-N-butyl formamide, N-propyl-N-heptyl formamide, N-propyl-N-octyl formamide, N,N-dibutyl formamide, N-butyl-N-pentyl formamide, N-butyl-N-heptyl formamide, N,N-dipentyl formamide, N-pentyl-N-hexyl formamide, etc.

In order to demonstrate the improvement in moderating the octane requirement increase obtained with the fuel compositions, a Multicylinder Octane Requirement Test was employed. The test uses a 350-CID Chevrolet V-8 engine. The test procedure involves engine operation for 100 hours on a prescribed load and speed schedule representative of typical vehicle driving conditions. The engine is stopped and force-cooled by circulation of cold water for a period of ½ hour during each 4 hours of operation. The test is started with clean combustion chambers and the engine octane requirement is measured at the start of the test and at daily intervals thereafter as combustion chamber deposits accumulate.

The engine ignition system is equipped with electronic circuitry, permitting the operator to retard the ignition timing of selected individual cylinders during the octane requirement measuring procedure, thus obtaining the requirement for each cylinder.

The gasoline employed in the test was representative of commercial unleaded fuel which, however, contained a relatively low concentration of heavy aromatic hydrocarbons. The formamides to be tested were added to the fuel at concentrations of 2,500 ppm. The data showed that the addition of dibutyl formamide to the fuel resulted in a substantial reduction in octane requirement increase compared with base fuel.

In addition to the reduction in ORI obtained with the formamides, an actual reduction in the octane requirement of an engine having octane-increasing deposits can be effected using generally larger quantities of the formamides. An automobile was operated for 264 miles on a typical nonleaded commercial fuel containing 10% by weight of N,N-dibutyl formamide. The automobile, which had 13,010 operating miles on its engine, had an OR of 91. At the end of the test, the OR was 90.

Polyoxyalkylene Aminocarbamates

It is highly desirable to provide fuel compositions which contain deposit control additives which effectively control deposits in intake systems (carburetor, valves, etc.) of engines operated with fuels containing them, but do not contribute to the combustion chamber deposits which cause increased octane requirements.

R.A. Lewis and L.R. Honnen; U.S. Patents 4,160,648; July 10, 1979; and 4,191,537; March 4, 1980; both assigned to Chevron Research Co. have provided additives which aid the fuel composition in maintaining engine intake cleanliness without contributing to combustion chamber deposits.

This fuel composition comprises a major amount of hydrocarbons boiling in the gasoline range, and from about 30 to about 2,000 ppm of a hydrocarbylpoly(oxyalkylene) aminocarbamate of molecular weight from about 600 to about 10,000 and having at least one basic nitrogen atom; wherein the poly(oxyalkylene) moiety is composed of oxyalkylene units selected from 2 to 5 carbon oxyalkylene units and containing at least sufficient branched chain oxyalkylene units to render the carbamate soluble in the fuel composition; and the hydrocarbyl group contains from 1 to about 30 carbon atoms.

The poly(oxyalkylene) aminocarbamate consists of an amine moiety and a poly(oxyalkylene) moiety comprising at least one hydrocarbyl-terminated poly(oxyalkylene) polymer bonded through a carbamate linkage, i.e., –OC(O)N<. The amine component of the carbamate and the poly(oxyalkylene) component of the carbamate are selected to provide solubility in the fuel composition and deposit control activity without octane requirement increase (ORI).

The polyether aminocarbamates will generally be employed in a hydrocarbon distillate fuel. The proper concentration of additive necessary in order to achieve the desired detergency and dispersancy varies depending upon the type of fuel employed, the presence of other detergents, dispersants and other additives, etc. Generally, however, from 30 to 2,000 weight parts per million, preferably from 100 to 500 ppm of polyether aminocarbamate per part of base fuel is needed to achieve the best results. When other detergents are present, a lesser amount of polyether aminocarbamate may be used. For performance as a carburetor detergent only, lower concentrations, for example, 30 to 70 ppm may be preferred.

Example 1: *Reaction of phosgene with a poly(oxypropylene) alcohol* – A 99 g (1.0 mol) portion of phosgene was condensed into 750 ml of toluene at 0°C. A 450 g (0.17 mol) portion of butylpoly(oxypropylene) alcohol, i.e., polypropoxylated butanol, having a molecular weight of about 2,400 was added in a slow stream to the phosgene-toluene mixture over a period of ½ hour while maintaining the temperature at 0° to 10°C. 200 ml of benzene were added. The temperature was raised to 80°C and excess phosgene and benzene were distilled from the product. A small sample was taken; toluene was evaporated from it. Infrared analysis showed a strong chloroformate absorption at 1,785 cm^{-1}.

Example 2: *Reaction of poly(oxypropylene) chloroformate with amine* – One half of the product from Example 1 (in toluene solution) was added at room temperature to 154 g (1.5 mol) of diethylenetriamine in 500 ml of toluene. Immediate precipitation of an amine hydrochloride occurred. The mixture was stirred for ½ hour, filtered and the toluene removed under reduced pressure. The residue was dissolved in 1½ volumes of hot n-butanol and extracted three times with 100 to 200 ml of hot water. The butanol was removed by vacuum providing 200 g of a product which contained 1.17% nitrogen and 0.80% basic nitrogen by ASTM D-2896. Infrared analysis revealed a typical carbamate absorption at 1,725 cm^{-1}. This product is designated Compound 1.

Example 3: Following the procedure of Examples 1 and 2, but using a butyl-capped poly(oxypropylene) glycol, i.e., butylpoly(oxypropylene) alcohol of about 1,800 molecular weight, the following carbamates were prepared.

Compound	Amine	Nitrogen (Wt %)
2	Ethylenediamine	0.83
3	Diethylenetriamine	1.36

A 35.4 g portion of Compound 3 was chromatographed on a silica gel column (7" long, 2" diameter), eluting with 1 ℓ of ethyl acetate; ethyl acetate:methanol, 4:1, and ethyl acetate:methanol:isopropyl amine, 7:2:1. The first fraction, 13.8 g, was found by infrared to be predominantly unreacted poly(oxypropylene). The latter two fractions, 20.7 g, were identified by the infrared carbonyl absorption as the desired carbamate. This material was designated Compound 3a.

Evaluation: In the following tests the polyether aminocarbamates were blended in gasoline and their deposit control capacity tested in an ASTM/CFR Single-Cylinder Engine Test.

In carrying out the tests, a Waukesha CFR single-cylinder engine is used. The run is carried out for 15 hours, at the end of which time the intake valve is removed, washed with hexane and weighed. The previously determined weight of the clean valve is subtracted from the weight of the valve. The difference between the two weights is the weight of the deposit with a lesser amount of deposit measured connoting a superior additive. The amount of carbonaceous deposit in milligrams on the intake valves is measured and reported in the following table.

The base fuel tested in the above test is a regular octane unleaded gasoline containing no fuel deposit control additive. The base fuel is admixed with varying amounts of the deposit control additives (in the table the additives are identified by compound numbers from the foregoing examples).

Intake Valve Deposit Tests*

Additive-Carrier Description	Amount (ppm)	Average Washed Deposit (mg) 11A Engine	12A Engine
Base fuel (control)	–	259**	102***
2	333	12	6
PPG-1800†	167		
2	200	33	18
PPG-1450††	300		

(continued)

Additive-Carrier Description	Amount (ppm)	Average Washed Deposit (mg) 11A Engine	12A Engine
3	500	6	15
3a	125	14	45
PPG-1800†	375		
3a	125	16	23
PPG-1450††	375		

*Single evaluations unless noted.
**Average of 8 runs.
***Average of 4 runs.
†The designation PPG-1800 refers to a monobutyl-capped poly(propylene glycol) of about 1,800 MW.
††Same as footnote †, but about 1,450 MW.

C.B. Campbell and R.J. Peyla; U.S. Patent 4,270,930; June 2, 1981; assigned to Chevron Research Co. have found that these hydrocarbyl polyoxyalkylene aminocarbamates, having molecular weights of from 600 to about 10,000, with at least one basic nitrogen, may be used in amounts of from 0.3 to 3 wt % based on the total fuel composition. These fuel compositions not only give the intake system deposit control which is shown at lower concentrations of the additive, but actually decrease the octane requirement of engines which have undergone octane requirement increases while operated with other fuel compositions.

In a continuation of the work reported in the three previous patents, *R.A. Lewis and L.R. Honnen; U.S. Patent 4,236,020; November 25, 1980; also assigned to Chevron Research Co.*, describe somewhat different additives which maintain cleanliness of intake systems without contributing to combustion chamber deposits. The additives are poly(oxyalkylene) carbamates comprising a hydrocarbyloxy-terminated poly(oxyalkylene) chain of 2 to 5 carbon oxyalkylene units bonded through an oxycarbonyl group to a nitrogen atom of ethylenediamine.

The preferred compounds may be described by the following general formula:

$$Z\text{-}(OC_gH_{2g})_j\text{-}O\text{-}\overset{\overset{\displaystyle O}{\|}}{C}\text{-}NH\text{-}CH_2\text{-}CH_2\text{-}NH_2$$

in which g is an integer 2 to 5, j is an integer such that the molecular weight of the compound is in the range of about 1,200 to about 5,000, Z is hydrocarbyl of 1 to 30 carbons. Sufficient of the oxyalkylene units in the compound are other than ethyleneoxy to render the compound soluble in hydrocarbon fuel boiling in the gasoline range.

The additives are usually prepared by the reaction of a suitable capped polyether alcohol with phosgene to form a chloroformate followed by reaction of the chloroformate with ethylenediamine to form the active carbamate.

Example 1: *Preparation of poly(oxypropylene) chloroformate* – Phosgene (298 g, 3.0 mols) was condensed into toluene (2.5 ℓ) at 0°C. n-Butoxy capped poly(oxypropylene)monool (5.0 kg, 2.78 mols) with a molecular weight of about 1,800 was added to the phosgene solution in a rapid stream, with stirring. The mixture was stirred an additional 30 minutes after completion of the addition, and excess phosgene was removed by purging with nitrogen while the temperature rose to ambient (about 2 hours). The product showed a strong chloroformate absorption at 1,790 cm^{-1}.

Example 2: *Reaction of poly(oxypropylene) chloroformate with ethylenediamine* – The chloroformate solution from Example 1 was divided in half, diluted with toluene (6 ℓ), and each half was added to ethylenediamine (527 g, 8.6 mol) in toluene (1 ℓ) at 0°C, with vigorous stirring. Immediate precipitation of ethylenediamine hydrochloride occurred. The reaction temperature was kept below 25°C and stirring was continued for one hour after addition. n-Butanol (5 ℓ) was added, and the mixture was extracted with hot water (approximately 15 ℓ). The two batches were combined and solvent was removed on a 5-gallon rotary evaporator. The product (5,050 g) contained 1.12% nitrogen and 0.46% basic nitrogen by ASTM D-2896. Infrared analysis revealed a typical carbamate absorption at 1,725 cm^{-1}.

The polyether amino carbamates were blended in gasoline and their deposit reducing capacity tested in an ASTM/CFR Single-Cylinder Engine Test, as described in U.S. Patent 4,191,537 above.

The amount of carbonaceous deposit in milligrams on the intake valves is measured and reported in the following table. The base fuel tested in this extended detergency test is a regular octane unleaded gasoline containing no fuel detergent. The base fuel is admixed with varying amounts of deposit control additives.

Intake Valve Deposit Tests

Additive-Carrier Description	Amount (ppm)	Average Washed Deposit (mg)	
		11A Engine	12A Engine
Base fuel (control)	–	259	102
PPG-1800* EDA Carbamate**	333	12	6
PPG-1800*	167		
PPG-1800* EDA Carbamate**	200	33	18
PPG-1450*	300		

*Refers to a monobutyl-capped poly(oxypropylene) glycol of about 1,800 MW; PPG-1450 is 1,450 MW.

**Poly(oxypropylene)ethylenediamine carbamate prepared as in Ex. 2.

These data show that these additives have excellent deposit control properties.

The compounds described in U.S. Patents 4,160,648 and 4,191,537 maintain engine intake system cleanliness without contributing to engine octane requirement increase (ORI). However, it has been found that carbamates containing certain poly(oxyalkylene) chains, when they are used in fuels employed in combination with certain lubricating oils, produce crankcase varnish. This does not occur when the poly(oxyalkylene) units of the aminocarbamates are derived from 1,2-epoxide monomers containing four or more carbon atoms. The C_4 or higher 1,2-epoxyalkanes provide oxyalkylene units with C_2 or higher alkyl side chains.

J.E. Lilburn; U.S. Patents 4,197,409; April 8, 1980; and 4,274,837; June 23, 1981; both assigned to Chevron Research Co. has therefore developed amino-carbamate compounds derived from certain poly(oxyalkylene) chains which have good deposit control characteristics in fuels, contribute little to engine ORI, and are compatible with most lubricating oil compositions. The compounds are hydrocarbylpoly(oxyalkylene) aminocarbamates, having at least one C_{1-30} hydrocarbyl-terminated poly(oxyalkylene) chain. The poly(oxyalkylene) chain comprises 1 to 5 branched oxyalkylene units, (e.g., oxy-1,2-alkylene units) each containing from 9 to 30 carbon atoms, as well as oxyalkylene units selected from 2 to 5 carbon oxyalkylene units which may be branched or linear. The aminocarbamates have molecular weights of from about 600 to 10,000, preferably from about 1,000 to 5,000. The polyamine moiety of the aminocarbamate will contain at least one basic nitrogen atom, i.e., a nitrogen titratable by a strong acid. It is preferably dimethylaminopropylenediamine, ethylenediamine, diethylenetriamine or triethylenetetramine.

These additives may be most conveniently prepared by reaction of phosgene with the monohydroxypoly(oxyalkylene) alcohol [itself prepared by reaction of a hydrocarbyl poly(lower oxyalkylene) alcohol with C_{9-30} epoxide] followed by reaction of the product with a suitable amine. The hydrocarbylpoly(lower oxyalkylene) alcohol is reacted with the C_{9-30} epoxide employing usually an excess of from 0.1 to 5 molar excess of the epoxide.

The presence of a basic catalyst is preferred. Usually from about 0.1 to 1.0 mol of catalyst such as an alkali metal or alkali metal hydroxide is used. The reaction is carried out at temperatures preferably from about 80° to 110°C. The reaction of the poly(oxyalkylene)monools thus prepared with phosgene is usually carried out on an essentially equimolar basis, at temperatures preferably in the range of 0° to 50°C. A solvent may be used in the chloroformylation reaction. Suitable solvents include benzene, toluene, etc.

The reaction of the resultant chloroformate with the amine may be carried out neat or, preferably, in solution. Temperatures of from –10° to 200°C may be utilized. The desired product may be obtained by water wash and stripping, usually by the aid of vacuum, of any residual solvent. The mol ratio of the basic amine nitrogen to the poly(oxyalkylene) chloroformates will generally be in the range from 5 to 15 mols of basic amine nitrogen per mol of chloroformate.

Example: Aminocarbamates were prepared from combinations of various capped poly(oxypropylenes) and epoxides. The epoxides included mixtures of C_{6-9} linear epoxides, C_{11-14} linear epoxides and C_{16} linear epoxide. The following compounds are carbamates of ethylenediamine (EDA).

	Poly(Oxypropylene)	Linear 1,2-Epoxide	Amine
Compound 1	1350 MW butyl-capped	C_{6-9}	EDA
Compound 2	1350 MW butyl-capped	C_{11-14}	EDA
Compound 3	1350 MW butyl-capped	C_{16}	EDA
Compound 4	1525 MW oleyl-capped	C_{11-14}	EDA
Compound 5	1800 MW butyl-capped	C_{16} 14	EDA

In the following tests these poly(oxypropylene) aminocarbamates were blended in gasoline and their deposit reducing capacity tested in an ASTM/CFR Single-Cylinder Engine Test.

In carrying out the tests, a Waukesha CFR single-cylinder engine is used. The run is carried out for 15 hours, at the end of which time the intake valve is removed, washed with hexane and weighed. The previously determined weight of the clean valve is subtracted from the weight of the valve. The difference between the two weights is the weight of the deposit with a lesser amount of deposit measured connoting a superior additive. The operating conditions of the test are as follows: water jacket temperature 100°C (212°F); manifold vacuum of 12 in Hg; intake mixture temperature of 50.2°C (125°F); air-fuel ratio of 12; ignition spark timing of 40°BTDC; engine speed is 1,800 rpm; the crankcase oil is a commerical 30W oil. The amount of carbonaceous deposit in milligrams on the intake valves is measured and reported in the following table.

The base fuel tested in the above test is a regular octane unleaded gasoline containing no fuel deposit control additive. The base fuel in each test run is admixed with 400 ppm of the deposit control additive and 200 ppm of poly(oxypropylene)-monobutyl ether (molecular weight about 1,450).

Intake Valve Deposit Tests

Additive Description	. Average Washed Deposit (mg). .	
	11A Engine	12A Engine
Base fuel alone	259	103
Compound 2	10	16
Compound 3	8	26
Compound 4	71	28

The above results show the significant reduction in valve deposits achieved by the test compounds compared with base fuel.

In a variation of the previous patent, *J.E. Lilburn; U.S. Patent 4,234,321; Nov. 18, 1980; assigned to Chevron Research Co.* provides deposit control additives made by reacting a hydrocarbylpoly(oxyalkylene) alcohol with excess phosgene and an excess amount of certain polyamines. The product comprises hydrocarbylpoly(oxyalkylene) ureylene carbamates.

In the process for making these additives the hydrocarbylpoly(oxyalkylene) alcohol is reacted with phosgene to produce a first product comprising a hydrocarbylpoly-(oxyalkylene) chloroformate as in the previous patent. Phosgene is preferably used in excess amounts within the limits of 1:1 to 1:5 mol ratio of alcohol to phosgene, preferably 1:1.1 to 1:2.5. Preferably without the removal of excess phosgene from the first product, the first product is reacted with polyamine in mol ratio alcohol:polyamine of 1:1.1 to 1:20, preferably 1:5 to 1:15. If sufficient excess phosgene is present from the first step, the excess polyamine is incorporated by the second step into the carbamate via the desired ureylene linkages. Optionally, the additional phosgene is added in a third step in such amount that the total phosgene reacted in the process is in mol ratio to the alcohol, as alcohol:phosgene, of about 1:1.1 to 1:5, preferably 1:1.1 to 1:2.5.

Example 1: *Preparation of poly(oxybutylene)* – The experiment was carried out under nitrogen using anhydrous conditions. Potassium (3.52 g, 0.09 mol) was added to 79 g (0.3 mol) of a phenol alkylated with propylene tetramer. The mixture was stirred and heated to 80°C for 20 hours, until the potassium was no longer visible. Freshly distilled 1,2-epoxybutane (646 ml, 7.5 mols) was added to this stirred solution and the mixture was refluxed for 64 hours. At the end of this time the pot temperature had reached 158°C. The product was extracted into 500 ml of n-butanol and washed with 400 ml of hot water. The solvent was removed under reduced pressure, yielding 560 g of a viscous liquid of molecular weight 1,550.

Example 2: *Preparation of poly(oxybutylene) chloroformate* – The reaction was carried out under nitrogen using anhydrous conditions. Phosgene (60 ml, 0.84 mol) was condensed into 500 ml of toluene at 5°C. The polymer from Example 1 (560 g, 0.36 ml) was added dropwise to the cooled phosgene solution. The reaction mixture was stirred without cooling for 1 hour after addition was complete. The IR spectrum of a sample of the product which had been treated with nitrogen to remove phosgene showed a strong carbonyl absorption at 1,785 cm^{-1}. Nitrogen was bubbled through the solution for an additional 2 hours to remove phosgene, but not all the excess phosgene was removed.

Example 3: *Preparation of product* – The reaction was carried out under nitrogen. Ethylenediamine (374 ml, 5.6 mols) was cooled in an ice bath. The product of Example 2 was diluted to 3,500 ml and added as quickly as possible to the rapidly stirred ethylenediamine while maintaining the pot temperature below 30°C. After the addition was complete, the ice bath was removed and the mixture was stirred for an additional hour. The product was extracted into 700 ml of n-butanol and washed four times with 500 ml aliquots of hot water. The solvent was removed under reduced pressure, yielding an orange-brown, gel-like material of average molecular weight 2,166, containing 0.29% basic nitrogen and 1.64% total nitrogen.

Example 4: Following the general procedure and starting materials of Examples 2 and 3, an alkylphenyl poly(oxypropylene) product was prepared from a starting polymer of about 1,300 molecular weight; the product had an average molecular weight of 2,197, 0.12% basic nitrogen and 1.68% total nitrogen (Product E).

Product E was compared to a product E' which was a poly(oxyalkylene) aminocarbamate such as those described in the previous patent using the ASTM/CFR Single-Cylinder Engine Test described in that patent. Product E showed superior deposit control to its corresponding aminocarbamate.

Hydrocarbylpolyoxyalkylene Aminoesters

In view of the explanation on increased octane requirements already given, it is obvious that it is highly desirable to provide fuel compositions which contain deposit control additives which effectively control deposits in intake systems (carburetor, valves, etc.) of engines operated with fuels containing them, but do not contribute to the combustion chamber deposits which cause increased octane requirements. While, in general, deposit control fuel additives are not believed to be useful dispersants for lubricating oil compositions, certain aminoesters are useful in this regard.

These additives, developed by *R.A. Lewis; U.S. Patent 4,198,306; April 15, 1980, assigned to Chevron Research Co.*, which are useful deposit control additives in hydrocarbonaceous fuel compositions and dispersants in lubricating oil compositions, are monoesters of an amino-substituted C_{2-20} monocarboxylic alkanoic acid and a hydrocarbylpoly(oxyalkylene) alcohol. The aminoesters have molecular weights of about 600 to 5,000. The amino-substituent contains from 1 to

12 amine nitrogen atoms, up to 40 carbon atoms and has a carbon:nitrogen ratio of up to about 10:1. The poly(oxyalkylene) moiety of the aminoester has a molecular weight from about 500 to about 5,000 and is composed of at least about 5 oxyalkylene units of from 2 to 5 carbon atoms each. At least a sufficient number of the oxyalkylene units are $C_{3\text{-}5}$ branched-chain oxyalkylene units to render the aminoester soluble in the fuel or lubricating oil composition. The hydrocarbyl group terminating the poly(oxyalkylene) chain contains about 1 to 30 carbon atoms.

The proper concentration of additive necessary in order to achieve the desired detergency and dispersancy varies depending upon the type of fuel employed, the presence of other detergents, dispersants and other additives, etc. Generally, however, from 30 to 2,000 weight ppm, preferably from 100 to 500 ppm of poly(oxyalkylene) aminoester per part of base fuel is needed to achieve the best results. When other detergents are present, a lesser amount of poly(oxyalkylene) aminoester may be used. For performance as a carburetor detergent only, lower concentrations, for example, 30 to 70 ppm may be preferred.

Example 1: *Preparation of butylpoly(oxypropylene) acrylate* – 500 g (0.27 mol) of a polypropylene glycol monobutyl ether, i.e., butylpoly(oxypropylene) alcohol, having a molecular weight of about 1,850 was combined with 350 ml of xylene, 31 g (0.43 mol) of acrylic acid, 10 g of p-toluenesulfonic acid and 1.0 g of hydroquinone in a 2-ℓ, 3-neck flask equipped with thermometer, stirrer, heater and a Dean-Stark trap. 5 ml of water was removed over about 4 hours. The reaction temperature reached 143°C. The mixture was cooled, filtered through diatomaceous earth and the solvent was partially removed on a rotary evaporator. The infrared spectrum of the product revealed a strong carbonyl absorption at 1,730 cm^{-1} and weak olefin bands at 1,630 and 810 cm^{-1}.

Example 2: *Preparation of butylpoly(oxyalkylene) N-(2-aminoethyl)-2-aminopropionate* – 15.7 g (0.26 mol) of ethylenediamine was combined with 100 g of the acrylate from Example 1 and 100 ml of xylene, and the mixture was heated for 4 hours at 120°C. After cooling, the mixture was diluted with 300 ml of n-butanol and washed 10 times with 100 ml portions of warm water. The solvents were removed under reduced pressure yielding 88 g of a brown oil. The product showed by infrared a carbonyl band at 1,740 cm^{-1} and no olefin absorption. The basic nitrogen content of the product was 0.49% by weight. This product is designated Compound A.

Example 3: *Preparation of poly(oxypropylene) aminoesters* – Using the apparatus and procedures of Examples 1 and 2, poly(oxyalkylene) aminoesters were prepared employing the acrylate precursor (product of Example 1), and diethylenetriamine (to form Compound B) and dimethylaminopropylamine (to form Compound C).

In order to demonstrate the capacity of the additives of this process to function in fuels for internal combustion engines without contributing significantly to engine ORI, the additives were subjected to thermogravimetric analysis (TGA). Deposit control additives showing low TGA values, i.e., more rapid thermal decomposition, have been found to show low ORI values in laboratory engine tests.

In thermogravimetric analysis a small weighed sample of the material to be analyzed is placed in the Analyzer and exposed to a flow of 60 ml of air per minute at the specified temperature and for a specified period. 30 minute exposures at 200°C and at 290° to 300°C were employed. The results of the test are set forth in the following table:

Thermogravimetric Analysis of Poly(Oxyalkylene) Aminoesters

Compound	.... Weight Remaining, %	
	(30 min at 290°-300°C)	(30 min at 200°C)
A	1	30
B	2	60
C	1	26
Z	50-60	100

Compound Z is a commercially available nitrogen-containing deposit control additive which has been found to yield higher ORI values. These data show that the poly(oxyalkylene) aminoesters have extremely low TGA values which correlate, in these deposit control additives, with low ORI values.

Hydrocarbylpolyoxyalkylene Polyamines

Still another additive which, when added to fuels, maintains the cleanliness of the engine intake system and which does not contribute to combustion chamber deposits that cause increased octane requirements have been provided by *L.R. Honnen; U.S. Patent 4,247,301; January 27, 1981; assigned to Chevron Research Co.*

The additives are hydrocarbylpoly(oxyalkylene) polyamines soluble in a hydrocarbon fuel boiling in the gasoline range. The poly(oxyalkylene) moiety comprises at least one hydrocarbyl-terminated poly(oxyalkylene) chain of 2 to 5 carbon atom oxyalkylene units which is bonded through a terminal carbon atom to a nitrogen atom of a polyamine having from 2 to about 12 amine nitrogens and from 2 to about 40 carbon atoms with a carbon-to-nitrogen ratio between about 1:1 and 10:1. The hydrocarbyl group will contain from 1 to 30 carbon atoms. The compounds have molecular weights in the range of about 500 to 10,000, preferably from 800 to 5,000.

Example 1: *Preparation of butyl-capped polyether-substituted diethylenetriamine* – Butylpoly(oxypropylene) alcohol, i.e., polypropylene glycol monobutyl ether (hydroxyl number 31 mg KOH/g, 500 g) and hydrochloric acid (37%, 400 g) were combined and heated at reflux under nitrogen for 20 hours. The reaction mixture was cooled, diluted with 2 parts of petroleum ether and washed repeatedly with water until no chloride was detected with silver nitrate. The organic phase was stripped on the rotary evaporator. The residue (416 g) had 0.38% chlorine; hydroxyl number, 26 mg KOH/g. Based on 1,810 molecular weight, the chlorine content was 20% of theory.

The chloride thus obtained (290 g, 106 mmols) was combined with diethylenetriamine (56 g, 540 mmols) and the mixture was heated to 160°C with stirring for 5 hours. The crude product was diluted with an equivalent volume of n-butanol and was washed 4 times with 200 ml of hot water. The organic phase was stripped of butanol on the rotary evaporator. The residue (197 g), a brown oil, had 0.23% N, 0.04% Cl. (Compound 1).

Example 2: *Preparation of hydroxy-terminated polyether-substituted diethylenetriamine* – Polypropylene glycol (hydroxyl number 56 mg KOH/g, 500 g, MW 2,000) was treated as above. The chloride contained 0.65% Cl and had a hydroxyl number of 59. 466 g were recovered. Based on a 2,000 MW for the starting material, the chlorine content was 36% of theory for the monochloromonohydroxypoly(oxypropylene). The chloride (200 g) was treated with diethylenetriamine as above, and provided a product with 0.44% nitrogen (21% of theory) and 0.09% chlorine. The MW was determined to be 1,408. This is entitled Compound 2.

Example 3: *Preparation of butyl-capped polyether-substituted diethylenetriamine* – Butylpoly(oxypropylene) alcohol, i.e., polypropylene monobutyl ether (hydroxyl number 31 mg KOH/g, calculated molecular weight 1,810; 2,200 g, 1.215 mols), and Celite (24 g) were combined in a 5-ℓ, 3-neck flask equipped with nitrogen inlet, mechanical stirrer, addition funnel, reflux condenser and thermometer. Thionyl chloride (217 g, 1.82 mols) was added in a slow stream with stirring. The mixture was heated to 80° to 90°C for 5 hours, and then allowed to cool under a nitrogen blanket. The reaction mixture was diluted with benzene (1 ℓ), filtered through Celite and was stripped of most of the benzene and excess thionyl chloride on the rotary evaporator. A small analytical sample was stripped of all solvent and was found to contain 1.46% chlorine, 0.28% sulfur, and had a molecular weight of 1,502.

The chloride prepared above (2,100 g, 1.15 mols based on starting material molecular weight) was added to diethylenetriamine (626 g, 6.08 mmols) and xylene (approximately 1 ℓ) and was heated to 160°C for 5 hours.

The mixture was cooled to 80°C, filtered from amine hydrochloride precipitate, and the filtrate was stripped of xylene on the rotary evaporator. n-Butanol (1 ℓ) was added and the product was washed 3 times with hot water (85°C, 1 ℓ). n-Butanol was removed on the rotary evaporator. Analysis revealed 0.41% chlorine and 0.54% nitrogen in the product.

The partially aminated product (2,000 g) was combined with diethylenetriamine (475 g) and xylene (500 ml) and heated to 165°C for an additional 5 hours. Work-up as before provided butylpoly(oxypropylene) diethylenetriamine (Compound 3); N: 0.88%; Cl: 0.04%. Based on the molecular weight of the starting material, conversion was 40%. This is entitled Compound 3.

In order to compare the surfactant properties of PPG amines capped with alkyl and hydroxy groups on the nonamine terminus, the aminated polymers were separated from the unreacted starting material and by-products. Compounds 1 through 3 (from Examples 1 through 3) were chromatographed on silica gel, eluting with (A) ethyl acetate, (B) 20% methanol in ethyl acetate, and (C) 10% isopropylamine, 20% methanol in ethyl acetate. The third eluate contained the active amine surfactant. This fraction was analyzed.

The additives were evaluated in a laboratory dispersancy test. The hexane-insoluble, chloroform-soluble portion of sludge scraped from the crankcase of high-mileage engines was added as a chloroform solution to a typical base gasoline containing varying amounts of the test additive. The concentration of additive necessary to prevent coagulation and precipitation of the sludge at 10 and 30 minutes was measured. The following table sets forth these results.

Compound No.	.... Dispersancy Cutoff (ppm) .... 10 Minutes	30 Minutes
1	175	300
2	>400	>400*
3	50	175

*Much higher.

The data in the table show that the butyl-terminated polyether amine prepared with HCl (as in Example 1) is decidedly superior in dispersancy to the hydroxy-terminated material prepared by the same method. The data also show that the butyl-terminated material prepared with thionyl chloride is superior to both other materials.

Reaction Product of a Hydrocarbylsuccinic Anhydride and an Aminotriazole

Many detergent additives for motor fuel compositions are known in the art and are eminently effective for this purpose. There is, however, a major drawback associated with the use of many of the known carburetor detergents including materials which are outstanding carburetor detergents. While not fully understood, many carburetor detergents cause an octane requirement increase in the engine of 3 or 4 or more octane units as measured by the Research Octane Number method. This phenomenon is believed to be due to an increase in deposits formation in the combustion zone of the engine leading to a higher than design compression ratio in the engine and a higher octane requirement and is a very serious problem.

R.L. Sung, W.P. Cullen and P. Dorn; U.S. Patent 4,257,779; March 24, 1981; assigned to Texaco Inc. have developed an effective oil-soluble, ashless detergent for adding to gasoline which does not increase the octane requirement in the engine. The additive comprises the reaction product of a hydrocarbylsuccinic anhydride with an aminotriazole.

The hydrocarbyl-substituted succinic anhydride reactant is represented by the formula:

$$\begin{array}{l} R\text{-}CH\text{—}C(=O) \\ \quad | \qquad\qquad \diagdown \\ \quad | \qquad\qquad\quad O \\ \quad | \qquad\qquad \diagup \\ CH_2\text{—}C(=O) \end{array}$$

in which R is a monovalent aliphatic hydrocarbon radical having preferably from 90 to 120 carbon atoms and may be straight or branched-chain, saturated or unsaturated. Suitable aminotriazoles include 3-amino-1,2,4-triazole and 5-amino-1,2,4-triazole. The following example illustrates the preparation of the reaction product.

Example: 295 g (0.22 mol) of polyisobutenylsuccinic anhydride in which the polyisobutenyl radical has a molecular weight of 1,320 and 21 g (0.25 mol) of 3-aminotriazole are dissolved in 50 ml of benzene. The temperature of the mixture is raised to the reflux temperature of benzene and the mixture is refluxed for about 8 hours. The mixture containing the reaction product is filtered and the benzene is removed by distillation under a vacuum.

In general, the additive is added to a motor fuel composition in a minor amount, i.e., an amount effective to provide detergency to the gasoline. The additive is effective as a detergent in a motor fuel composition in amounts preferably ranging from about 0.02 to 0.10 wt % in gasoline.

A motor fuel composition containing the reaction product of the example was tested for its effectiveness as a carburetor detergent in the Chevrolet Carburetor Detergency Test. The Base Fuel employed in these examples was a premium grade gasoline having a Research Octane Number of about 100 and contained 3 cc of tetraethyl lead per gallon.

The results of the Chevrolet Carburetor Detergency Test are set forth in the table below. The commercial detergent fuel contained approximately 173 PTB of a carburetor detergent additive.

Run	Fuel	Deposit Removed (%)
1	Commercial Detergent Fuel	87
2	Base Fuel + 100 PTB of Example	73

A motor fuel composition containing the prescribed reaction product was also tested for its effect on the ORI or octane requirement increase in an engine. The Base Fuel employed in this was a premium grade gasoline essentially unleaded with less than 0.05 g of tetraethyl lead (TEL) per gallon having a Research Octane Number of about 91.

The Octane Requirement Increase Test was run on a Waukesha RDH engine loaded by a cradled electric dynamometer. The Base Fuel employed was the unleaded gasoline described above. One run was conducted using the detergent additive of the example in the unleaded gasoline at a concentration of 100 PTB. The second run was conducted using a commercial carburetor detergent additive.

The motor fuel containing the detergent additive of the example gave the same ORI as the Base Fuel. The motor fuel composition containing the commercial carburetor detergent additive gave an ORI that was 4 units higher than the Base Fuel.

These tests demonstrate that this carburetor detergent did not aggravate the ORI of the engine and, in this regard, is a considerable improvement over the motor fuel with the commercial detergent additive fuel composition which caused a marked increase in the ORI of the engine.

Cerium(III or IV) 2-Ethylhexanoate

C. Bello, R.J. Hartle and G.M. Singerman; U.S. Patent 4,264,335; April 28, 1981; assigned to Gulf Research & Development Co. describe a procedure which suppresses the tendency of a gasoline-fired engine to require an increase in octane rating of the gasoline after use. This is accomplished by incorporating in the gasoline preferably from about 0.05 to 2.0 g of cerium(III or IV) 2-ethylhexanoate per gallon of gasoline.

It was found that cerous 2-ethylhexanoate is soluble in gasoline and that although this compound has no effect on the antiknock characteristics of a motor gasoline, it does have a positive effect on the octane requirement increase of an automobile engine when incorporated in the gasoline combusted in that engine. For example, when cerous 2-ethylhexanoate is incorporated in a motor gasoline that causes knocking in a particular engine, the cerium compound will not reduce the knock-producing characteristics of the gasoline such as would result with an equivalent quantity of tetraethyl lead. However, if cerous 2-ethylhexanoate is introduced into a motor gasoline that does not cause knocking in a particular engine, the tendency of that engine to knock with prolonged use of the fuel containing this cerium carboxylate additive is substantially reduced.

This result occurs because this particular cerium compound partially suppresses the inherent tendency of an automotive engine to require a higher octane gasoline after a substantial period of use. This inhibition of the octane requirement increase of an automotive engine is also obtained herein by the use of ceric 2-ethylhexanoate as an additive in the gasoline.

It is not necessary that the cerium carboxylate modified gasoline be used in an engine when it is new in order to obtain the desirable suppression of the requirement for octane increase. If an automotive engine has been used over a substantial period of time with an unmodified fuel such that a substantial increase in the octane requirement of the engine has resulted, then this octane requirement increase can be significantly decreased after regular use of the cerium carboxylate modified fuel. For this reason, this cerium carboxylate additive can be used either periodically as needed, or it can be used continuously.

In the following experiments the research method octane number (RON) was determined by ASTM D2700 and the motor method number (MON) was determined by ASTM D2699.

The experiments were carried out on two different 1976 Chevrolet engines having 305 in^3 (5 ℓ) displacement. The engine was taken apart and cleaned before the start of each experiment. The octane requirement of the engine was then determined by the use of a series of reference motor fuels of slightly different octane ratings. In determining the octane requirement, the engine was run at 1,400, 1,800 and 2,200 rpm using a reference gasoline with gradual loading up to wide open throttle. The incipient knocking under these test conditions with a particular reference gasoline established the octane requirements of the engine.

The engine was then run under simulated normal, nonknocking conditions for an extended period such as 24 hours before the octane requirements of the engine were again determined as just described. This simulated normal driving involved a series of 10-minute cycles continually repeated between octane determinations over the entire period of each experiment. Each 10-minute cycle involved three phases. The first phase was a 1-minute idle at 600 to 650 rpm and no load. The second phase representing 25 mph (40 kmph) was for 4½ minutes at 1,200 rpm and 20 to 25 lb ft (27.1 to 33.9 Nm) of load. The third phase representing 45 mph (72 kmph), also for 4½ minutes, was carried out at 1,750 rpm and 45 to 50 lb ft (61.0 to 67.8 Nm) of load. Each experiment was carried out until the engine's octane requirements stabilized as illustrated by the following data.

All of these experiments were carried out with a standard commercially available lead-free motor gasoline as the test fuel except for the use of the reference fuels which were used in determining the engine octane requirement, as described.

Example 1: This experiment, carried out in engine 1, was a blank run with no additive in the test gasoline. The octane requirement of the engine was initially 93 RON. After 24 hours the octane requirement was 94, after 48 hours it had increased to 98 and at 72 hours it was still 98. At 122 hours the octane requirement was found to be 100 and this requirement of 100 octane was repeated at 170 hours and 194 hours giving an average stabilized octane requirement of 100 RON for the last three octane determinations.

Example 2: Engine 1 was again used but in this experiment the test fuel contained 0.4 g of cerous 2-ethylhexanoate per gallon. The octane requirement of the engine was initially 91 RON. At 24 hours it was 92, at 48 hours 94, at 72 hours 95, at 122 hours 96, at 170 hours 97, at 190 hours 98, and 97 after 200 hours, giving an average octane rating of 97.3 RON for the last three octane determinations. This experiment shows that the cerium 2-ethylhexanoate additive provides a significant suppression in the octane requirement increase.

-2-

DETERGENTS FOR FUELS AND LUBRICATING OILS

MULTIPURPOSE FUEL AND LUBRICANT DETERGENTS

One of the many problems associated with internal combustion engine operation relates to the crankcase lubricant and its use. Crankcase lubricating oils are inevitably contaminated with foreign substances such as dirt, water and decomposition products from the combustion process or from the breakdown of the lubricating oil itself. Significant amounts of sludge can be produced in the crankcase of an engine as a result of the presence of foreign matter and this sludge tends to adhere to the walls and passages in the engine.

A particularly serious problem arises when the sludge in the crankcase lubricant deposits in the small passageways of the engine, thereby restricting the flow of the lubricating oil to bearings and valves in the engine. In the more serious instances, the oil flow through the oil passageways tends to be completely restricted resulting in a failure of the system to lubricate critical engine bearing surfaces. This condition leads to excessive cam shaft wear and ultimately to reduced engine life. An effective detergent in the crankcase lubricating oil serves to keep the foreign substances dispersed in the oil and also improves the effectiveness of the oil filter to remove a substantial amount of the foreign matter from the oil.

Reaction Product of a Phenol plus Aldehyde plus Active Hydrogen Compound

J.F. Pindar and J.M. Cohen; U.S. Patent 4,147,643; April 3, 1979; assigned to The Lubrizol Corporation describe compositions useful as fuel and lubricant additives which can be made by reacting:

(a) An aromatic compound having an OH or SH group attached to an aromatic nucleus and an aromatic hydrogen atom (e.g., a substituted phenol wherein the substituent is an alkyl group of at least 50 carbon atoms);

(b) An aldehyde or reactive equivalent thereof (e.g., formaldehyde);

(c) A non-amino-hydrogen, active hydrogen compound (e.g., a

phenol, N,N-dimethyl aniline, etc.); and optionally

(d) An aliphatic alkylating agent of at least 12 carbon atoms (e.g., a polyisobutene of about 50 carbon atoms).

In the following table, examples 1 through 13 are carried out according to the following general procedure. A mixture of the polybutenyl-substituted phenol, mineral oil, n-butanol, sodium hydroxide and paraformaldehyde is heated at 82° to 87°C for 3 hr. Glacial acetic acid is then added to neutralize the hydroxide catalyst. Distillate is removed as the mixture is heated to 125°C under nitrogen.

The active hydrogen compound is added and the mixture heated at 175° to 185°C for 3 hr. The reaction product is then stripped at 190° to 200°C under vacuum and filtered to yield an oil solution of the desired product containing about 40 wt % of mineral oil.

In the following table, examples 14 through 17 are carried out by the same procedure with the exception that the active hydrogen compound is added immediately after the glacial acetic acid addition.

Ex. No.	Polybutenyl Phenol $\bar{M}_n$*	Polybutenyl Phenol (pbw)	$(CH_2O)_x$ (pbw)	NaOH equiv.	Solvent n-Butanol (pbw)	Solvent Oil (pbw)	Acetic Acid (equiv.)	Active Hydrogen Compound	(pbw)
1	a	4,000	165	0.25	165	2,863	0.25	Phenol	250
2	b	960	39.6	0.06	40	701	0.06	o-chlorophenol	77.2
3	c	1,800	99	0.15	99	1,318	0.15	Phenol	150
4	a	960	39.6	0.06	40	693	0.06	Anisole	64.8
5	b	960	36	0.06	36	709	0.06	α-naphthol	87
6	a	1,120	46	0.07	46	809	0.07	p-cresol	73
7	a	960	33	0.06	40	698	0.06	N,N-dimethyl-aniline	72.6
8	b	1,120	46	0.07	46	828	0.07	o-t-butyl phenol	105
9	b	715	33	0.05	33	519	0.05	o-cresol	52
10	a	4,160	172	0.26	172	3,005	0.26	Catechol	286
11	b	1,120	46	0.07	46	845	0.07	Bisphenol A**	106.5
12	a	960	39.6	0.06	40	699	0.06	Pyrogallol	75.6
13	b	800	33	0.05	33	578	0.05	Hydroquinone	55
14	a	1,360	56	0.085	56	976	0.085	Mercapto acetic acid	78
15	b	1,040	43	0.065	43	762	0.065	Thiophenol	82
16	d	664	165	0.35	165	851	0.35	Thiophenol	550
17	a	1,760	72.6	0.11	73	1,271	0.11	Resorcinol	121

*$\bar{M}_n$ of phenol by VPO (Vapor pressure osmometry); a = 1,300; b = 1,340; c = 920; and d = 266
**4,4'-isopropylidenebisbenzenol

These compositions are useful as additives for lubricants, in which they function primarily as sludge dispersants and detergents. Such dispersants and detergents disperse and remove from surfaces sludge which forms in the lubricant during use.

These additives may also be used in fuel, such as gasoline, to impart dispersant and detergent properties to the fuel, usually in an amount of 1 to 10,000 pbw, preferably 4 to 1,000 pbw of the reaction product per million parts by weight of fuel. The preferred gasoline-based fuel compositions generally exhibit excellent engine sludge dispersancy and detergency properties. In addition, they exhibit antirust and carburetor/fuel lines deposit-removing and deposit-inhibiting properties.

Polyolefinic Copolymers

Internal combustion engine sludge is produced by the oxidative degradation of lubricating oils and by the partial oxidation of motor fuels and by-products of motor fuel combustion. The partially oxidized by-products of motor fuel combustion contain reactive intermediates such as aldehydes, acids and hydroxy acids which undergo complex condensation reactions to form insoluble resinous materials known as sludge and varnish. Accordingly, unless the components are dispersed relatively soon after formation, they will settle out of the lubricant, causing filter plugging and sticking of moving parts.

Previously, lubricating oils and hydrocarbon motor fuels have been formulated with several additives to provide a plurality of properties, including improved viscosity-temperature characteristics (viscosity index or VI), pour point depressancy, oxidation inhibition, antirust and detergency. However, multiple additives add substantially to the cost of a lubricating oil or motor fuel and cause problems of incompatibility and interaction of the additives.

The graft copolymer additives provided by *R.L. Stambaugh and R.A. Galluccio; U.S. Patent 4,160,739; July 10, 1979; assigned to Rohm and Haas Company* combine several of these properties in a single material and, therefore, provide a significant improvement over mixtures of additives.

The described graft copolymers combine the efficient thickening properties of polyolefinic viscosity index improvers and the dispersancy provided by nitrogen-containing materials by the grafting of a monomer system comprising maleic acid or anhydride and at least one other (different) monomer which is addition copolymerizable therewith, the grafted monomer system then being post-reacted with a polyamine.

The copolymerizable monomers are selected for their reactivity with maleic acid or anhydride so that more maleic acid or anhydride may be incorporated into the polymer than would occur in the absence of the comonomers.

It has also been found that the graft copolymers are efficiently produced with little or no wasteful by-product by forming an intimate admixture of backbone polymer, copolymerizable monomer system, and a free radical initiator, wherein the temperature of the mixture, at least during the time that the initiator is being uniformly dispersed therein, is maintained below the decomposition temperature of the initiator. Thereafter, the temperature is increased to or above the temperature at which the initiator decomposes, preferably while continuing agitation of the reaction mixture, to form a graft copolymer, followed by post-reaction with a polyamine.

In still another aspect, the substrate backbone polymer has a relatively high molecular weight, of the order of 100,000 to 200,000 viscosity average molecular weight, and the backbone polymer is not degraded prior to the grafting. The molecular weight of the final graft copolymer is reduced by homogenization or other degrading technique to provide the desired balance of thickening capability, viscosity index improvement, shear stability, detergency and other properties in lubricating oils or motor fuels. The backbone polymer is usually degraded before grafting in prior art practices.

The graft copolymers of this process have the significant advantage over known nitrogen-containing dispersant improvers of lower production cost. Moreover, in many cases they exhibit activity substantially equivalent to that of known nitrogen-containing graft copolymers but at lower additive concentrations, thus further enhancing economic advantage.

The backbone or substrate polymers are any substantially linear, substantially saturated, rubbery, olefinic hydrocarbon polymers which are oil soluble before or after grafting of the copolymerizable monomers thereon. By "substantially saturated" is meant less than 4 mol % olefinic unsaturation, preferably 2 mol % or less. Suitable backbone polymers include ethylene/propylene copolymers, ethylene/propylene/diene modified terpolymers, hydrogenated styrene-butadiene copolymers, styrene-isoprene copolymers and atactic polypropylene.

Backbone polymers which can be rendered oil soluble by the grafting or after grafting include low density polyethylene, and the like. The backbone polymers may have a wide range of molecular weight, usually 100,000 to 150,000.

The monomers copolymerizable with maleic acid or anhydride (preferably maleic anhydride) are any α,β-monoethylenically unsaturated monomers which are sufficiently soluble in the reaction medium and reactive towards maleic acid or anhydride so that substantially higher amounts of maleic acid or anhydride can be incorporated into the grafted polymeric product than is obtainable using maleic acid or anhydride alone. Suitable monomers include the esters, amides and nitriles of acrylic and methacrylic acid, and other monomers containing no free acid groups. Representative of these classes are the methyl, ethyl, propyl, butyl, pentyl, hexyl, cyclohexyl, 2-ethylhexyl, and isodecyl esters of acrylic or methacrylic acid.

Other useful monomers are styrene, α-methyl styrene, C_{1-4} alkyl and alkoxy ring substituted styrenes such as p-methyl styrene, p-sec-butyl styrene, p-methoxy styrene, and C_{4-12} alpha olefins such as isobutylene, and the like. Other types of monomers are the vinyl esters such as vinyl acetate, propionate and butyrate; vinyl ketones such as methyl and ethyl vinyl ketone; and other vinyl and vinylidene monomers such as vinyl chloride and vinylidene chloride.

The graft copolymers are prepared in accordance with conventional free radical copolymerization techniques except for those aspects relating to formation of an intimate admixture of the reactants with the free radical initiator at a temperature below the decomposition temperature of the initiator, and subsequent increase of reaction temperature above the decomposition temperature of the initiator.

Typically, in terms of an ethylene/propylene copolymer as the substrate polymer backbone material, the backbone polymer is dissolved to a concentration of 20 to 30 wt % in a suitable inert solvent such as a halogenated aromatic hydrocarbon at a temperature of about 80° to 150°C. Dissolution of the polymer is promoted by suitable agitation such as magnetic or mechanical stirring. Graft monomer is then blended into the solution, usually at a lower temperature such as about 80°C, preferably also while the reaction mixture is being agitated. The graft monomers are added to the mixture in a total amount of 2 to 30 wt % based on the ethylene/propylene copolymer, preferably 5 to 20 wt %. If necessary, the temperature of the mixture is again lowered below the decomposition temperature of the initiator, and the initiator is added and uniformly admixed into the solution.

Conditions of temperature, agitation, sequence and rate of addition are carefully selected to minimize homopolymerization and graft copolymerization at this point. The initiator is added in an amount of 0.5 to 2 wt % based on the ethylene/propylene copolymer. After the initiator has been uniformly admixed, the temperature is gradually raised to or above the decomposition temperature of the initiator.

In the case of the preferred t-butylperbenzoate initiator, the temperature is 120° to 140°C. This temperature is maintained until the reaction is substantially complete. About 1 to 2 hr reaction time is usually sufficient. The reaction product typically will contain 2 to 25 wt % of the graft monomers, preferably 4 to 20 wt %.

The carboxyl groups provided by the maleic acid or anhydride grafts are postreacted with a nonpolymerizable polyamine to form a structure which for convenience is called a "carboxyl-polyamine adduct." The polyamine is added in a sufficient amount to react with substantially all of the carboxyl functionality.

The polyamines may be characterized as aliphatic, cycloaliphatic, aromatic or heterocyclic, and may have mixed character. The polyamines may carry other functional groups, such as hydroxyl, provided such groups do not cause interfering reactions. Examples of suitable polyamines are: N,N-dimethylaminopropylamine; N-(2-aminoethyl)morpholine; N-(3-aminopropyl)morpholine; N-(5-aminopentyl)-2,5-dimethyl pyrrole; N-phenylurea; 2-(2-aminoethyl)pyridine; N-(3-aminopropyl)-N'-methylpiperazine; and 2-aminobenzothiazole.

In similar work, *R.L. Stambaugh and R.A. Galluccio; U.S. Patent 4,161,452; July 17, 1979; assigned to Rohm and Haas Company* describe graft copolymers with the same backbone polymer as that described in the previous patent and the grafted units are the residues of an addition copolymerizable monomer system comprising an unsaturated polycarboxylic acid or anhydride and at least one other addition monomer.

Preferably, the monomer residues are postreacted with an alcohol or an amino alcohol. The graft copolymers are prepared with a free radical initiator which is blended with the reactants at a temperature below the initiator decomposition temperature, followed by heating to the decomposition temperature, as described in the previous patent. The graft copolymers also impart combined detergent, viscosity index improvement and other useful properties to lubricating oils and hydrocarbon motor fuels.

The unsaturated polycarboxylic acids or anhydrides comprising Component 1 of the grafted monomer units include any such acids or anhydrides which graft by free radical addition polymerizable onto polymeric backbones. Representative of such compounds are maleic acid, fumaric acid, citraconic acid, mesaconic acid, itaconic acid, methylenemalonic acid, acetylenedicarboxylic acid, aconitic acid, the anhydrides of any of the foregoing acids, and similar acids and anhydrides containing 4 to 12 carbon atoms. Maleic acid or maleic anhydride is preferred. Mixtures of any of the acids or anhydrides may be used.

Component 2 of the graft monomer system comprises any one or a mixture of monomers different from Component 1 and which contain only one copolymerizable double bond and are copolymerizable with Component 1. Typically, such monomers do not contain free carboxylic acid groups and are esters containing α,β-ethylenic unsaturation in the acid or alcohol portion; hydrocarbons, both aliphatic and aromatic, containing α,β-ethylenic unsaturation, such as the C_{4-12} alpha olefins, styrenes, and vinyl monomers.

The components of the graft copolymerizable system are used in a ratio of monomer Component 1 to monomer Component 2 of 1:4-4:1, preferably 1:2-2:1 by weight. After the graft polymerization the product preferably is reacted with an alcohol or an amino alcohol, containing up to 40 carbon atoms.

Suitable amino alcohols are the mono- and polyalkanol monoamines containing 2 to 20 carbon atoms and 1 to 6 hydroxy groups such as monoethanolamine, diethanolamine, triethanolamine, diethylaminoethanol, diethanolaminoethanol, the corresponding propanol amines, and the like, including mixtures.

The alcohols include any of the saturated alcohols described with reference to the copolymerizable monomers which contain ester groups. Sufficient alcohol or amino alcohol is used to react with substantially all of the carboxylic acid groups of the carboxylic acid monomer Component 1. Preferably, the alcohol or amino alcohol is used in excess (about 5 to 10% excess) over the stoichiometric amount.

Example 1: A 5 ℓ, 3 neck flask is equipped with a thermometer fitted via a flexible rubber mount attached to a ground glass adaptor, a C-type stirrer mounted in a ground glass adaptor via a Teflon insert, and a "Y" tube containing a pressure-equalizer addition funnel and a water-jacketed condenser. Atop the condenser is fitted an inlet tube to provide a nitrogen atmosphere throughout the reaction zone. A Variac-controlled heating mantle is used as a heat source.

To the reaction flask is added 750 g of o-dichlorobenzene. The solvent is heated with stirring to 120°C and 225 g of a 50/50 ethylene-propylene copolymer ("Epcar" 506, Goodrich Chemical Company) is added. The temperature is maintained at 150°C for about 5 hr to complete solubilization. The temperature is then reduced to 80°C and 22.5 g of methyl acrylate and 11.25 g of maleic anhydride is added over a period of 15 min. After an additional hour, 1.32 g of t-butyl perbenzoate is added.

The solution is then stirred at 80°C for 30 min, followed by rapid heating to 130°C. An additional 1.32 g of t-butyl perbenzoate is then added and stirring is continued at 140°C for 1 hr. The solution is then diluted to 25% theoretical solids (based on oil) and is stripped at 1.5 mm Hg and 140°C. A final dilution with 100 neutral oil is then made to produce a concentrate containing 9.4% of graft copolymer.

Shear stability (typically about 35% as measured by ASTM test procedure D-2603) can be improved, to give a product which is essentially stable in engine oil use, by homogenizing the polymer to a viscosity average molecular weight of about 30,000.

Example 2: The procedure of Example 1 is followed in all essential respects except that styrene, 22.5 g, is used in place of the methyl acrylate, and the initiator is cumene hydroperoxide. A graft copolymer is obtained which may be post-reacted with an alcohol or amino alcohol to obtain polymers exhibiting good viscosity and dispersancy in lubricating oils. Shear stability is improved by homogenization.

Example 3: The procedure of Example 1 is repeated in all essential respects except that a commercial ethylene/propylene/diene modified (EPDM) rubber containing 2.5 wt % hexadiene is used as the substrate polymer. A graft copolymer is obtained having good detergency and viscosity index properties. Shear stability can be improved by homogenization.

N,N'-Substituted Diamines

B.A.O. Alink and N.E.S. Thompson; U.S. Patent 4,163,646; August 7, 1979; assigned to Petrolite Corporation have developed a process for preparing linear N,N'-substituted diamines. This is accomplished by the hydrogenation of tetrahydropyrimidines to yield linear N-substituted diamines which can be reacted with carbonyls to form imines, which imines can be reduced to N,N'-substituted diamines as illustrated by N,N'-substituted 2,4-diamino-2-substituted pentanes.

Example 1: *N-Cyclohexyl-2,4-diamino-2-methylpentane* – In a 1 ℓ stirred autoclave was placed 95 g of 2,2-pentamethylene-4,4,6-trimethyl-2,3,4,5-tetrahydropyrimidine, 200 cc of methanol and 6 g of 5% Pt/C catalyst. The autoclave was pressurized with 400 psi of hydrogen gas and the mixture heated for 75 min at 75° to 80°C while a pressure of 400 to 800 psi was maintained.

The reaction product was removed through an internal filter leaving the catalyst behind. After removal of the solvent, there was isolated 95 g of N-cyclohexyl 2,4-diamino-2-methylpentane.

Example 2: *N-Cyclohexyl-N'-isobutyl-2,4-diamino-2-methylpentane* – A sample of 50 g of N-cyclohexyl-2,4-diamino-2-methylpentane and 18.7 g of isobutyraldehyde in 50 cc of benzene was refluxed under azeotropical conditions for 1 hr. The benzene was removed under diminished pressure to yield 67 g of product. The product was dissolved in 200 cc of methanol and 9.6 g of sodium borohydride was slowly added with stirring.

After the reaction was completed, the solvent was removed under diminished pressure. To the resulting product was added water and the organic layer was separated.

The aqueous layer was extracted with ether and the ethereal solution combined with the organic layer. After removal of the ether under diminished pressure the product was distilled under diminished pressure to yield 60 g of N-cyclohexyl-N'-isobutyl-2,4-diamino-2-methylpentane, $BP_{0.6}$ = 98°C. The analytical data were consistent with the assigned structure as follows.

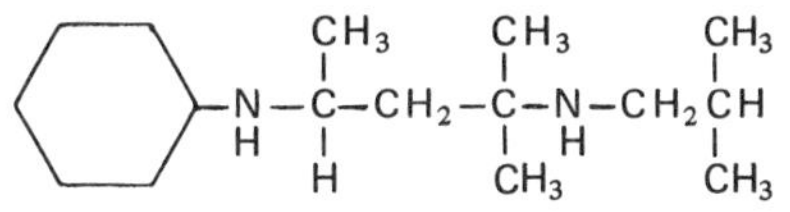

$$
\begin{array}{ccccccc}
 & & CH_3 & & CH_3 & & CH_3 \\
 & & | & & | & & | \\
\text{(cyclohexyl)}- & N- & C- & CH_2- & C- & N-CH_2 & CH \\
 & H & | & & | & H & | \\
 & & H & & CH_3 & & CH_3
\end{array}
$$

The substituted 2,4-diamino-2-methylpentanes have been found to be useful fuel additives to inhibit the formation of color and sludge in the oil. This is particularly true for No. 1, 2 and 3 fuel oils used for heating and for diesel fuel. They are preferably used in amounts of 3 to 300 ppm, based on weight of oil.

Alkenyl-Substituted Oxa-Amines

G. Soula; U.S. Patent 4,168,242; September 18, 1979; assigned to Orogil, France has developed a class of alkenyl amines which impart to lubricating oils and fuels particularly interesting detergent-dispersant qualities.

These additives contain at least one alkenyl amine and they are prepared by the reaction of a halogenated polyolefin with at least one polyamine having an ether group, such as tris(6-amino-3-oxa-hexyl)amine, N,N-bis(6-amino-3-oxa-hexyl)ethanol amine, N-ethyl-N,N-bis(6-amino-3-oxa-hexyl)amine, etc.

The process also relates to lubricating oils and fuels improved by the addition of a small percentage of their weight of the described compositions, imparting excellent detergent-dispersant properties to them. The additives, furthermore, impart antirusting and antifoaming properties to the lubricating oils.

The quantity of additive to be added depends on the intended use to be made of the lubricating oil. Thus, for gasoline motor oil the amount of additive to be added will be from 1 to 10%, while for an oil for a diesel engine, it will be from 4 to 10 wt %.

The fuels which can be used are hydrocarbons of the petroleum distillate type, such as automobile gasoline, aviation gasoline, fuels for diesel engines, fuel oil, etc. The quantity of additive to be added is preferably from 40 to 2,000 ppm.

Example 1: The apparatus employed consists of a 4 ℓ, 3 necked, round bottom flask provided with a mechanical agitator, a dip tube and a distillation system followed by two flasks, the first containing water and the second containing a 20% aqueous solution of sodium hydroxide.

2 kg (about 2 mols) of polyisobutylene of a number average molecular weight of 960 is poured into the round bottom flask. It is heated to 110°C and gaseous chlorine is introduced for 5 hr at that temperature. Analysis shows a chlorine content of the resultant chlorinated polyisobutylene of 4.8 wt %.

Example 2: Into a 1 ℓ, three necked, round bottom flask provided with a mechanical agitator, a vertex, and a reflux condenser there are introduced, in succession: (a) 207 g (i.e., 0.2 mol) of chlorinated polyisobutylene prepared in accordance with Example 1, (b) 15 g of sodium carbonate, and (c) 21.5 g (i.e., 0.67 mol) of tris(6-amino-3-oxa-hexyl)amine, which corresponds to a molar ratio of amine to chlorinated polyisobutylene of 0.33.

The mixture is heated at 160°C for 15 hr. It is cooled to 90°C and 250 cc of a 10% aqueous sodium hydroxide solution is then added. It is refluxed for 1 hr, whereupon the mixture is transferred into a separating funnel. The two phases are separated after cooling. 350 cc of xylene are added to the organic phase in order to favor the following separations. Six successive washings are effected with 250 cc of water. 80 cc of isopropanol may be added in order to promote the separation.

After filtration, the solvents are distilled off at 140°C under a pressure of 20 mm of mercury. A viscous washed product is obtained. Analysis of this product shows a nitrogen content of 1.5%, and a residual chlorine content of 0.5%.

Example 3: Into a 2 ℓ, three necked, round bottom flask provided with a mechanical agitator, a vertex, and a reflux condenser there are introduced in succession: (a) 1,007 g (about 1 mol) of the chlorinated polyisobutylene prepared in Example 1, (b) 15 g of sodium carbonate, and (c) 152 g of a mixture of amines containing 75% of tris(6-amino-3-oxa-hexyl)amine and 25% of N,N-bis-(6-amino-3-oxa-hexyl)ethanolamine, which corresponds to a molar ratio of amine to chlorinated polybutylene of about 0.5.

The mixture is heated at 140°C for 15 hr. It is cooled to 90°C and thereupon 200 cc of a 10% aqueous sodium hydroxide solution are added. The reaction mixture is then treated in the same manner as in Example 2. At the end of the treatment, there is a light product obtained, analysis of which shows a content of 2.1% nitrogen, and 0.6% residual chlorine.

These compounds, upon testing, showed good dispersing, antirust and antifoam properties.

Mannich Condensation Products

According to *R.E. Malec; U.S. Patent 4,186,102; January 29, 1980; assigned to Ethyl Corporation*, Mannich condensation products having excellent dispersant-detergent properties in lubricating oil and liquid hydrocarbon fuels and improved corrosion properties may be obtained by reacting: (a) high molecular weight (over 650) hydrocarbon-substituted phenols, (b) aldehydes, (c) ammonia or an amine having a reactive hydrogen atom, and (d) alkylene oxides.

A preferred embodiment composition made by the process comprising reacting: (a) 1 mol part of an aliphatic hydrocarbon-substituted phenol having an average molecular weight of 650 to 5,000, (b) 1 to 10 mol parts of a C_{1-4} aldehyde, (c) 1 to 10 mol parts of a nitrogen compound selected from the group consisting of ammonia and amines containing at least 1 $HN<$ group and containing 1 to 20 carbon atoms, and (d) 0.1 to 50 mol parts of an alkylene oxide containing 2 to 6 carbon atoms.

The most useful aliphatic hydrocarbon substituent is a polymer of a C_{2-4} olefin, and the most useful aldehyde is formaldehyde.

A highly preferred class of amine reactants are the alkylene polyamines which have the formula $H_2N{-}(R_1{-}NH{-})_nH$ wherein n is an integer from 1 to 6 and R_1 is a divalent hydrocarbon group containing 2 to 4 carbon atoms. These include the series ethylene diamine, diethylene triamine, triethylenetetramine, tetraethylene, and the like.

Of these alkylene polyamines, a most preferred reactant is tetraethylenepentamine or a mixture containing mainly tetraethylenepentamine or having an average composition corresponding to tetraethylenepentamine. Such a material is commercially available as Polyamine H.

Example 1: In a reaction vessel was placed 169 g of phenol, 500 g of heptane and 920 g of polybutene ($\overline{M}_n$ 920). This was stirred and heated to 40°C and 40 g of BF_3·2-phenol added. Stirring was continued for 1 hr at 50°C and then the mixture was washed with a solution of 14.3 vol % concentrated ammonium hydroxide, 28.6 vol % water and 57.1 vol % methanol.

The mixture was washed a second time with 57.1 vol % methanol in water. The washed polybutene-substituted phenol was heated to 35°C and 300 g of tetraethylenepentamine and 72 g of paraformaldehyde added. This was stirred for 6.75 hr while gradually heating to 175°C. Full aspirator vacuum was applied during the last 2 hr after the mixture reached 130°C to remove volatiles.

A concentrate was prepared by blending 600 g of the above product with 300 g of mineral oil.

A 798 g portion of the concentrate was placed in a reaction vessel and, while stirring, ethylene oxide was bubbled through it for 2.5 hr at 90° to 120°C. A weight increase of 48 g was obtained. The final product was water washed and volatiles removed by heating under vacuum. The product was an effective dispersant-detergent.

Example 2: In a reaction vessel was placed 920 g of polybutene ($\overline{M}_n$ 920), 169 g of phenol and 500 g of heptane. The mixture was stirred and heated under nitrogen to 45°C. Then, 40 g of BF_2·2-phenol complex was added and the mixture stirred at 50°C for 1 hr. The polybutene-substituted phenol product was then washed in the manner of Example 1.

The product was transferred to a second reaction vessel and 50 g of diethylenetriamine was added at 35°C. The mixture was stirred and then 36 g of paraformaldehyde was added. While stirring, the resultant mixture was gradually heated to a temperature of 180°C over a 7 hr period. Aspirator vacuum was applied after 5 hr when the mixture attained 130°C. The product weighed 1,051 g and was diluted with 526 g of mineral oil to give a 66.6% active concentrate.

A 1,035 g portion of the above concentrate was placed in a reaction vessel and heated to 65°C. Ethylene oxide was bubbled through this for 3 hr while raising the temperature gradually to 120°C. A 44 g weight increase was observed. One-half of this product was removed and 11 g of diluent mineral oil added to it to maintain a 66.6% active concentrate.

The additives are useful as ashless dispersants in a broad range of lubricating oils, both synthetic and mineral.

The amount of dispersant added should be an amount sufficient to impart the required degree of dispersancy. A preferred range is from 1 to 5 wt %. The additives are also useful in liquid hydrocarbon fuels such as distillate fuel oil, diesel fuels and gasoline. They help in suppressing sludge formation in distillate fuels during storage. They also are beneficial in preventing diesel injector plugging. In gasoline they help maintain a clean carburetor and clean intake valves.

A highly preferred embodiment is a liquid hydrocarbon fuel of gasoline boiling range of 80° to 430°F containing in addition to the detergent additive a small amount of a mineral oil. This embodiment is particularly advantageous in promoting the cleaning of intake valves and stems. The amount of oil added can be any amount from 0.05 to 0.5 vol %, based on the final gasoline. Although the oil adjuvant can be any of the well-known mineral oils including those obtained from Pennsylvania, midcontinent, Gulfcoast, or California crudes, the more preferred are the naphthenic mineral oils. The viscosity of the mineral oil can vary from 100 to 2,000 SUS at 100°F.

Example 3: In a blending vessel place 10,000 gal of Gasoline E, 2.5 lb of detergent of Example 1, and 50 lb of a neutral mineral oil (viscosity 100 SUS at 100°F). Stir the mixture, resulting in an unleaded gasoline having good detergent properties.

Haloalkyl Phenol/Unsaturated Compound Condensation Product

D.E. Ripple; U.S. Patent 4,194,886; March 25, 1980; assigned to The Lubrizol Corporation has devised a process for making a class of compounds which are effective as fuel additives for antiicing, carburetor cleaning and the prevention of screen clogging. The additives are made by reacting at a temperature ranging from 100° to 300°C the following:

(A) At least one alpha haloalkyl aromatic compound of the general formula:

$$\text{(1)} \qquad \left[\begin{matrix} R_m \\ R_2'C(X) \end{matrix} \right] Ar - (OH)_n$$

wherein Ar is a hydrocarbyl aromatic nucleus of 6 to 30 carbon atoms, preferably a benzene nucleus, or a substituted analog of such an aromatic nucleus substituted with one or more, up to three each of lower alkoxy, lower alkylthio, chloro, or nitro substituents, each R is a nonfused hydrocarbyl group of 25 to 700 carbon atoms, X is a halogen atom (preferably a Cl or Br atom), each R' is independently a hydrogen atom, an alkyl group of 1 to 36 carbon atoms, or a halogen substituted alkyl group of 1 to 36 carbon atoms, n is 1 to 3, and m is 1 to 5 with the provisos that the total number of carbon atoms in both the R' groups does not exceed 36, and where m exceeds 1, one of the R groups can also be a $R_2'C(X)-$ group with

(B) At least one alpha-beta olefinically unsaturated compound selected from the group consisting of C_{2-40} hydrocarbyl nitriles, C_{2-40} hydrocarbyl carboxylic acids and anhydrides, esters, amides and ammonium and metal salts of the C_{2-40} carboxylic acids.

The ratio of (A) and (B) is between about 0.5:1-2:1.

The reaction of (A) with (B) results in the formation of a carbon-to-carbon bond, including the carbon of at least one $R_2'C(X)-$ group.

Particularly preferred hydroxy-aromatic precursors used as sources of the halo-alkyl hydroxy-aromatic compounds are monosubstituted phenols and naphthols, particularly the monosubstituted phenols (i.e., where Ar is phenyl and n and m are both one in Formula 1). In such monosubstituted phenols, R can be a relatively high molecular weight long-chain group containing about 25 to 250 carbon atoms. Typical of such groups are alkyl and alkenyl groups made from homo- and interpolymers of ethylene, propylene, butylenes and isobutylene.

The olefinically unsaturated reactant (B) is preferably an acid or anhydride, e.g., acrylic, methacrylic, cinnamic, crotonic, maleic, fumaric, mesaconic, and itaconic acid or their derivatives, such as maleic anhydride.

The following examples will illustrate the process for preparing these additives. All parts and percentages are by weight unless otherwise stated.

Example 1: A mixture of 1,412 parts of phenol and 1,090 parts of benzene is heated to 50° to 55°C; then 283 parts of a boron trifluoride phenol complex (BF_3·2 phenol) is added over 20 min. Following this addition, 5,000 parts of a polyisobutene having a $\overline{M}_n$ of about 1,000 is added. The mixture is stirred for 2 hr at 55° to 60°C and then 645 parts of ammonium hydroxide is added. Stirring is continued for an additional hour. The mixture is heated to 160°C for 4 hr while an azeotrope of phenol, water and benzene distills from it. Stripping the mixture to 220°C/10 mm Hg and filtering it through diatomaceous earth provides the desired alkylated phenol which has a $\overline{M}_n$ of 1,047 and an infrared spectrum consistent with its structure as an alkylated phenol.

Example 2: A mixture of 4,549 parts of the alkylated phenol described in Example 1, 540 parts of paraformaldehyde and 2,500 parts of petroleum naphtha boiling between 96° and 102°C is heated to 55° to 60°C for 2 hr to effect homogenization. Gaseous hydrogen chloride is then bubbled into the reaction mixture at a rate of 2 cfh through a glass tube whose orifice is located below the mixture's surface for a total of 12 hr. The mixture is stirred for an additional 3.5 hr and blown with nitrogen at a rate of 1.5 to 2 cfh for an additional 8 hr. The mixture is filtered through filter aid and the filtrate stripped to 90°C/25 mm Hg to provide the final product which has a chlorine content of 2.1%, and an infrared spectrum consistent with its structure as a chloromethylated alkylated phenol.

Example 3: A mixture of 4,302 parts of the chloromethyl alkylated phenol described in Example 2 and 274 parts of maleic anhydride is heated to 210° to 215°C for 7.5 hr while being nitrogen blown. Excess maleic anhydride is removed by stripping the mixture to 210°C/10 mm and the reaction mixture diluted with 2,973 parts of a diluent oil. Filtration with filter aid provides a 40% solution of the desired product.

Examples 4 through 8: These examples are all carried out in substantially the same fashion. Details as to the reactants and proportions used in Examples 4 through 8 are summarized in the following table.

Ex. No.	α-Haloalkylphenol	(mols)	Unsaturated Reactant	(mols)
4	Example 2	0.245	2-Ethylhexyl acrylate	0.25
5	Example 2	0.245	Di(n-butyl) furmarate	0.25
6	Example 2	0.32	Itaconic acid	0.32
7	Example 2	0.2	4-Cyclohexene-1,2-carboxylic acid anhydride	0.2

(continued)

Ex. No.	α-Haloalkylphenol	(mols)	Unsaturated Reactant	(mols)
8	Alkyl-substituted chloromethyl phenol*	2.55	Maleic anhydride	2.8

*Made by alkylating phenol with a mixture of C_{15-18} 1-olefins according to general procedure described in Example 1.

The fuel compositions of this process can contain about 0.001 to 5% (based on the weight of the final composition) preferably 0.001 to 1%, of the abovedescribed products. The presence of these products can impart many desirable characteristics to the fuel composition depending upon the particular composition and fuel mixture selected. Thus, in gasolines they may improve the overall composition ability to retard corrosion of metal parts with which it may come in contact or improve the fuel's ability to clean carburetors and reduce carburetor icing. On the other hand, these products can be used in fuel oil compositions and other normally liquid petroleum distillate fuel compositions to impart antiscreen clogging and demulsifying properties to the fuel.

Hydroxyalkyl/Hydroxy Aromatic Condensation Product

In related work, *C.P. Bryant; U.S. Patent 4,205,960; June 3, 1980; assigned to The Lubrizol Corporation* describes additive compounds which have the same general formula as Formula 1 in the previous patent except that the halogen atom (X) is replaced by a hydroxy group. The formula then, is as follows, with Ar, m, n, R and R' having the same meanings as those given in the previous patent:

$$R_m\text{—}(Ar)\text{—}(OH)_n \quad \text{with} \quad R'_2C(OH)\text{—}$$

Example 1: A mixture of 1,040 parts of a polyisobutyl phenol (polyisobutyl group $\overline{M}_n$ 885) and 326 parts of diluent oil at 55°C is treated with four parts of saturated aqueous sodium hydroxide. The treatment of the mixture is then increased to 75°C and 72 parts of paraformaldehyde is added. The mixture is held at 75°C for two additional hours and then stored at 30°C for 16 hr. The hydroxide catalyst is neutralized with four parts of acetic acid and an additional 480 parts of diluent oil is added to the reaction mixture before it is stripped at 40° to 75°C/8 torr. Infrared analysis of the residual oil solution establishes formation of the desired hydroxymethyl intermediate.

Example 2: A mixture of 1,918 parts of the oil solution described in Example 1 and 150 parts of maleic anhydride is heated in slow stages to 194°C over 3 hr and held at 190° to 195°C for 2 hr. It is then stripped to 190°C/10 torr for 1 hr and filtered through filter aid to provide an oil solution of the desired condensation product.

Example 3: To 1,453 parts of the oil solution described in Example 2 at 85°C is slowly added 63.5 parts of a commercial polyethylene polyamide mixture having 3 to 7 amino groups per molecule. The reaction mixture is then heated to 165°C and held at this temperature for 3 hr while being blown with nitrogen. After storage overnight at room temperature, it is heated to 175°C with nitrogen blowing for an additional 3 hr and stripped at 170°C/8 torr for 1 hr. Filtration through filter aid provides an oil solution of the desired product.

Gasoline compositions containing from 0.001 to 1% of the above additives retard corrosion, clean carburetors and reduce carburetor icing. In fuel oil compositions the additives impart demulsifying properties and retard screen clogging.

Aminoalkylalkanolamine Reaction Product with Acids

B.R. Bonazza and S. Schiff; U.S. Patent 4,230,588; October 28, 1980; assigned to Phillips Petroleum Company have developed a hydrocarbon-soluble detergent suitable for use in the fuel of an internal combustion engine. It is made by reacting fatty acid with aminoalkylalkanolamine.

In another embodiment a hydrocarbon-soluble detergent having improved water tolerance is provided by reacting with arylsulfonic acid the reaction products described above in which all of the reactant nitrogen has not been amidated.

In the reaction of fatty acid with aminoalkylalkanolamine, depending upon the amine chosen and the molar ratio of acid to amine, the additive product will be an amide amine alcohol, a multiamide-alcohol, or a multiamide-ester.

These will hereinafter be referred to as amide product. In the fatty acid, RCOOH, R is a hydrocarbyl radical of 7 to 99 carbon atoms, preferably 12 to 33 carbon atoms. These hydrocarbyl radicals include alkyl, alkenyl, cycloalkyl, and aralkyl groups. Suitable examples are capric acid, myristic acid, stearic acid, oleic acid, phenylstearic acid, naphthyllauric acid, and the like.

Reactant aminoalkylalkanolamines have the general formula:

$$H_{3-a}N(CH_2CH_2NR'CH_2CH_2OH)_a$$

where R' is H or $CH_2CH_2NH_2$ and a is 1, 2 or 3. Suitable examples are N-(2-aminoethyl)-2-aminoethanol, N,N-bis(2-aminoethyl)-2-aminoethanol, tris-[N-(2-hydroxyethyl)aminoethyl] amine, and the like.

Each primary nitrogen, secondary nitrogen, and carbinol group in the reactant amine can be reacted with the fatty acid(s) to form amide, amide and ester, respectively. Since amidation and esterification are effected by elimination of a mol of water, determination of the quantity of water evolved from the reactants provides a criterion to determine how far the reaction has proceeded. It is presumed that, when the quantity of fatty acid is insufficient to react with all available functional groups, i.e., primary and secondary amines and carbinols, the amines are substantially completely amidated before esterification begins.

The amide product, provided that not all of the reactant nitrogen has been amidated, may be reacted with an arylsulfonic acid to improve its water tolerance. Suitable sulfonic acids have the general formula $R''SO_3H$ where R'' is an aryl or an alkaryl group with 6 to 100 carbon atoms. Benzenesulfonic acid, dodecylbenzenesulfonic acid, or the acid oil product made by treating lubricating stock with sulfur trioxide are examples of suitable sulfonic acids. The ratio of sulfonic acid to amide product to prepare the finished additive can be determined by titrating them separately with standard base and standard acid, respectively, using a glass electrode pH meter. Samples are dissolved in titration solvent (equal volumes of benzene and isopropanol plus 0.5 vol % water) for the titration. The quantity of sulfonic acid added to the amide product should produce material having pH of about 7 to 8. The reaction is effected by combining predetermined quantities of sulfonic acid and the amide product, warming to 50° to 70°C for 15 to 30 min, with stirring to produce a homogeneous phase. Viscous reactants may conveniently be thinned by dilution with lubrication stock or other solvents to facilitate their mixing.

These amide products or their derivatives that have been neutralized (with sulfonic acid) are detergent additives that are added to motor fuel in the concentration range of 1 to 100 lb/1,000 barrels, preferably 5 to 30 lb/1,000 barrels, to prevent harmful carburetor and fuel intake system deposits.

Example 1: The amide amine alcohol resulting from reaction between equimolar quantities of phenylstearic acid and N-(2-aminoethyl)-2-aminoethanol was prepared by combining 74.6 g (0.20 mol) of the former with 20.8 g (0.20 mol) of the latter plus about 20 ml of toluene. These were heated to reflux (about 130°C) in a flask fitted with a condenser, a Barrett water trap, a thermowell for temperature observation and controlling, and a magnetic stirrer.

A slow stream of nitrogen, introduced via a tube in the condenser, maintained an inert atmosphere in the reactor. During 2 hr of refluxing, the stoichiometric quantity of water was collected. The phenylstearamide of N-(2-aminoethyl)-2-aminoethanol produced, dissolved in toluene, was cooled. Analysis of the solution by potentiometric titration with standard acid showed it to contain 0.239 equivalents of basic nitrogen. A portion of this product was freed from solvent in a rotary evaporator for subsequent evaluation as an additive.

Example 2: A portion representing 30% (0.072 equivalents) of the solution made in Example 1 was reacted with 20 g of acid oil which was made by sulfonating lubricating base stock with sulfur trioxide, the equivalent weight of the acid oil was 1,290. This quantity of acid oil was chosen because its product with the amide from Example 1, when diluted with titration solvent, had a pH of 7.5. After removal of solvent, the product was evaluated as an additive.

Example 3: Additives whose preparation is described in the preceding examples were subjected to a series of tests in gasoline at the concentrations listed.

(a) Falcon engine test: 10 lb/1,000 barrels. (All additives containing acid oil were tested at 20 lb/1,000 barrels.)

(b) Thin layer chromatography (TLC) test for detergency: 6.3 wt % additive.

(c) Spray gum deposit: 0.07 wt % additive or 175 lb/1,000 barrels.

The additive of Example 2 was markedly superior to the unreacted product.

In similar work, *H.D. Holtz and B.R. Bonazza; U.S. Patent 4,249,912; February 10, 1981; assigned to Phillips Petroleum Company* prepared detergent additives by reacting a polycarboxylic amino acid with alkylamines and, similarly, further combines the aminoamide with a sulfonic acid to improve its water tolerance.

The acids have the general formula $R_2N[(CH_2)_aNR]_bR$ where a is 1 to 6, b is 0 to 5, and R is $-CH_2COOH$. Suitable examples are N-nitrilotriacetic acid, ethylenediaminetetraacetic acid, diethylenetriaminepentaacetic acid, and the like. Primary amines used in the synthesis have the formula $R'NH_2$ where R' is a hydrocarbyl radical of 8 to 100 carbon atoms, preferably 12 to 25 carbon atoms. These hydrocarbyl radicals include alkyl, alkenyl, cycloalkyl, cycloalkenyl, aryl, and combinations such as alkaryl, aralkyl, alkylcycloalkyl, arylcycloalkyl, aralkenyl, and arylcycloalkenyl. Suitable examples are caprylamine, lauryl amine, palmityl amine, oleyl amine; also any isomers or mixtures of isomers of isostearyl amine, phenylstearyl amine, cyclohexylaniline, butylaniline, and the like.

Synthesis of the additives of this process requires essentially complete amidation of all carboxyl groups on the reactant amino acid(s). Since a mol of water is made for each amide group formed, measurement of water produced in the reaction is a convenient way to follow its progress. The reaction produces a compound of the formula $R''_2N[(CH_2)_aNR'']_bR''$ where a is 1 to 6, b is 0 to 5, and R'' is $-CH_2CONHR'$ with R' as defined above.

Example 1: Additive 1 was made by reaction of 38.5 g (0.14 mol) of tallow amine with 10.2 g (0.035 mol) of ethylenediaminetetraacetic acid (EDTA). The amine, Armeen T, (Armak Co.) is made by amination of the fatty acids contained in tallow. The hydrocarbyl portion of the amine ranges from C_{14-18}, and is distributed approximately 50% paraffinic and 50% olefinic. These materials together with about 100 cc of mixed xylenes were placed in a 250 cc flask fitted with a Barrett water trap and its associated water-cooled condenser to collect evolved water, and a thermometer-containing thermowell, and a magnetic stirrer.

The flask was heated, and during about 6 hr refluxing the theoretical volume (0.14 mol) of water was collected. Solvent was removed by warming the preparation under reduced pressure, leaving a brown, solid product. Then, to 21.2 g (0.0152 mol) of the tetraamide of EDTA, 9.9 g (0.0304 mol) of dodecylbenzenesulfonic acid (DBSA) was added to neutralize the two amine nitrogens. The neutralization was effected by combining the components, each dissolved in about 30 cc of toluene, and warming to about 90°C until all solids were dissolved. Finally, solvent was removed by warming at reduced pressure.

Example 2: Additive 2 was made by reaction of 75.6 g (0.36 mol) of coco amine with 26.3 g (0.09 mol) of EDTA in xylene solution. The amine, Armeen C, (Armak Co.) is made by the amination of the fatty acids contained in coconut oil. It contains C_{8-18} chains that are mostly paraffinic. Reaction conditions were essentially identical to those detailed for Additive 1. After 4 hr of reflux the theoretical quantity (0.36 mol) of water had been collected. Half of the resulting tetraamide (0.045 mol) was combined with 29.3 g (0.09 mol) of dodecylbenzenesulfonic acid in about 200 cc of xylene solution. After warming to complete the neutralization, and to dissolve the acid, solvent was removed by evaporation.

Example 3: Additive 3 was made by reaction of 83.3 g (0.24 mol) of phenylstearyl amine with 15.3 g (0.08 mol) of N-nitrilotriacetic acid in about 80 cc of xylenes. Experimental arrangement and conditions were essentially the same as stated for Additive 1. The theoretical volume of water (0.24 mol) was collected in 3.5 hr while refluxing at 150°C reactor temperature. This solution of triamide was divided into two equal parts. The solvent was removed from one part of reduced pressure on a rotary evaporator.

Example 4: Additive 4 consisted of reacting the remaining half of Additive 3 with 12.4 g (0.038 mol) of dodecylbenzenesulfonic acid. Solvent was removed from the product by warming at reduced pressure in a rotary evaporator, and the residue was reserved for testing.

Tests similar to those performed on the additives of the two previous patents showed them to have good detergency; the Falcon engine test demonstrated that they reduced carburetor deposits.

Imidazoline and Its Reaction Product with a Sulfonic Acid

Detergent additives as described by *B.R. Bonazza and H.D. Holtz; U.S. Patent 4,247,300; January 27, 1981; assigned to Phillips Petroleum Company* are made by reaction of a polyamine with a carboxylic acid to produce an imidazoline. In the carboxylic acid, RCOOH, R is a hydrocarbyl radical of 7 to 99 carbon atoms, preferably 11 to 25 carbon atoms. These hydrocarbyl radicals include alkyl, alkenyl, cycloalkyl, cycloalkenyl, aryl and combinations such as alkaryl, aralkyl, alkylcycloalkyl, arylcycloalkyl, aralkenyl, and arylcycloalkenyl.

Suitable examples are capric acid, myristic acid, stearic acid, oleic acid; and also any isomers or mixtures of isomers of isostearic acid, phenylstearic acid, naphthyl lauric acid, phenylcyclohexylcarboxylic acid, cyclohexylbenzoic acid, and the like.

Polyamines required to make imidazolines have the general formula:

$$H_2NCH_2CH_2NH(CH_2CH_2NR')_aH$$

where R' can be hydrogen, alkyl or $(CH_2CH_2NH)_bH$. The sum of a and b should not exceed 10; a can have values between 0 and 10; b can range between 0 and 8. Suitable examples of polyamines include ethylenediamine, diethylenetriamine, tetraethylenepentamine, hexaethyleneheptamine, and polyamines such as these in which internal (secondary) nitrogens bear the $(CH_2CH_2NH)_bH$ group.

When desired, and to improve its water tolerance, the imidazoline additive is reacted with an arylsulfonic acid. A basic additive is preferred, so the amount of sulfonic acid used is not greater than the stoichiometric quantity necessary completely to neutralize the amine nitrogen. Suitable sulfonic acids have the general formula $R''SO_3H$ where R'' is an aryl or an alkaryl group with 6 to 100 carbon atoms. Benzenesulfonic acid, isopropylbenzenesulfonic acid, cyclohexylbenzenesulfonic acid, dodecylbenzenesulfonic acid, and dioctylbenzenesulfonic acid are suitable examples. The ratio of sulfonic acid to imidazoline required to prepare the finished additive may be determined by titrating them separately with standard base and standard acid, respectively, using a glass electrode pH meter.

Samples are dissolved in titration solvent (equal volumes of benzene and isopropanol plus 0.5 vol % water) for the titration. The quantity of sulfonic acid added to the imidazoline is chosen to make a product having pH of 7 to 9.

The imadazolines, or the products of their reaction with sulfonic acid, are detergent additives which can be added to motor fuel in the concentration range of preferably 5 to 30 PTB, to prevent harmful carburetor and fuel intake system deposits.

Example 1: *Reaction of isostearic acid and tetraethylenepentamine* – To a 500 ml round bottom flask fitted with a Barrett water trap with associated water-cooled condenser, and with a thermowell containing a thermometer were added 79.5 g (0.28 mol) isostearic acid, 26.5 g (0.14 mol) tetraethylenepentamine, and about 50 cc of toluene.

Reactants were blanketed with nitrogen that was admitted through a tube via the Barrett trap. The flask was heated to reflux (135°C); in about 30 min 2.5 cc of water had been collected. Using a stopcock that permitted draining condensate from the Barrett trap, sufficient toluene was removed to raise the boiling point to about 170°C. In about 3 hr a total of 8.4 cc (0.47 mol) of water was collected.

In 90 min of additional refluxing no more water was liberated. The amount of water collected represents complete reaction of the isostearic acid, equivalent to formation of about 70% imidazoline and 30% amide. By titration with standard HCl the equivalent weight of this additive was determined to be 568.

Example 2: *Reaction with acid oil* – To 15.3 g (0.0269 eq) of the additive from Example 1 was added 34.7 g (0.0270 eq) of acid oil made by sulfonating a 250 weight lubricating base stock with sulfur trioxide. The product of this neutralization, as a 1% solution in titration solvent (described above), had a pH of 8.6.

Example 3: *Reaction with dodecylbenzenesulfonic acid* – To 31.8 g (0.0560 eq) of the additive from Example 1 was added 18.1 g (0.0559 eq) of dodecylbenzenesulfonic acid, Sulframin 98 Hard Acid (Witco Chemical). The product of this neutralization, as a 1% solution in titration solvent, had a pH of 8.2.

These additives were evaluated by the Falcon engine test, the thin layer chromatography test and the spray gum deposit test, all of which showed them to have good or good-excellent ratings for detergency; in the Falcon engine test all three additives reduced carburetor deposits by at least 80%. All additives gave excellent results in the spray gum test, leaving essentially no residue. The diisostearylimidazoline of tetraethylenepentamine (Example 1) failed the water tolerance test. However, reaction with either acid oil or dodecylbenzenesulfonic acid produced an additive that readily passed the test.

Nitrophenol-Amine Condensates

K.E. Davis; U.S. Patent 4,231,757; November 4, 1980; assigned to The Lubrizol Corporation has found that nitrogen-containing compositions made by condensing a nitrophenol with an organic amino compound having at least one hydrogen atom directly bonded to a nitrogen or oxygen atom are useful as additives for lubricants and fuels. A specific embodiment comprises compositions made by reacting nitrophenols having hydrocarbyl substituents of up to 750 aliphatic carbon atoms with alkylene polyamines.

In the following examples all parts and percentages are by weight, and likewise all temperatures are in degrees Celsius, unless expressly stated to the contrary.

Example 1: An alkylated phenol is prepared by reacting phenol with a polybutene having a number average molecular weight of approximately 1,000 (Vapor Phase Osmometry) in the presence of a boron trifluoride-phenol complex catalyst. Stripping of the product thus formed, first to 230°/760 torr (vapor temperature) and then 205°/50 torr (vapor temperature) provides polybutene-substituted phenol.

Example 2: A mixture of 4,578 parts of the polybutene-substituted phenol, 3,052 parts of diluent mineral oil and 725 parts of textile spirits is heated with agitation to 60°C to achieve homogeneity. After cooling to 30°C, a mixture of 319.5 parts of 16 molar nitric acid and 600 parts of water is slowly added into the mixture which is kept below 40°C by external cooling.

After stirring the mixture for an additional 2 hr, 3,710 parts are transferred to a second reaction vessel. The remaining material is stripped to 150°/43 torr (vapor temperature), cooled to 110°C and filtered through diatomaceous earth to provide as a filtrate the desired nitrophenol. This material has a nitrogen content of 0.58%.

Example 3: To 1,353 parts of the nitrophenol described in Example 1 under nitrogen atmosphere, at 49°C is added 61.5 parts of a commercial polyethylene polyamine mixture containing 33.5% nitrogen and substantially corresponding in emperical formula to tetraethylene pentamine. The reaction mixture thus formed is heated to 80°C for 1.5 hr and then stored for 16 hr at 25°C. It is then heated to 130° to 160°C for a total of 15 hr and finally stripped to 160°/30 torr (vapor temperature). The residue was filtered through diatomaceous earth to give the desired product as a filtrate. This product contains 1.5% nitrogen.

Example 4: A mixture of 1,600 parts of a polybutene-substituted phenol prepared in essentially the same fashion as that described in Example 1 from a polybutene having a number average molecular weight of 1,400 (Gel Permeation Chromatography), 10 parts of aqueous hydrochloric acid and 33 parts of paraformaldehyde is heated to 90°C under a nitrogen atmosphere for a total of 20 hr with intermittent storage at room temperature. 500 parts of textile spirits are then added, followed by 91.3 parts of concentrated nitric acid in 100 parts water.

During the nitric acid addition the reaction temperature was maintained at 31° to 38°C by external cooling. The reaction mixture is then stirred for 2 hr at room temperature and 61.5 parts of the polyethylene polyamide described in Example 3 is added slowly. The reaction mixture is heated to 160°C for 7 hr and then stripped at 160°/30 torr (vapor temperature) and the residue filtered through diatomaceous earth. The filtrate is the desired product and has a nitrogen content of 0.88%.

A fuel composition suitable for use in an automobile spark-ignited engine equipped with a carburetor is prepared by adding the product of Example 4 at a dosage rate equivalent to 26.2 lb/1,000 barrels to a commercially available gasoline. These fuel additives function effectively as detergents and exhibit engine sludge dispersancy.

Overbased Manganese Salts of Organic Acids

Overbased manganese salts of organic acids, which are compounds in which manganese is present in excess of the stoichiometric amount required to react with the acidic groups of the organic acids, are widely used as additives in liquid hydrocarbon fuels and in lubricants for internal combustion engines. The basicity of these additives counteracts the corrosive acidic compounds that are formed during the operation of the engines and inhibits the formation of deposits of soot, lacquer, and sludge in the engines. In addition, these additives act as smoke suppressants in the fuels and improve the detergency of the lubricants.

A. Ali; U.S. Patent 4,252,659; February 24, 1981; assigned to Tenneco Chemicals, Inc. has developed an improved process for the production of overbased manganese salts and of solutions of overbased manganese salts that contain at least 13 wt % of manganese.

In this process, a reaction mixture that comprises manganous oxide, an oil-soluble organic acid, a solvent system, a promoter, a copromoter, and a third promoter that is a lower alkanoic acid is contacted with carbon dioxide while it is maintained at a temperature in the range of 70° to 120°C at a pressure in the range of 1 to 10 atm until the carbonation reaction has been completed and there is obtained a clear, fluid solution of an overbased manganese salt that contains 13 wt % or more of manganese.

The organic acids that are used in the process are organic carboxylic acids and organic sulfonic acids that are oil soluble and that form manganese salts that are oil soluble. They are preferably aliphatic and cycloaliphatic monocarboxylic acids having 4 to 10 carbon atoms, aromatic monocarboxylic acids having 7 to 12 carbon atoms, and mixtures thereof. In most cases a 5 to 100% molar excess of manganous oxide is used.

The promoters include ammonium halides, ammonium nitrate, mono-, di-, and trialkylamine hydrochlorides, ammonium sulfide and ammonium peroxydisulfate. Among the useful copromoters are alkaline earth metal halides, aluminum chloride, and ferric chloride. Excellent results have been obtained using 2 to 4% ammonium chloride as the promoter and 4 to 8% of calcium chloride or barium chloride as the copromoter, based on the weight of manganous oxide in the reaction mixture.

The solvent system preferably contains 25 to 40 wt % of an alcohol and 60 to 75 wt % of a hydrocarbon. Best results have been obtained when the alcohol was methoxyethanol or methoxyethoxyethanol and the inert organic solvent was hi-flash naphtha that contained 9% C_8 alkylbenzenes, 83% C_9 alkylbenzenes, and 8% C_{10} alkylbenzenes.

The process is further illustrated by the following examples in which all parts and all percentages are by weight.

Example 1: To a stainless steel autoclave were charged 59.0 parts of n-heptanoic acid, 69.0 parts of manganous oxide (77.2% $\overline{M}_n$), 3.3 parts of 90% formic acid, 2.0 parts of ammonium chloride, 3.4 parts of calcium chloride, 152.0 parts of hi-flash naphtha, and 59.0 parts of 2-methoxyethoxyethanol.

The autoclave was sealed and purged twice with carbon dioxide to remove air from it. The reaction mixture was stirred and heated to 96° to 102°C while the pressure in the autoclave was maintained at 2 atm by the addition of carbon dioxide as needed. After 2 hr, a vigorous reaction occurred as shown by a sharp exotherm and by rapid absorption of carbon dioxide.

The reaction mixture was maintained at 96° to 102°C and carbon dioxide was added to it at a rate sufficient to maintain the pressure at 2 atm for 3 hr after the start of the exotherm.

The system was vented to the atmosphere, and a solution of 0.066 part of an antifoam agent (SAG-47) in 1.0 part of the product was added to the reaction mixture. The reaction mixture was then sparged with nitrogen at 93° to 99°C under subatmospheric pressure (15" Hg absolute) for 30 min cooled to 65°C, treated with filter-aid, and filtered. There was obtained 344.3 parts of an overbased manganese heptanoate solution that contained 13.5% of manganese.

The autoclave and filter cake were washed with 40.0 parts of hi-flash naphtha. The concentrated product was then diluted with a portion of the wash liquor to obtain 373 parts of an overbased manganese heptanoate solution that contained 13.0% of manganese and that had a Gardner-Holdt viscosity of <A-5 at 25°C and A-4 at –15°C.

Example 2: A mixture of 45 parts of n-heptanoic acid, 81.5 parts of methoxyethanol, 138 parts of hi-flash naphtha, 2.3 parts of glacial acetic acid, 2.6 parts of ammonium chloride, 5.2 parts of calcium chloride, and 65 parts of manganous oxide was stirred and heated in an autoclave at 95° to 100°C under a carbon dioxide pressure of 1.5 atm. After the reaction mixture had been maintained under these conditions for 90 min, the carbonation reaction began, as was shown by a sharp exotherm and by the rapid absorption of carbon dioxide.

The reaction mixture was maintained at 95° to 100°C and carbon dioxide was added to it at a rate that maintained the pressure at 1.5 atm until there was no further absorption of carbon dioxide. The reaction mixture was cooled, treated with filter-aid, and filtered.

There was obtained an overbased manganese heptanoate solution that contained 13.0% of manganese. The yield was 93.5%, based on the weight of manganous oxide charged, and the overbasing was 347%.

Substituted Phenol/Epichlorohydrin/Amine Adducts

Detergent motor fuel and lubricating oil additives available generally suffer from one or more deficiencies. Either they are used at very high concentrations, for example, of the order of 4,000 ppm, or if used at lower, more economical levels, their detergency and other desirable properties are substantially diminished or lost.

W.H. Machleder and J.M. Bollinger; U.S. Patent 4,259,086; March 31, 1981; assigned to Rohm and Haas Company have demonstrated that the reaction products of certain substituted phenols, epichlorohydrin and amines show excellent carburetor, induction system and combustion chamber detergency, and in addition provide effective rust inhibition when used in hydrocarbon fuels at low concentrations, i.e., between 20 to 600 ppm, and more preferably, between 60 to 400 ppm.

In addition to their activity as fuel additives, these compounds are also ashless rust inhibitors and dispersants for use in lubricating oils at concentrations of 0.1 to 10 wt %, preferably 0.5 to 8 wt %. The products may also be described as, for example, the reaction products of an alkyl phenol with epichlorohydrin followed by amination with an amine such as ethylene diamine or other primary or secondary mono- or polyamine. As hydrocarbon motor fuel (such as gasoline or diesel fuel) additives, the adducts of the process act to control spark plug fouling and thus help to keep the spark plugs relatively clean and free of deposits.

Although there are many carburetor detergents on the market, it is believed that only one, Chevron F-310, can be classified as a true second generation additive possessing the broad based activity achieved with the described amine adducts of this process. However, F-310 is recommended at a high treating level of 4,000 ppm, and that may exceed the industry's handling or economic capabilities.

The multipurpose additives described herein are the reaction products of: (a) a glycidyl ether compound of the formula:

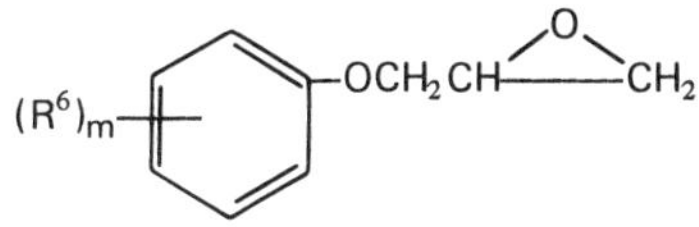

$$(R^6)_m\text{-}C_6H_{5-m}\text{-}OCH_2CH\overset{O}{\frown}CH_2$$

where R^6 is an aliphatic hydrocarbon group having at least 8 carbon atoms and m is 1 to 3, and (b) an amine having at least one amino group, the amine having two or more active hydrogen atoms, and wherein the mol ratio of (a) to (b) is at least 2:1, the amine being selected from aliphatic monoamines, cyclic amines, heterocyclic amines, aromatic amines, and aliphatic polyamines of the formula:

$$R'\underset{\displaystyle R^2}{N}-(R^4\underset{\displaystyle R^5}{N})_n-R^4-\underset{\displaystyle R^3}{NH}$$

wherein R^1, R^2 and R^3 independently are hydrogen, C_{1-6} alkyl substituted by $-NH_2$ or $-OH$, R^4 is a C_{1-6} divalent hydrocarbon radical, R^5 is hydrogen or C_{1-6} alkyl, and n is 0 to 5.

The mol ratio of the glycidyl ether compound to the amine is preferably 2:1-4:1. Suitable amines are N-(3-aminopropyl) morpholine, 3-aminoethyl pyridine, 2-aminoethylaminoethanol, N,N-dimethylethylenediamine, aniline and hexylamine.

A preferred product is N,N'-bis[3-(p-H35-polyisobutylphenoxy)-2-hydroxypropyl]-ethylene diamine, shown by the structural Formula 1 as follows, where R^6 is hydrogen or PIB_{H35}:

$$\text{(1)}\quad \text{Ar}-O-CH_2\underset{}{\overset{OH}{C}}HCH_2NHCH_2CH_2NHCH_2\overset{OH}{C}HCH_2-O-\text{Ar}$$

where each Ar is a benzene ring bearing $-R^6$ ortho to the oxygen and PIB_{H35} para to the oxygen.

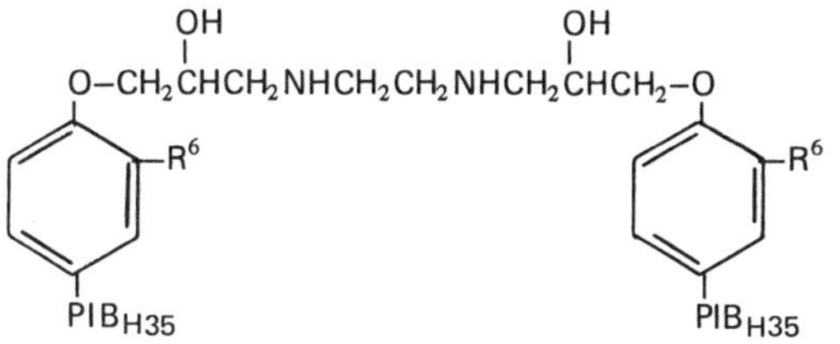

PIB is an abbreviation for a polyisobutene generically of any molecular weight. H35 is the commercial designation for Amoco Chemical Company's polyisobutene having a number average molecular weight of about 670.

More generally, the PIB component may have a number average molecular weight of 500 to 2,000, preferably 600 to 1,500. Optionally, some of the polyisobutene may be in the ortho position (R^6). R^6 may, therefore, simply be the same as PIB or R^6 may be hydrogen.

As indicated in the general description above, the preferred product can be a mixture of Formula 1 and Formula 2 set forth below or it can be Formula 1 or Formula 2 taken singly.

In other words, on a parts per 100 parts basis, Formula 1 can vary from 1 to 99 parts and Formula 2 can vary from 99 to 1 part; or there can be 100 parts of Formula 1 or 100 parts of Formula 2, all parts being on a weight basis,

(2)

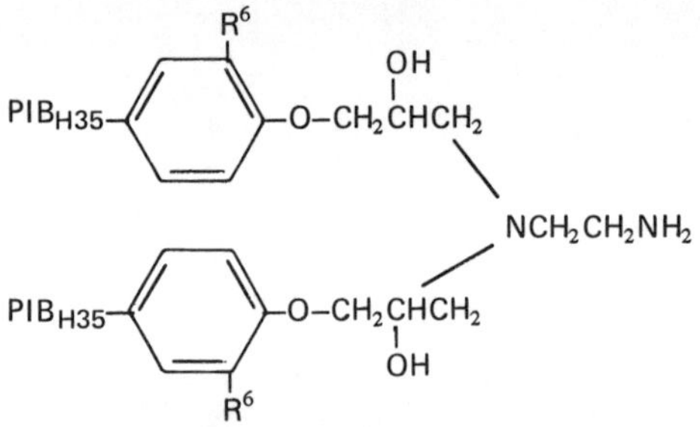

R^6

OH

PIB_{H35}–(benzene ring)–O–CH_2CHCH_2

$NCH_2CH_2NH_2$

PIB_{H35}–(benzene ring)–O–CH_2CHCH_2

OH

R^6

where R^6 is as defined in Formula 1.

The overall amount for use of the amine adduct(s) remains the same no matter what may be the proportions of isomers in the products.

The preferred chemical gasoline additive compound of this process is prepared by the following reaction sequence where R^6 is PIB_{H35}:

(a) Phenol is alkylated with polyisobutene, i.e., polyisobutylene, of molecular weight of about 670 (Amoco H35) using an acid catalyst.

(b) The polyisobutylphenol is converted to the sodium phenoxide using sodium hydroxide and then reacted with epichlorohydrin.

(c) 2 mols of the epichlorohydrin adduct are reacted with 1 mol of ethylene diamine to form the desired product.

The table below presents data comparing the preferred product of this case (Formulas 1 or 2) with Chevron F-310. The essential component in Chevron F-310 is believed to be a polybutene amine as described in U.S. Patent 3,438,757.

The data which indicate the percent reduction in deposits versus untreated gasoline shows that the preferred product greatly improves the performance of untreated gasoline and provides performance comparable to F-310 at a much reduced treating level.

Additive	Recommended Treating Level, (PTB)	ASTM-D665 Rust Test % Area Rusted	Carburetor Detergency Blowby Test	Induction System Test Single Cylinder
			...... % Deposit	Reduction ......
Control base gasoline	–	100	0	0
Chevron F-310	1,000	0	96	99
Described amine adduct	75	0–5	95	94

Example: *Part A* – Polyisobutene H35 phenol: To a 5 ℓ 3 necked flask equipped with a thermometer, mechanical stirrer, and reflux condenser with Dean-Stark trap was charged 1,920 g (2.9 mols) of polyisobutene H35 (Amoco) of about 660 molecular weight, 564 g (6 mols) of phenol, 200 g of Amberlyst 15 acid catalyst, and 550 ml of hexane. The stirred mixture was heated at reflux (pot temperature 100° to 107°C) under a nitrogen atmosphere for 24 hr, during which time 5.4 ml of water had separated.

After cooling to 60° to 80°C, the mixture was filtered to remove the resin beads, the latter being washed with hexane, and the filtrate subjected to vacuum concentration with a pot temperature of 160°C. There was obtained 1,971.4 g of product residue having an oxygen content of 2.92% (theoretical: 2.12%) and molecular weight of about 754. The product is actually a mixture of alkylated phenols with an average molecular weight of 548 based upon the oxygen and 556 calculated from ultraviolet spectral parameters, i.e., in some of the products R^6 is hydrogen and in others R^6 is PIB_{H35}.

Part B – 1,2-Epoxy-3-[p-(H35-polyisobutyl)phenoxy] propane: To a 5 ℓ 3 necked flask fitted with a thermometer, mechanical stirrer, addition funnel and reflux condenser was charged 973 g (1.75 mols based upon 2.92% oxygen) of the polyisobutene H35 phenol obtained in Part A, 72 g (1.75 mols based upon 97.4% assay) of sodium hydroxide pellets, 450 ml of toluene. The stirred mixture was heated under a nitrogen atmosphere at 84° to 90°C for 1 hr to effect the dissolution of the base.

Epichlorohydrin (161.9 g, 1.75 mols) was then added dropwise at 60°C during 2.5 hr, followed by a hold period at 70°C. The reaction mixture was then cooled, filtered, and the salt (107 g dry) washed with toluene. The filtrate was stripped (100°C/15 mm) to give 1,075.3 g of product residue having a molecular weight of about 810.

Part C – N,N'-Bis[3-(p-H35-polyisobutylphenoxy)-2-hydroxypropyl] ethylene diamine: A mixture of 1,018.4 g of the epoxide product of Part B, 122.6 g (2.04 mols) of ethylene diamine, and xylene (700 ml) was heated at reflux (131° to 136°C) with stirring under a nitrogen atmosphere for 18 hr. After vacuum stripping (18 mm, pot temperature of 120°C), there was obtained 1,053.4 g of turbid residue which was filtered through a bed of Celite 545 in a steam-heated Buchner funnel to give clear, yellow viscous product of molecular weight about 1,680.

The mixture obtained in this synthesis may contain the N,N' diadduct, the N,N diadduct and some N or N' monoadduct. The product prepared in this way had 1.26% basic nitrogen (1.67% theory) and 5.26% oxygen (3.81% theory).

In related work, *W.H. Machleder and J.M. Bollinger; U.S. Patent 4,134,846; January 16, 1979; assigned to Rohm and Haas Company* use a mixture of the previously described Component 1, substituted phenol/epichlorohydrin/amine adduct with Component 2, a polyalkylene phenol as a low concentration additive for fuels and lubricating oils.

The polyalkylene phenol Component 2 is used primarily for cost considerations and lowers treating costs. However, Component 2 also promotes deposit reduction as shown in the Induction System Test described in the patent.

Component 2 is any of the polyalkylene phenols used as reactants for the preparation of the glycidyl ether compounds, such as polyisopropylene phenol or polyisobutylene phenol having a number average molecular weight of about 500 to 3,000, preferably about 600 to 1,500.

The polyalkylene phenols may also be mixtures of two or more of the mono-, di- and trialkylated phenols and any straight or branched chain products resulting from the alkylation of phenol. Component 2 may be present in the additive mixture as material remaining unreacted from the preparation of Component 1, or Component 2 may be added separately.

The following table presents data on the use of an additive mixture of the process containing an adduct of Formula 1 or 2 and the polyisobutylene phenol prepared in the above Example, Part A of the previous patent. The comparision primarily is with Chevron F-310. As can be seen from the following table, the mixture of the amine adduct Component 1 and the polyisobutene phenol Component 2 compares very favorably with Component 1 and with the Chevron F-310, when the latter two additives are used alone.

It should also be noted that the Chevron F-310 is used at a much higher level (1,000 lb/1,000 barrels of gasoline) than is either the amine adduct component or the mixture of Component 1 and Component 2.

Performance of Carburetor Blends or Mixtures

Additive	Treating Level (PTB)	Carburetor Detergency Blowby Test*	Induction System Test** Single Cylinder
Control base gasoline	–	0	0
Chevron F-310	1,000	96	~99
Component (1)	75	95	94
Additive mixture Component (1)/ Component (2)	25/50	90, 89, 86	53

*% deposit reduction leaded gasoline (MS-08)
**% deposit reduction nonleaded gasoline (Phillips "J" reference fuel)

In the third patent of this series, *W.H. Machleder and J.M. Bollinger; U.S. Patent 4,147,641; April 3, 1979; assigned to Rohm and Haas Company* use a mixture of Component 1, the substituted phenol/epichlorohydrin/amine adduct already described and Component 2, a polycarboxylic acid ester for the gasoline or lubricating oil additive.

The polycarboxylic acid ester Component 2 is used primarily for cost considerations and lowers treating costs. However, Component 2 also promotes deposit reduction as shown in the Induction System Test described in the patent.

Component 2 preferably is a mixed ester of a di- or tricarboxylic acid, although polycarboxylic acid esters, wherein the ester groups are the same, are also useful. An especially preferred polycarboxylic acid ester is the mixed adipate diester comprising the monoisodecyl, monooctyl phenoxy polyethoxy ethanol (containing an average of 5 mols of condensed ethylene oxide) mixed ester of adipic acid made by a conventional acid esterification process.

The proportions or amounts of the polycarboxylic acid esters apply whether the polycarboxylic acid is completely esterified with the same alcohol or whether the polycarboxylic acid is esterified with different alcohols, in which case it is referred to as a mixed acid ester or a mixed polycarboxylic acid ester, e.g., the mixed dicarboxylic acid ester for which test results are reported in the following table.

In this table, the comparison primarily is with Chevron F-310. As can be seen, the mixture of the amine adduct Component 1 and the polycarboxylic acid ester Component 2 compares very favorably with Component 1 and with the Chevron F-310, when the latter two additives are used alone. It should also be noted that the Chevron F-310 is used at a much higher level (1,000 lb/1,000 barrels of gasoline) than is either the amine adduct component or the mixture of Component 1 and Component 2.

Performance of Carburetor Blends or Mixtures

Additive	Treating Level (PTB)	Carburetor Detergency Blowby Test*	Induction System Test** Single Cylinder
Control base gasoline	–	0	0
Chevron F-310	1,000	96	~99
Component (1)	75	95	94
Additive mixture***	25/50	84, 90, 83	~81
Additive mixture***	25/75	94	~98

*% deposit reduction leaded gasoline (MS-08)

**% deposit reduction nonleaded gasoline (Phillips "J" reference fuel)

***Mixture of component (1)/component (2) as follows: component (1), adduct of formula 1 or 2 where R^6 is H or PIB_{H35}; component (2), monoisodecyl, monooctyl phenoxy polyethoxy ethanol (containing an average of 5 mols of condensed ethylene oxide) mixed ester of adipic acid.

LUBRICANT AND DIESEL FUEL ADDITIVES

Chelate Derived from Hydroxyalkylated Benzotriazoles

S. Chibnik; U.S. Patent 4,212,754; July 15, 1980; assigned to Mobil Oil Corporation provides Werner coordination complexes. These are prepared by: (a) reacting a benzotriazole with a monoepoxide to form a mixture of 1- and 2-hydroxyalkylbenzotriazoles; (b) esterifying the mixture or one of the isomers thereof with alkenylsuccinic anhydride to form a monoester; (c) converting the monoester to a salt of a metal selected from Groups I-B, II-B, IV-B and VIII of the Periodic Table; and (d) complexing it with a ligand containing amine, hydroxyl, oxazoline, or imidazoline groups to form a chelate.

In preparing these Werner complexes, the benzotriazole reactant is benzotriazole (azimidobenzene) or a lower alkylbenzotriazole, such as methylbenzotriazole (azimidotoluene or toluoltriazole). The monoepoxide reactant can be a 1,2-epoxy compound having the formula:

$$\overset{\quad O}{R\overset{/}{C}H-\overset{\backslash}{C}H_2}$$

wherein R is hydrogen or an alkyl radical having between 1 and 30 carbon atoms, or it can be an internal oxirane or an epoxidized ring. Such compounds include ethylene oxide (oxirane); propylene oxide (2-methyloxirane); butyl glycidylether; 2,3-epoxyoctane; 2,3-epoxydecane, styrene oxide; cresyl glycidyl ether; epoxidized natural oils, such as soybean oil; epoxycyclohexane; and epoxidized mixtures of C_{10-28} alpha-olefins. A wide variety of monoepoxides are available commercially.

The reaction between the benzotriazole and monoepoxide in step (a) is preferably carried out in a suitable solvent including tert-butanol, benzene, toluene, xylene, and halogenated hydrocarbons. In general, the reaction is carried out at a temperature between 25° and 200°C for 1 to 24 hr.

The alkenylsuccinic anhydrides are well known and are available commercially. They are prepared by known techniques from an olefin or polyolefin and maleic anhydride.

The reaction between the hydroxyalkylbenzotriazole and alkenylsuccinic anhydride is carried out under mild conditions to form a monoester, leaving a free carboxyl group. Generally, the reaction is carried out at a temperature between 40° and 130°C for 0.5 to 8 hr.

The resulting acid ester of the hydroxyalkylbenzotriazole is neutralized with a polyvalent metal, especially divalent metals, using known techniques of salt formation. The acceptable metals include zinc, tin, nickel, copper, cobalt, cadmium, chromium, and lead. Of these, the most preferred are zinc and nickel.

The metal salts produced as described are reacted with a ligand to form a coordinated complex or chelate. The preferable ligands are compounds containing at least two amine, hydroxyl, oxazoline, or imidazoline groups which have the capability to complex metal salts to form chelates, such as triethylenetetramine, tetraethylenepentamine, di(methylethylene) triamine, and hexapropyleneheptamine.

The amount of this chelate that is added to a lubricating oil or a hydrocarbon fuel is a minor amount sufficient to impart detergency properties thereto. In general, the amount used can be between 0.05 and 2.5 wt % thereof.

The compounds were evaluated in a diesel oil test developed to produce deposits from the oxidation of lubricating oil under conditions which closely approximate those found in the piston zone of a diesel engine. The test comprises an aluminum cylinder heated by radiation from an external source. The surface temperature of the cylinder is maintained at 575°F during the 140 min test period. The shaft turns slowly (2 rpm) and dips into an oil sump where it picks up a thin film of oil. This thin film is carried into the oxidation zone where heated gases (moist air at 350°F is typically employed, but nitrogen oxides, sulfur oxides and the like may be used) form oxidation deposits. These deposits can be affected by the detergent as the test cylinder rotates into the sump. The efficiency of the detergent is rated by the color and intensity of the deposits on the shaft at the end of the test.

Specifically, 20 parts of each chelate (on a nonoil basis) were compounded with 947 parts of a solvent refined SAE 30 grade lubricating oil, 16 parts of calcium sulfonate, 4 parts of calcium phenate, 10 parts of zinc organodithiophosphate and 1 part of an acrylic ester polymer VI improver. Parts are by weight. The results are as follows. In the test, 100 is clean. Without detergent, the rating is <50. With the chelate, ratings ranged from 67 to 81.

Aroyl Derivative of an Alkenylsuccinic Anhydride

S. Chibnik; U.S. Patent 4,219,431; August 26, 1980; assigned to Mobil Oil Corporation has provided lubricant and fuel compositions comprising a major amount of a liquid hydrocarbon fuel or a lubricating oil and a detergent amount of a product selected from the group consisting of:

(a) The product of reaction between an alkenylsuccinic acid, ester or anhydride and a hydroxyaromatic compound;

(b) The product of reaction between (a) and an amine selected from the group consisting of an alkanepolyol such as an amino alkanediol and a polyalkylene polyamine;

(c) The reaction product of (b) and an aldehyde;

(d) The reaction product of (c) with a metal salt capable of forming a stable complex with amines;

(e) The reaction product of (b) with a metal salt; and

(f) The product of reaction between an alkenylsuccinic acid or anhydride and an alkyl phenyl ether.

The alkenylsuccinic compound is one wherein the alkenyl group is a hydrocarbon containing a double bond and containing preferably 50 to 100 carbon atoms. These are produced by known techniques from an olefin or polyolefin and maleic anhydride.

The alkenylsuccinic esters include the mono- and diesters and may be represented by the formula:

$$\begin{array}{l} R-CH-COOR_1 \\ \quad\;\; | \\ \quad\;\; CH_2-COOR_2 \end{array}$$

wherein R is the alkenyl group defined hereinabove and R_1 and R_2 are hydrogen or an alkyl having 1 to 18 carbon atoms. While both R_1 and R_2 may be a hydrocarbyl group, either the same or different, only one of them may be hydrogen. In other words, at least one of R_1 and R_2 must be a hydrocarbyl group.

The term "hydroxyaromatic compound" is meant to include phenol, naphthol, anthrol, hydroquinone, catechol, resorcinol and the like, as well as alkyl-substituted high molecular weight hydroxyaromatic compounds. The alkyl group preferably contains 1 to 50 carbons and may be a simple alkene or a polymer of such an alkene.

Reaction (a) may advantageously be carried out with from 0.5 to 2.0 mols of the hydroxyaromatic compound per mol alkenylsuccinic compound in the presence of such solvents as benzene or toluene and a catalyst, preferably BF_3.

In reaction (b) the amine may be an amino alkanepolyol such as an amine alkanediol. An example of such an alkanediol is 2-amino-2-(hydroxymethyl)-1,3-propanediol. The amine may also be a polyalkanepolyamine of the structure:

$$H_2N(RNH)_xH$$

wherein R is a 2 to 5 carbon alkylene group and x is from 1 to 10. The formula includes triethylenetetramine, tetraethylenepentamine, di(methylethylene)-triamine, hexapropyleneheptamine and the like.

The amine may be present to the extent of from 0.1 to 3.0 mols/mol of the reaction product of (a). As is with the case of reaction (a), a solvent may be used if desired.

For reaction (c), the usable aldehydes preferably include formaldehyde and paraformaldehyde, and will be used to the extent of from 0.1 to 5.0 mols of aldehyde per mol of (b).

Reaction (d) is the reaction product of (c) with, preferably zinc methanesulfonate, using 0.1 to 3.0 mols salt per mol of (c).

Reaction (e) is the reaction of (a) with a polyalkylene polyamine [such as that outlined for reaction (b)] with zinc methanesulfonate.

The alkyl phenyl ethers used for reaction (f), have the formula:

R'—C_6H_4—O—R

wherein R is a hydrocarbyl of from 1 to 18 carbon atoms, and R' is hydrogen or the same as R. When R and R' are hydrocarbyls, they may be the same or different. Preferably R is an alkyl of from 1 to 8 carbon atoms and R' is hydrogen.

Example 1: The basic reaction between an alkenylsuccinic anhydride and a phenolic material is as follows. A reactor was charged with 3,200 parts polybutenylsuccinic anhydride (prepared from 1,300 MW polybutene and maleic anhydride), 240 parts of a 1:2 molar boron trifluoride phenol complex and 4,395 parts benzene solvent. The mixture was stirred for 1 hr at 85°C and then stripped of solvent, catalyst and unreacted phenol by heating to 105°C for 2 hr at 2 mm of Hg. Infrared analysis indicated that the product was a mixture but the main absorptions were due to presence of o- and p-substituted aromatics, phenolic hydroxyl and anhydride moieties.

Example 2: A solution of zinc methanesulfonate was prepared from 7 parts zinc oxide, 22.6 parts of 70% aqueous solution of methanesulfonic acid and 41.4 parts water. To this solution was added 15.5 parts tetraethylenepentamine over a 1 hr period. A solution of 375.5 parts of Example 1 product in 176.7 parts Promor oil was added, and the materials were reacted at 150°C for 2 hr under a vacuum of 2 mm of Hg. The product, after dilution with 160.7 parts Promor oil, and filtration, contained 0.40% nitrogen, 0.45% zinc and 0.36% sulfur.

The product of Example 2 was evaluated in a 1-G Caterpillar engine test. This test is performed to evaluate detergency characteristics of an additive. The formulation used was comprised of 1.6% calcium sulfonate, 0.4% calcium phenate, 1% zinc isopropylethylhexylphosphorodithioate and 2% (nonoil basis) experimental detergent in solvent-refined paraffinic neutral and bright oils blended to provide a 62 to 64 SUS viscosity grade. The engine is periodically dismantled and inspected for carbon and lacquer deposits with the following results after 120 hr of operation.

	Piston Rating	Lacquer Deposits	Percent Packing
Example 2	68.3	19.4	94
Control	47	37.4	51

This test indicates the excellent dispersant properties of this product.

Process for Preparing Highly Basic Magnesium Sulfonates

Highly basic magnesium sulfonates are known to be particularly effective additives for internal combustion engine lubricating oils and for diesel fuels. In this context, the term "highly basic" refers to sulfonates containing an amount of magnesium substantially in excess of that required to form a neutral sulfonate. Other terms which have been used to describe such sulfonates include overbased, superbasic and hyperbasic.

Many processes are known for preparing fluid, highly basic magnesium sulfonates. While many of these prior art processes and patents claim to be suitable for commerical scale production of highly basic magnesium sulfonates, few highly basic magnesium sulfonates are available on the market.

Accordingly, there is still a need for improved, commercially viable processes for the preparation of commercial scale batches of fluid, highly basic magnesium sulfonates, especially processes which have a high utilization of magnesium and which produce sulfonates that have a high total base number (TBN), of about 385 or more, a low viscosity, an absence of haze and sediment, and good stability under engine use conditions.

J.D. Arnold, H.J. Fair, L.V. Fair and J.D. Berry; U.S. Patent 4,225,446; September 30, 1980; assigned to Calumet Industries, Inc. developed a process for treating a neutralized, fluid, oil-soluble, magnesium sulfonate with magnesium oxide and a unique combination and sequence of promoters and processing conditions to produce a fluid, highly basic magnesium sulfonate. The highly basic magnesium sulfonate. The highly basic magnesium sulfonate produced by this process possesses an excellent combination of properties, including a high base number, a low viscosity, an absence of sediment and haze, and excellent engine stability; and it can be made economically, on a commercial scale.

In a highly advantageous embodiment of the process, an oil-soluble sulfonic acid is mixed with a previously prepared overbased magnesium sulfonate having a total base number (TBN) in excess of 200, in the presence of a suitable volatile solvent. Preferably, a minor amount of a lower alkanol, such as methanol, is added to the mixture. This mixture is then heated to a temperature between 180° and 260°F or more, preferably 200° and 220°F, to neutralize the sulfonic acid.

After the neutralization has been completed, the solution is cooled, and a naphthenic acid, a lower alkanol, such as methanol, and magnesium oxide are added to the neutralized solution to form a second admixture. The magnesium oxide preferably has the following specific characteristics: a bulk density between 13 and 20 lb/ft^3, an iodine number between 35 and 60 meq/100 g, and a particle size of 200 mesh or smaller. Preferably, a nonvolatile diluent oil is added at this point in the process.

This second admixture is then carbonated, preferably with gaseous or liquid carbon dioxide, at a temperature below 120°F, preferably between 100° and 120°F. Water and ammonia or an ammonium compound are added as promoters during the initial phase of carbonation.

Substantially all of the water and the ammonia or ammonium compound must be added after carbonation has begun and before the addition of the first 25% of the carbon dioxide has been completed. Preferably, these carbonation promoters are added slowly during the addition of the first 15 to 20% of the carbon dioxide. After carbonation has been completed, the solution is heated to a temperature in excess of the boiling point of the water and the lower alkanol, advantageously to 250°F or more, to remove those substances and to eliminate or minimize the haze in the final product.

After heating, the solution is cooled somewhat, to a temperature between 180° and 220°F at which temperature additional carbon dioxide is added for a short time, about 10 to 30 min. Next, the solution is heated to an elevated temperature, advantageously about 250°F or more, and then it is cooled to a temperature below 220°F. Preferably, additional volatile solvent is then added. Next, the solution is clarified, preferably by centrifugation and/or filtration, to remove suspended solids and sediment, after which the volatile solvents advantageously are removed, by heating the solution, preferably under a vacuum.

A fluid, highly basic magnesium sulfonate made in accordance with this process has a high TBN of 385 (mg KOH/g) or more, preferably 400 to 500, a relatively low viscosity, an absence of haze and sediment, and good stability under engine use conditions and in compatibility tests. Such magnesium sulfonates also normally consist of 25 to 35% magnesium sulfonate and from about 9 to 10.5% magnesium.

LUBRICANTS FOR TWO-CYCLE ENGINES

Aminophenols plus Certain Other Detergent/Dispersants

The increasing use of two-cycle engines, coupled with the increasing severity of the conditions under which they have been operated and the need to maximize usage of petroleum-derived materials in the face of increasing shortages, has led to an increasing demand for oils and fuels which adequately lubricate such engines (it is a common practice to add the oils used to lubricate such engines to the fuel).

Among the problems associated with the lubrication of two-cycle engines are piston ring sticking, rusting, lubrication failure of connecting rod and main bearings, and deposit formation. The formation of varnish is a particularly vexatious problem since the build-up of varnish on piston and cylinder walls can cause loss of compression through seal failing. This is particularly damaging in two-cycle engines since they depend on suction to draw the new fuel charge into the exhausted cylinder.

D.L. Clason, J.F. Pindar and J.M. Cohen; U.S. Patent 4,200,545; April 29, 1980; assigned to The Lubrizol Corporation describe additive combinations of amino phenols with certain detergent/dispersants and to oils and fuels containing same which are especially useful in two-cycle engines.

This additive combination comprises (a) at least one aminophenol having a substantially saturated hydrocarbon substituent of at least 10 aliphatic carbon atoms; and (b) one or more detergent/dispersants selected from the group consisting of: (1) neutral or basic metal salts of an organic sulfur acid, phenol or carboxylic acid; (2) hydrocarbyl-substituted amines wherein the hydrocarbyl substituent is substantially aliphatic and contains at least 12 carbon atoms; (3) acylated nitrogen-containing compounds having a substituent of at least 10 aliphatic carbon atoms; and (4) nitrogen-containing condensates of a phenol, aldehyde and amino compound.

In a preferred embodiment, the amino phenol is of the formula:

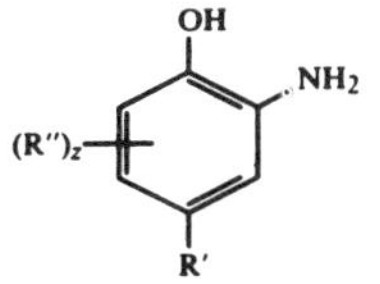

OH
NH_2
$(R'')_z$
R'

wherein R' is derived from homopolymerized or interpolymerized C_{2-10} 1-olefins and has an average of 30 to 400 aliphatic carbon atoms and is usually derived from ethylene, propylene, butylene and mixtures thereof. Typically, it is derived from polymerized isobutene. R'' is a lower alkyl, lower alkoxy, nitro group or halogen atom and Z is 0 or 1.

In general the detergent/dispersants (b) used in these combinations are materials known to those skilled in the art and they have been described in numerous books, articles and patents.

The following specific illustrative examples describe how to make the amino phenols and detergent/dispersants which comprise the additives of this process.

In these examples all percentages, parts and ratios are by weight, unless otherwise expressly stated to the contrary, and temperatures are in degrees centigrade.

Example 1: A mixture of 4,578 parts of a polyisobutene-substituted phenol prepared by boron trifluoride-phenol catalyzed alkylation of phenol with a polyisobutene having a number average molecular weight of approximately 1,000 (vapor phase osmometry), 3,052 parts of diluent mineral oil and 725 parts of textile spirits is heated to 60° to achieve homogeneity. After cooling to 30°, 319.5 parts of 16 molar nitric acid in 600 parts of water is added to the mixture.

Cooling is necessary to keep the mixture's temperature below 40°. After the reaction mixture is stirred for an additional 2 hr, an aliquot of 3,710 parts is transferred to a second reaction vessel. This second portion is treated with an additional 127.8 parts of 16 molar nitric acid in 130 parts of water at 25° to 30°. The reaction mixture is stirred for 1.5 hr and then stripped to 220°/30 torr. Filtration provides an oil solution of the desired intermediate.

Example 2: A mixture of 810 parts of the oil solution of the intermediate described in Example 1, 405 parts of isopropyl alcohol and 405 parts of toluene is charged to an appropriately sized autoclave. Platinum oxide catalyst (0.81 part) is added and the autoclave is evacuated and purged with nitrogen four times to remove any residual air.

Hydrogen is fed to the autoclave at a pressure of 29 to 55 psig while the content is stirred and heated to 27° to 92° for a total of 13 hr. Residual excess hydrogen is removed from the reaction mixture by evacuation and purging with nitrogen four times. The reaction mixture is then filtered through diatomaceous earth and the filtrate stripped to provide an oil solution of the desired amino phenol. This solution contains 0.578% nitrogen.

Example 3: A mixture of 906 parts of an oil solution of an alkyl phenyl sulfonic acid (having an average molecular weight of 450, vapor phase osmometry), 564 parts mineral oil, 600 parts toluene, 98.7 parts magnesium oxide and 120 parts water is blown with carbon dioxide at a temperature of 78° to 85° for 7 hr at a rate of about 3 ft^3 of carbon dioxide per hour. The reaction mixture is constantly agitated throughout the carbonation. After carbonation, the reaction mixture is stripped to 165°/20 torr and the residue filtered. The filtrate is an oil solution of the desired overbased magnesium sulfonate having a metal ratio of about 3.

Example 4: A polyisobutenyl succinic anhydride is prepared by reacting a chlorinated polyisobutene (having an average chlorine content of 4.3% and an average of 82 carbon atoms) with maleic anhydride at about 200°. The resulting polyisobutenyl succinic anhydride has a saponification number of 90. To a mixture of 1,246 parts of this succinic anhydride and 1,000 parts of toluene there is added at 25° 76.6 parts of barium oxide. The mixture is heated to 115° and 125 parts of water is added drop-wise over a period of 1 hr. The mixture is then allowed to reflux at 150° until all the barium oxide is reacted. Stripping and filtration provides a filtrate having a barium content of 4.71%.

Example 5: To 1,133 parts of commercial diethylene triamine heated at 110° to 150° is slowly added 6,820 parts of isostearic acid over a period of 2 hr. The mixture is held at 150° for 1 hr and then heated to 180° over an additional hour.

Finally, the mixture is heated to 205° over 0.5 hr; throughout this heating, the mixture is blown with nitrogen to remove volatiles. The mixture is held at 205° to 230° for a total of 11.5 hr and then stripped at 230°/20 torr to provide the desired acylated polyamine as a residue containing 6.2% nitrogen.

As noted hereinabove, these nitrogen-containing compositions are particularly useful in formulating lubricating oils for use in two-cycle engines. In general, the two-cycle engine lubricating oil compositions contain about 98 to 50% oil or mixture of oils of lubricating viscosity. Typical compositions contain about 90 to 60% oil. The preferred oils are mineral oils and mineral oil-synthetic polymer and/or synthetic ester oil mixtures.

Polybutenes of molecular weights of 250 to 1,000 (as measured by vapor phase osmometry) and fatty acid ester oils of polyols such as pentaerythritol and trimethylol propane are typical synthetic oils used in preparing these two-cycle oils.

These oil compositions contain 2 to 30%, typically 5 to 20% of at least one amino phenol as described hereinabove and 1 to 30%, typically 2 to 20% of at least one detergent/dispersant. The ratio (by weight) of amino phenol to detergent/dispersant in these oils varies between 1:10-10:1. Other additives such as viscosity index (VI) improvers, lubricity agents, antioxidants, coupling agents, pour point depressing agents, extreme pressure agent, color stabilizers and antifoam agents can also be present.

Tertiary Carbinamine-Modified Mannich Compositions

Considerable interest exists in additives that prevent wear and the formation of deposits in engines that operate at high temperatures (e.g.) two-cycle, rotary engines, etc. Engines operating at high temperatures are commonly used in recreation and industry. The increasing use of these engines results in an increasing demand for improved high temperature lubricants.

Mannich additives are commonly made by the condensation reaction of an amine, generally a polyamine, a formaldehyde-yielding reagent and a substantially hydrocarbon compound having at least one acidic or active hydrogen.

The conventional Mannich additive compositions are highly effective in reducing the formation of harmful deposits in four-cycle gasoline and diesel engines. However, lubricants containing conventional Mannich additives suffer the drawbacks that at high temperatures the antiwear and antideposit properties of the lubricant are reduced. That Mannich additives have heretofore not been useful for two-cycle high temperature engine operation is unfortunate in view of the fact that Mannich additives are generally less expensive than the polyamine derivatives of isostearic acid commonly used as the standard two-cycle engine additive.

The general objective of the formulation by *P.J. Cahill and J.H. Udelhofen; U.S. Patent 4,202,784; May 13, 1980; assigned to Standard Oil Company (Indiana)* is to improve lubricating properties of synthetic Mannich additives for high engine temperatures and to improve the ability of Mannich additives to prevent wear and deposits at high temperature in two-cycle engine lubrication. These objectives are accomplished by producing tertiary alkyl carbinamine (tertiary alkyl azomethine) derivatives of Mannich products. The high temperature activity of the tertiary carbinamine derived Mannich additives apparently results from the presence of tertiary alkyl groups linked to the Mannich product at various sites through an aminomethylene bridge.

Briefly the high temperature modified Mannich additive is made by the reaction of a tertiary alkyl primary carbinamine, a substantially hydrocarbon compound with at least one active or acidic hydrogen, formaldehyde, and a polyamine.

Tertiary carbinamines useful in this formulation have the general formula:

$$R{-}N{=}CH_2$$

where R is preferably a tertiary substantially aliphatic group having from 4 to 20 carbon atoms.

Mannich additives suitable for reaction with tertiary alkyl carbinamine are made by the condensation of a polyamine, a formaldehyde-yielding agent, and a substantially hydrocarbon compound containing at least one active or acidic hydrogen; for example, a p-alkyl phenol.

The modified Mannich products disclosed herein can be used in lubricating oil at a concentration of 1.0 to 50.0 wt %; preferably 5 to 15 wt % of the oil composition for highest performance at lowest cost. The lubricant containing the Mannich is dissolved in gasoline for use in two-cycle engines. About 1 part lubricant is used per 1 to 100 parts by weight gasoline. The two-cycle engines in use require 1 part lubricant per 35 to 60 parts gasoline depending on the engine.

In high temperature engine operation the Mannich additive made with the tertiary alkyl carbinamine can be used with a variety of other additives other than the hindered phenol compounds. One useful coadditive is the reaction product of an alkyl succinic anhydride and a polyamine used as an antirust agent. The alkyl succinic anhydride can have a molecular weight from 300 to 6,000, and various polyamines can be used. From 0.001 to 1 part of this product can be used with 1 part of the Mannich additive made with the tertiary alkyl primary amine. Various other additives can be beneficially used and are well known for activity in high temperature engine operation.

Example: A tertiary alkyl primary carbinamine was prepared as follows. 630 g (2 mols) of Primene JM-T, a mixture of about C_{20} tertiary alkyl primary amines, was charged into a 2 ℓ flask equipped with a stirrer and thermometer. 168 g of 37% aqueous formalin (2.07 mols) was added dropwise over a 30 min period and the temperature was maintained at about 30°C. The mixture was stirred for 30 min and 20 g of potassium hydroxide was added to the mixture. The mixture was stirred until the potassium hydroxide was fully dissolved and two phases separated.

The mixture was poured into a separatory funnel and the aqueous phase was discarded. The organic phase was dried over potassium hydroxide and was vacuum distilled. The final product was recovered by the distillation of the organic phase between 66° and 150°C at 0.2 mm/Hg. The tertiary carbinamine modified Mannich was prepared as follows. To a 3 ℓ, three-necked flask equipped with a reflux condenser, thermostat, and thermometer was charged 1,875 g (1.14 mols) of 60% active polybutyl phenol in oil, molecular weight about 1,000, 190 g of

5W oil and 34.3 g (0.57 mol) of ethylenediamine. The contents of the flask were mixed well, heated to about 85°C, and 92.5 g (1.14 mols) of 37% aqueous formalin were added to the flask over a 60 min period. The temperature was maintained during this period and for another 30 min after the addition was complete. The mixture was heated to 150°C and volatiles including water were stripped with a nitrogen stream for 4 hr. To this mixture were added dropwise over a 60 min period 186.5 g of tertiary carbinamine prepared above. The flask was stirred and maintained at 150°C and stripped of volatiles with a heated nitrogen for 1 hr. The mixture was cooled and was ready for use. Nitrogen content of the product was 1.04%.

The additive of the example was evaluated and compared with industry standard additive Oronite 340K in a standard Yamaha and hot tube test. In terms of deposit prevention, the two were essentially equivalent. However, there was a significant increase in the prevention of piston varnish and ring sticking in the tertiary carbinamine modified Mannich blended with a small amount of di-tert-butyl-p-dodecyl-phenol compound.

In the hot tube test, performance of the modified Mannich in the 425° and 495°F tests was superior due to the hindered phenol.

Mannich Additives Modified by Ditertiaryalkyl Phenols

Another method of modifying Mannich compounds for high temperature lubrication uses is by making them from a polyamine, formaldehyde, a substantially hydrocarbon compound having at least one active or acidic hydrogen and, as a substitute for the tertiary carbinamine described in the previous patent, a ditertiary alkyl phenol. This process is described by *J.B. Hanson; U.S. Patent 4,242,212; December 30, 1980; assigned to Standard Oil Company (Indiana).*

The preferred ditertiary alkyl phenol compounds, including phenol compounds commonly called "hindered" phenols, useful in this process are 2,4-ditertiary alkyl phenols, and 2,6-ditertiary alkyl phenols, wherein each tertiary alkyl group can contain from 4 to 1,000 carbon atoms. These phenols are commercially available and economical to use.

Any alkyl group having size or steric hindrance equivalent to or greater than that of a tertiary butyl group can be used to produce the improved Mannich additives. However, as the size or steric hindrance of the group increases, the Mannich reaction rate of the phenol decreases. Thus, the preferred tertiary alkyl groups that provide both acceptable reaction rates and antiwear and antideposit activity are the tertiary butyl, amyl, hexyl, heptyl, octyl, nonyl and decyl groups.

These modified Mannich products can be used in lubricating oil at a concentration of about 1.0 to 50.0 wt %, preferably 5 to 15 wt % based on the oil for highest performance at lowest cost. The lubricant containing the Mannich is dissolved in gasoline for use in two-cycle engines. About 1 part lubricant is used per 1 to 200 parts gasoline. The two-cycle engines in use require 1 part lubricant to 25 to 150 parts gasoline depending on the engine. As a coadditive, the reaction product of an alkyl succinic anhydride and a polyamine, as was used in the previous patent, is useful.

The hot tube test is commonly used to evaluate the antideposit properties of additives. In the hot tube test, a test oil formation commonly containing 13.0 wt % of the modified Mannich product and heated NO_x are passed upward through a 2 mm capillary tube heated in an aluminum block. The test oil is consumed in the test, and the ability of the additive to prevent deposits is rated by the amount of dark colored deposits left on the tube at the end of the test. The deposits are rated from 10 (clear and colorless) to 1 (dark, black and opaque).

Example 1: Into a 500 ml 3-neck distilling flask equipped with a heating mantle, stirrer, Dean-Stark trap, thermometer, and nitrogen atmosphere was charged 200 g (0.137 mol) of a polybutenyl phenol ($\overline{M}_n$ = 936) 64% active in oil solution, 20 g 5W oil, 6.17 g (0.013 mol) ethylenediamine, and 2 drops of an antifoam agent.

The mixture was stirred and heated to 100°C (212°F) and 11.1 g (0.317 mol) 37.0 wt % aqueous formalin were added. The temperature was maintained at 100°C (212°F) for 30 min, and then raised to 155°C (311°F). A stream of nitrogen was passed through the mixture for 2 hr to strip volatiles. To the stripped reaction mixture was added dropwise 14.1 g (0.0685 mol) of 2,6-ditertiary butyl phenol and 5.6 g (0.0685 mol) of 37% aqueous formalin. The mixture was maintained at 155°C (311°F) for 1 hr as volatiles were again stripped with a nitrogen stream. The product had an activity of 60.1%, a nitrogen content of 0.864 wt % and a viscosity at 100°C (212°F) of 2,536 SSU.

Example 2: Into a 500 ml 3-neck flask equipped with a heating mantle, stirrer, Dean-Stark trap, thermometer and nitrogen atmosphere were charged 200 g (0.137 mol) of a polybutyl phenol molecular weight ($\overline{M}_n$ of 936) 64% active in 5W oil, 20 g 5W oil, 5.64 g (0.027 mol) 2,6-ditert-butyl phenol, 4.91 g (0.0818 mol) of ethylene diamine and 2 drops of an antifoam agent.

The mixture was heated to 100°C (212°F) for 30 min and 13.3 g (0.164 mol) of 37% aqueous formalin were added dropwise. The mixture was sparged with nitrogen at 155°C (311°F) to remove volatiles for 3 hr. The product was 60% active, contained 0.618 wt % nitrogen and had a viscosity of 3,636 SSU at 100°C (212°F).

Example 3: Example 2 was repeated with 28.2 g (0.14 mol) of 2,4-ditertiary butyl phenol, 8.22 g (0.137 mol) ethylene diamine and 11.1 g (0.137 mol) of 37% formalin in place of the 5.64 g (0.027 mol) of the 2,6-ditert-butyl phenol, 4.91 g (0.0825 mol) ethylenediamine and 13.3 g (0.164 mol) of aqueous formalin.

The results of the hot tube test show that the Mannich products having hindered phenol functional groups (Examples 1, 2 and 3) have improved properties over the Mannich compound without the hindered phenol functional groups.

ADDITIVES FOR USE IN CAT-CRACKED GASOLINE

According to *S.P. Thomas; U.S. Patent 4,225,319; September 30, 1980; assigned to Phillips Petroleum Company* carburetor deposit formation may be suppressed in fuel compositions containing cat-cracked gasoline by blending with the fuel some adsorbent-treated cat-cracked gasoline.

It is well known in the art to crack catalytically many different heavy streams resulting from operations to refine crude materials. For example, heavy naphtha, gas oil, or heavier streams are employed as feed stocks to catalytic crackers. It is known in the art that heavier feeds contain more oxygen- and nitrogen-containing species which ultimately appear in the cat-cracked gasoline.

As is well known, the effluent from a catalytic cracker is fractionated to give a variety of products. Those fractions generally boiling in the same range as useful gasolines are employed as fuel-blending stocks. The art indicates that oxygen- and nitrogen-containing materials in the cat-cracked gasoline which contribute to carburetor deposit formation include substituted phenols and substituted aromatic amines such as anilines, pyridines, indoles and higher molecular weight materials.

Adsorbents which are useful in this process for treating the cat-cracked gasoline include many of the well known adsorbents such as silica, alumina, silica-alumina, charcoal, carbon black, magnesium silicate, aluminum silicate, zeolites, clay, fuller's earth, magnesia, and the like. A wide variety of such materials is commercially available in a wide variety of particle sizes and surface areas. Particle size and surface area are not deemed to be critical, but can be selected by the practitioner in accordance with current availability and desired operating conditions.

Although the addition of any finite amount of adsorbent-treated cat-cracked gasoline to a fuel composition containing untreated cat-cracked gasoline improves deposition problems, the amount of adsorbent-treated cat-cracked gasoline generally employed can be stated to be up to 50 wt % of the total composition, and preferably in the range of 5 to 25 wt %.

Example: Silica gel-treated cat-cracked gasoline was prepared by passing cat-cracked gasoline through a glass chromatographic column packed with silica gel (Grade 11 or 12 from W.R. Grace & Co.) at the maximum flow rate obtained by gravity flow and at room temperature (about 25°C).

In the following runs, treated and untreated cat-cracked gasolines were blended with alkylate. The resulting blends were employed as the fuel for a 170 in^3 displacement 6 cylinder automobile engine with a tared removable carburetor throat insert. Operation of the engine was for 23 continuous hours at 1,800 rpm and 11.4 brake horsepower. After these conditions of operation the insert was removed, washed with n-heptane, subsequently dried, and weighed to give the weight (in milligrams) of deposit formed.

In the following table, the results of the abovedescribed test employing as the motor fuel the alkylate, alkylate/untreated cat-cracked gasoline blends, alkylate/treated cat-cracked gasoline blends, and alkylate/treated and untreated cat-cracked gasoline blend.

Run Number	Alkylate	UCCG*	TCCG**	Deposits (mg)
1	100	0	0	2.3
2	90	10	0	7.3
3	80	20	0	10.3

(continued)

Run Number	Alkylate	UCCG*	TCCG**	Deposits (mg)
4	90	0	10	2.3
5	80	0	20	3.6
6	80	10	10	4.5

*Untreated cat-cracked gasoline
**Silica gel-treated cat-cracked gasoline

The results of Runs 1, 2 and 3 in the above table demonstrate that the addition of untreated cat-cracked gasoline to alkylate results in a substantial increase in carburetor deposits with increasing amount of untreated cat-cracked gasoline.

The results of Runs 4 and 5 demonstrate that the use of 10 vol % of treated cat-cracked gasoline in alkylate gives no detectable increase in carburetor deposits and that 20 vol % treated cat-cracked gasoline in alkylate gives only a slight increase in carburetor deposits. The result of Run 6 demonstrates that the mixture of 10 vol % untreated cat-cracked gasoline and 10 vol % treated cat-cracked gasoline in alkylate gives substantially less carburetor deposits than the use of not only 20% untreated cat-cracked gasoline, but also 10% untreated cat-cracked gasoline.

The result of Run 6 demonstrates a deposit level lower than would be expected by simply diluting a fuel blend containing 10% untreated cat-cracked gasoline with a clean fuel such as alkylate or a clean cat-cracked gasoline.

-3-

SLUDGE DISPERSANTS AND SEDIMENT INHIBITORS

SLUDGE DISPERSANTS FOR LUBRICANTS AND FUELS

Haze-Free Grafted Ethylene Copolymer

The process developed by *R.L. Elliott and J.B. Gardiner; U.S. Patent 4,144,181; March 13, 1979; assigned to Exxon Research & Engineering Co.* relates to haze-free polymeric dispersant additives which may also be useful as viscosity index improvers for lubricating oils.

These additives are haze-free solutions of substantially saturated polymers comprising ethylene and one or more C_{3-28} alpha-olefins, preferably propylene, which have been grafted in the presence of a free-radical initiator with an ethylenically-unsaturated carboxylic acid material preferably at an elevated temperature and in an inert atmosphere, and then reacted first with polyfunctional material reactive with carboxy groups, such as polyamine, a polyol or a hydroxy amine, or mixtures thereof, to form multifunctional polymeric reaction products and thereafter reacted with an oil-soluble acid.

In a preferred method, haze is eliminated and/or prevented in oil additive compositions by the step of treating the composition with a polymethylene substituted benzene sulfonic acid having a MW of 350, in an amount of from 0.01 to 8 wt % at a temperature within the range of 150° to 200°C for a period from 0.1 to 20 hr, e.g., for 1 hr at 190°C. This method results in an additive oil composition which has no visually perceptible haze. The antihazing agent appears to convert at least part, i.e., the visually perceptible part, of the oil-insoluble hazing substances to an oil-soluble material.

Example 1: 23.4 gal (180 lb) of a 15 wt % solution of an ethylene-propylene copolymer concentrate [made by the Ziegler-Natta process using H_2-moderated $VOCl_3$/aluminum sesquichloride catalyst; having a crystallinity of less than 25%; containing about 45 wt % ethylene and 56 wt % propylene; and having a TE of 2.11 (MW is 56,000)] in S100N (Solvent 100 Neutral Mineral Oil) was heated to 154°C under a nitrogen blanket. To this was added with stirring 0.87 lb (0.33 wt %) of 2,5-dimethyl hex-3-yne-2,5-bis-tert-butyl peroxide and 3.42 lb (1.29

wt %, 0.0132 mol %) of maleic anhydride. After about 2.5 hr at 154°C the system was sparged with nitrogen for 1.5 hr to remove all of the untreated maleic anhydride. To this system at 154°C, 3.6 lb (1.36 wt %, 0.0096 mol %) of N-aminopropyl morpholine was added. All weight percent values are based on the weight of the initial oil solution. The reaction was conducted for 1 hr at 154°C followed by vacuum stripping (84 kPa) at 154°C for 5 hr. To the final product was added 77 lb of S100N. This resulted in a nitrogen-containing grafted copolymer which contains 0.46 wt % nitrogen and a final oil concentration of about 10.5 wt % graft copolymer whose haze reading was >230 nephelos (off-scale) as measured on a Model 9 nephelometer from Coleman Industries.

Example 2: The same general procedure of Example 1 was followed modified by changes to the charges below given as weight percent of total charge.

Ingredient	Weight Percent
Maleic anhydride	0.995
2,5-dimethyl-hex-3-yne-2,5-bis-tert-butyl peroxide	0.198
N-aminopropyl morpholine	1.003
Ethylene-propylene polymer concentrate of Example 1	97.9

The resulting grafted copolymer contained 0.21 wt % nitrogen and the haze reading of the final oil concentrate was again >230 nephelos.

Example 3: 300 g of the oil concentrate of Example 2 was placed into a 4-necked 1 ℓ flask under a nitrogen blanket. The solution was heated to 205°C with good agitation and thereafter treated with 3 g of SA-119 (SA-119 is a commercial alkaryl sulfonic acid sold by Esso SA France). The SA-119 is a 90% active oil concentrate of primarily C_{24} (average) alkylbenzene sulfonic acid having $\overline{M}n$ of ~500. The reaction was continued for 15 min at about 205°C and then the clear solution cooled and bottled. Haze reading of the final oil concentrate was 48 nephelos as measured on the nephelometer.

When the polymers were tested for their sludge dispersant activity, it was found that at a concentration of 0.5 or 1 wt % the sulfonium amine salt of the morpholine derivative of a maleic acid graft copolymer had superior dispersant activity to the morpholine derivative.

Polyolefin Graft Polymers

Gasoline and middle distillate fuels, such as home heating oils, diesel fuels and jet fuels, tend to deteriorate oxidatively upon standing and form gummy deposits. These deposits can foul screens, burners, or fuel injectors. In the case of gasoline, such gummy residues are deposited in the carburetor, making control of air-fuel ratio impossible.

R.L. Stambaugh and R.A. Galluccio; U.S. Patent 4,146,489; March 27, 1979; assigned to Rohm and Haas Company have developed additives for lubricants and fuels which will disperse such deposits, preventing deterioration in the fuel quality during storage. These additives, which are copolymers resulting from grafting polar nitrogen-containing monomers selected from C-vinylpyridines and N-vinylpyrrolidone into an oil-soluble substantially linear, rubbery ethylene/propylene backbone polymer, also impart detergency and viscosity index (VI) improvement to lubricating oils. They may be used at very low concentrations when used in hydrocarbon motor fuels, where primary use is made of their superior dispersancy, typically from 0.001 to 0.1 wt %. In lubricants the additives are typically used in amounts of 0.6 to 1.5 wt %.

Example: *(A) Preparation of a 2-vinylpyridine-ethylene/propylene graft copolymer* – 675 g of o-dichlorobenzene was added to a clean, nitrogen flushed 5 ℓ flask and heated to 100°C under nitrogen. A sample of a 60/40 mol % ethylene/propylene copolymer was cut into small pieces and added to the solvent. A total of 225 g of the hydrocarbon polymer was added to the dichlorobenzene. After a homogeneous solution was obtained, the contents of the flask were cooled to 80°C and 22.5 g of commercially available 2-vinylpyridine was added to the flask.

When mixing was complete, 1.32 g of 85% t-butyl perbenzoate was added to the flask. The temperature was held at 80°C for one-half hour to complete mixing and then raised to 140°C over a one-half hour period. After a 40 min holding period an additional 1.32 g of perester was added. The solution was held at 140°C for 1 hr and then 1,720 g of 100 neutral solvent refined mineral oil was added. The product solution was then vacuum stripped of solvent and unreacted monomer with final conditions of 0.5 mm Hg and 150°C being held for 1 hr. The product was then further diluted with 305 g of 100 neutral oil, producing a concentrate containing 10.4% graft copolymer.

A sample of the graft copolymer was isolated by dialysis and found to contain 0.41% nitrogen by Kjeldahl analysis. The titratable nitrogen content was found to be 0.40%.

(B) Blending data – A crankcase oil blend was prepared using the Formulation B as follows.

Ingredient	Percent
Oil concentrate of the graft copolymer product of (A)	8.10
Pour point depressant	0.50
Polybutene succinimide ashless dispersant	2.00
Overbased magnesium sulfonate (400 TBN)	2.00
Zinc dialkyldithiophosphate	1.50
100 neutral oil	38.65
200 neutral oil	47.24
Silicone antifoam solution	0.01

The viscometric properties of the above blend are compared below with those of the additive-treated base stock without viscosity index improver and demonstrate that the graft copolymer of (A) can be used to formulate a quality 10W40 crankcase oil.

(C) Dispersancy tests – Asphaltenes test: In the standard asphaltenes test, 0.0625% of the graft copolymer of (A) dispersed 0.4% of asphaltenes at 150°C. The starting ethylene/propylene copolymer will not disperse asphaltenes even at 2.0% polymer concentration.

Sequence V-C Test: This test was also carried out on the Formulation B. The sequence V-C test is an engine test procedure that evaluates crankcase motor oil with respect to sludge and varnish deposits produced by engine operation under a combination of low and midrange temperatures. This test also indicates a capacity of the oil to keep positive crankcase ventilation (PCV) valves clean and functioning properly. The test results are given below along with the specifications required to meet the American Petroleum Institute's SE service classification.

The second column shows results for a widely-used dispersant methacrylate copolymer. Note that the dispersant activity of the graft copolymer of (A) used in Formulation B (above) is substantially superior to that of the commercial polymethacrylate even though graft copolymer (A) is used at only 25% of the treating level of the commercial product.

Sequence V-C Test Results, 192 hr

	Formulation B	Commercial Polymethacrylate Dispersant	SE Spec
Polymer, %	0.84	3.3	
Average sludge rating*	9.3	7.1	8.5, min
Average varnish rating*	8.4	7.9	8.0, min
Piston skirt rating* Varnish	8.2	7.4	7.9, min
Oil ring clogging, %	0	0	5, max
Oil screen clogging, %	0	0	5, max

*10 is clean

Acylated Polyamine-Substituted Cycloaliphatic Compounds

According to *E.W. Kluger, J.W. Miley and T.K. Su; U.S. Patent 4,153,567; May 8, 1979; assigned to Milliken Research Corporation*, fuel and lubricant additive compositions may be produced by the process comprising simultaneously reacting a polyamine substituted cycloaliphatic compound having the general structure:

(1) NH—$CH_2CH_2CH_2NH_2$; NHR

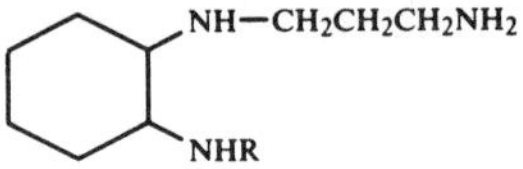

wherein R is hydrogen or $CH_2CH_2CH_2NH_2$, with at least one substituted aliphatic polycarboxylic acid acylating agent selected from the group consisting of carboxylic acids and their corresponding halides, anhydrides, and esters, having at least 12 carbon atoms per molecule. More specifically, the oil-dispersible and/or oil-soluble additives have the general structure:

(2) NH—$CH_2CH_2CH_2$—N ; NHR ; O ; O ; R_1

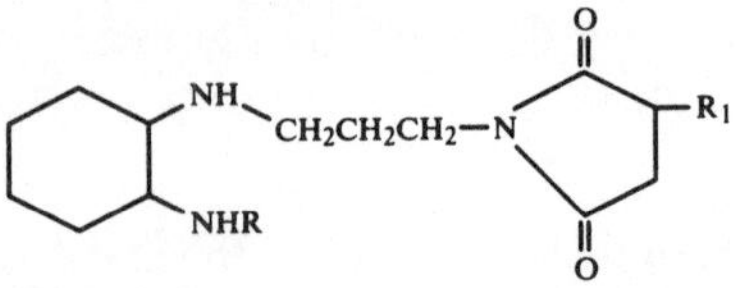

wherein R is H, $-CH_2CH_2CH_2NH_2$ or

(3) 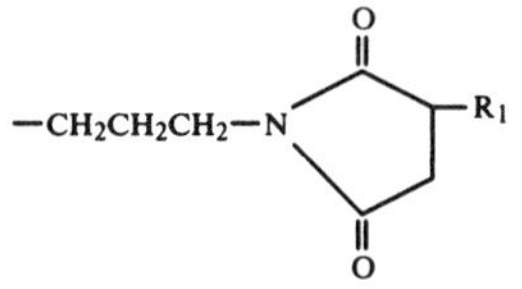 $-CH_2CH_2CH_2-N$ (succinimide ring bearing R_1, with two C=O groups)

and R_1 is an alkyl moiety containing at least 8 carbon atoms or a substantially saturated olefin polymer having an average molecular weight of 200 to 5,000.

The polyamine substituted cycloaliphatic compounds of Formula 1 can readily be prepared by cyanoethylation of 1,2-diaminocyclohexane to produce the cyanoethylated 1,2-diaminocyclohexane which can be reduced with ammonia to provide the amine derivative.

Example 1: In a 250 cc three-necked flask equipped with a magnetic stir bar, Dean-Stark trap, and heating mantle was placed 31 g (0.087 mol) of alkenylsuccinic anhydride having an average molecular weight of 356 in 50 cc of toluene. To this solution was added 19.9 g (0.087 mol) of the tetramine, e.g., N,N'-di-(3-aminopropyl)-1,2-diaminocyclohexane.

An exothermic reaction occurred on mixing the solutions. The reaction mixture was then refluxed for 14 hr and the resulting water was collected in the Dean-Stark trap. The toluene was removed by evaporation to give 97% yield of the crude viscous adduct. An infrared spectrum indicated that no anhydride was present. The compound produced from the above reaction can be represented structurally as:

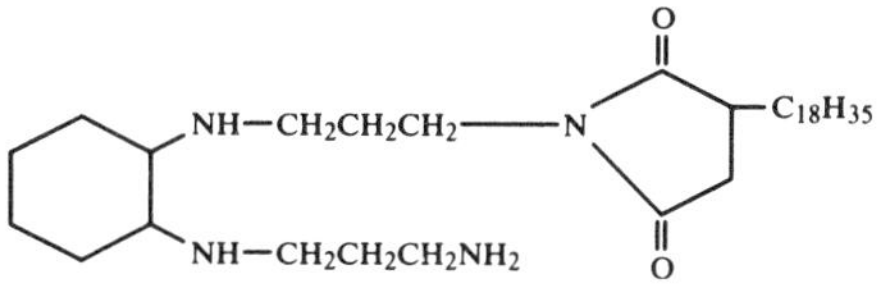

Cyclohexane ring bearing $NH-CH_2CH_2CH_2-N$ (succinimide ring bearing $C_{18}H_{35}$) and $NH-CH_2CH_2CH_2NH_2$

Example 2: A series of admixtures were formulated using varying amounts of the product produced in Example 1 and a mineral oil having a kinematic viscosity of 7.7 cs at 100°F. Each admixture contained a total weight of 80 g. Into each admixture was added 1 g of carbon black having a particle size of 24 mm and 15 g of glass beads. Each admixture, which had been placed in a transparent container, was shaken using a Red Devil paint mixer for 10 min and thereafter centrifuged for 30 sec at a speed of 3,000 rpm. Each admixture was then removed from the centrifuge and the admixtures were evaluated to determine the effect of the product of Example 1 as an oil dispersant. Tabulated below are the results of such experiments.

Product of Example 1 in Oil, (%)	Observation of Carbon Black in Oil
0	Substantially complete precipitation of carbon black
0.01	Substantially complete precipitation of carbon black

(continued)

Product of Example 1 in Oil, (%)	Observation of Carbon Black in Oil
0.10	Partial precipitation of carbon black
1.00	Substantially uniform dispersion of carbon black in oil, e.g., substantially no carbon black precipitation

The data clearly indicate that such compounds can effectively disperse carbon black in mineral oil, and thus would indicate that such compounds have the desired dispersing properties for use as dispersing agents. Further, the solubility of such compounds in the mineral oil is indicative of their use as a lubricating or fuel dispersing agent or additive.

Alkyl-Guanidino Heterocyclic Compound

One of the main problems concerning crankcase oils is that the unavoidable presence therein of metal particles, carbonaceous products, decomposition products resulting from the degradation of fuel and oil, and water, leads to the formation of sludges, which are particularly detrimental to a good operation of the engine.

It is known that the use of detergent additives such as metal phenates and sulfonates makes it possible to maintain the particles in suspension, and consequently to avoid the formation of sludges. But these "basic" detergents are efficient mainly at high temperatures and are not so well adapted to the operating conditions of the engine when the car is used over short distances (particularly in town traffic, for the so-called "stop-and-go" driving), which are insufficient for bringing the oil to a high temperature.

A major purpose of the process developed by *C. Cohen and B. Sillion; U.S. Patent 4,159,898; July 3, 1979; assigned to Institut Francais du Petrole, France* is to provide efficient detergent and dispersing lubricating compositions for use in engines operated at either low or high temperatures and/or, alternately low and high temperatures.

The alkyl-guanidino-heterocyclic compounds found useful as fuel additives are defined as resulting, as a general rule from the alkylation, with a halogenated hydrocarbon RX_n, of guanidino-heterocyclic compounds of the following general formula.

(A)

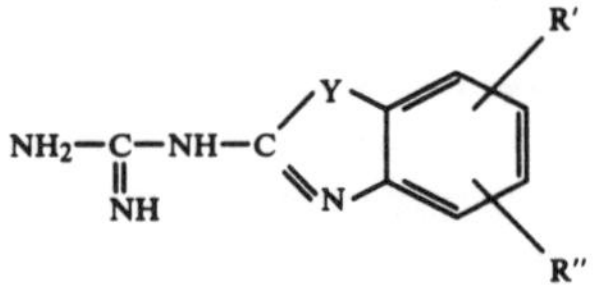

R′

Y

$NH_2-C(=NH)-NH-C$

N

R″

In the formulation RX_n of the halogenated hydrocarbon, R is a saturated or unsaturated substantially hydrocarbon radical containing 10 to 200 carbon atoms, and which may further contain polar atoms or groups in proportions which do not alter its hydrocarbon nature, for example, oxygen atoms –O–, carbonyl groups or halogen atoms; X is a halogen atom, for example, chlorine, bromine or iodine, and n is an integer, generally 1 or 2.

In the formula of the guanidino-heterocyclic Compound (A), Y may be the following.

An oxygen atom –O–: in this case the heterocycle is a benzoxazole;

A sulfur atom –S–: in this case the heterocycle is a benzothiazole;

A –NH– group: in this case the heterocycle is a benzimidazole;

A –CONH– group in which the carbon atom is directly bound to the benzene ring: in this case the heterocycle is a 4-quinazolone;

R' and R'' may be a hydrogen atom, or an alkyl radical having 1 to 10 carbon atoms; when they are in ortho positions, R' and R'' may also form together a ring fused to the benzene ring, this ring being either aliphatic or aromatic; R' and R'' may also represent a –Z–Ar group in which Z is an oxygen atom, a sulfur atom or a divalent aliphatic group, such as: $-CH_2-$, $-(CH_2)_4-$ or $-C(CH_3)_2-$ and Ar is a simple aromatic radical such as a phenyl radical or a substituted aromatic radical such as an alkyl-phenyl radical in which the alkyl group contains 1 to 10 carbon atoms.

The halogenated hydrocarbons RX_n involved in the process are preferably derived from substantially saturated petroleum cuts, or from polymers of α-olefins or of internal olefins.

The guanidino-heterocyclic compounds of the general Formula (A) may be obtained in a known manner by reacting dicyandiamide with hydrochlorides of the corresponding substituted anilines.

The ratio of the two reactants used to prepare the additives may vary, for example, from 0.7 to 2.5 mols of Compound (A) per halogen atom of the halogenated hydrocarbon; preferably, a ratio close to 1:1 is used.

The reaction is performed at a temperature from 100° to 250°C, generally over a period of a few minutes to several hours, in most cases 15 min to 6 hr.

Example 1: In a glass reactor, provided with a stirrer, shaped as an anchor, and heated in an oil bath, various compounds were prepared (designated in the following table as Ia to Ig), by reacting, under nitrogen scavenging, chlorinated poly-isobutenes with 2-guanidino benzimidazole (A) in the presence of sodium carbonate, at the conditions reported in the table below.

	Chlorinated Polyisobutene			Compound		Reaction		Final Product	
Ref.	M*	Cl (wt %)	Weight (g)	(A) (g)	Carbonate (g)	Duration (hr)	Temperature (°C)	N (wt %)	Cl (wt %)
Ia	950	3.66	10	4.37	1	4	190	2.7	0.4
Ib	1,280	2.7	12.8	4.37	1	4	200	2.3	0.45
Ic	480	8.04	18	8.74	2	4	200	5.1	0.6
Id	980	4.3	10	4.37	1	4	200	3.2	0.4
Ie	950	3.66	10	2.2	1	8	200	3	0.35
If	950	3.66	2	0.84	0.2	2	250	3	0.4
Ig	960	0.4	460	200	48	4	200	1.4	0.2

*M is the molecular weight of the chlorinated polyisobutene.

Products Ia to If have been diluted with hexane after reaction, filtered with a filtration aid and then evaporated.

Product Ig has been diluted with 470 g of a 100 N oil at the end of the reaction and then filtered. The N and Cl contents mentioned for Ig have been determined on the obtained solution.

Example 2: By operating as in Example 1, there is reacted 255 g of chlorinated polyisobutene having a molecular weight of 950 and containing 3.66 wt % of chlorine, with 112 g of 2-guanidino benzoxazole and 25 g of sodium carbonate for 4 hr at 200°C. The reaction medium is diluted with hexane, filtered and the hexane is evaporated therefrom. The resulting product (Additive II) contains 3.4 wt % of nitrogen and 0.5 wt % of chlorine.

The dispersing power of lubricating compositions containing certain additives prepared as described in the preceding examples have been tested by examination of spots obtained on a filter paper after deposition thereon of a drop of used mineral oil containing 2 wt % of additive. In strictly identical operating conditions was determined, in each case, the ratio of the average diameter of the black spot to the average diameter of the oil aureole. The higher this ratio, the better the dispersing power of the additive. The results obtained are reported in the table below, in which are also mentioned the results of identical tests conducted without additive and with 2% of an additive available in the trade (alkenylsuccinimide).

	Additive					
(°C)	None	Commercial	Ib	Ic	Ig	II
25	0.4	0.60	0.67	0.54	0.65	0.56
150	0.30	0.48	0.54	0.46	0.58	0.43

Oxazoline-Linked Dicarboxylic Acids and Certain Amino Acids

The process described by *J. Ryer, J. Zielinski, H.N. Miller and S.J. Brois; U.S. Patent 4,195,976; April 1, 1980; assigned to Exxon Research & Engineering Co.* is based upon the discovery that reaction of long chain hydrocarbyl dicarboxylic acid material, i.e., acid, or anhydride, or ester, with certain classes of amino alcohols, under certain conditions, will result in a different type of linkage, namely an oxazoline linkage, and that materials with this oxazoline linkage appear very effective as detergents or dispersants for oleaginous compositions such as lube oil and gasoline.

The long chain hydrocarbyl substituted dicarboxylic acid material, i.e., acid or anhydride, or ester, used in the process includes alpha or beta unsaturated C_{4-10} dicarboxylic acids, or anhydrides or esters thereof, such as fumaric acid, itaconic acid, maleic acid, maleic anhydride, chloromaleic acid, dimethyl fumarate, etc., which are substituted with a long hydrocarbon chain, generally an olefin polymer chain.

The hydrocarbyl portion should average at least 50 aliphatic carbon atoms per dicarboxylic acid group and be substantially saturated.

Particularly useful olefin polymers have number average molecular weights within the range of 900 to 3,000 with approximately one terminal double bond per polymer chain. An especially valuable starting material for a highly potent dispersant additive made by this process is polyisobutylene.

Examples of amino alcohols used to make these dispersants include 2-amino-2-methyl-1,3-propanediol, 2-amino-2-(hydroxymethyl)-1,3-propanediol (also known as tris-hydroxyaminomethane or THAM), 2-amino-2-ethyl-1,3-propanediol, etc. Because of its effectiveness, availability, and cost, the THAM is particularly preferred.

The formation of the oxazoline dispersants, in a fairly high yield, can be effected by adding about 1 to 2 mol equivalents of the aforesaid 2,2-disubstituted-2-amino-1-alkanol per mol equivalent of the dicarboxylic acid material, with or without an inert diluent, and heating the mixture at 140° to 240°C, preferably 180° to 220°C for one-half to twenty-four, more usually two to eight, hours.

Although not necessary, the presence of small amounts, such as 0.01 to 2 wt %, preferably 0.1 to 1 wt %, based on the weight of the reactants, of a metal salt can be used in the reaction mixture as catalyst to shorten the reaction time.

When these products are used as detergents or dispersants in fuels such as gasoline, kerosene, diesel fuels, No. 2 fuel oil and other middle distillates, a concentration of the additive in the fuel in the range of 0.001 to 0.5 wt %, based on the weight of the total composition, will usually be employed.

Example: *Part A* – 350 g (0.328 mol) of polyisobutenylsuccinic anhydride having an ASTM saponification number of about 105 and a molecular weight of about 1,067 was reacted with 79.4 g (0.656 mol) of 2-amino-2-(hydroxymethyl)-1,3-propanediol (THAM) as follows. The polyisobutenylsuccinic anhydride and THAM were dissolved in 250 ml xylene in a 2 ℓ 4-neck flask equipped with thermometer, stirrer, dropping funnel, a condenser including a Dean-Starke water trap, and having a nitrogen bleed to provide a nitrogen blanket.

The heat was raised to 290° to 306°F and over a period of 1 hr and 45 min, about 9 cc water collected in the trap. The heat was then turned off overnight and the following day the reaction mixture was refluxed another hour at 308°F whereupon 10 cc of water had now collected in the trap.

The reaction mixture was then sparged with nitrogen to evaporate the xylene during which time the temperature rose to 380°F. Then the system was connected to a vacuum pump and heated to 360° to 380°F for 3 hr under a pressure of 20 to 30 mm Hg. The heat was then turned off overnight. The following day the mixture, which was not free of xylene, was warmed to 200°F and 489 cc (419 g) xylene was added to make a concentrate containing 50 wt % of the bis-oxazoline reaction product dissolved in the xylene to give a concentrate having a nitrogen content of 1.23 wt % against a calculated nitrogen content of 1.1 wt %.

Part B – A sample of the concentrate product of Part A above was added to a gasoline in an amount equivalent to 25 lb of the concentrate (50% ai) per 1,000 barrels of gasoline. This additive treated gasoline was then tested for its effectiveness in a carburetor detergent and the carburetor cleanup indicated that the additive was very effective as a carburetor detergent when added to gasoline.

Part C – A Sludge Inhibition Bench (SIB) test was carried out on the product of Part A to determine its effectiveness as a sludge dispersant in lubricating oil.

This test data showed that it is an extremely effective sludge dispersant, particularly at about the 0.75 wt % concentration level, i.e., about 0.375 wt % active ingredient.

Oxazoline-Containing Dispersants Stabilized Against Oxidation with Sulfur

Dispersants which are mono- or bisoxazolines of a dicarboxylic acid moiety that have been prepared by reaction of hydrocarbyl-substituted C_{4-10} dicarboxylic acid or anhydride or ester thereof that has at least 50 carbon atoms in the hydrocarbyl group, with a 2,2-disubstituted 2-amino-1-alkanol that has from 2 to 3 hydroxy groups and that contains from 4 to 8 total carbon atoms, such as those described in the previous patent and in U.S. Patents 4,102,798, 4,113,639, 4,049,564 and U.K. Patent 1,483,681 may be stabilized against oxidation.

The process for stabilizing these dispersants is related by *R. O'Halloran and E.D. Winans; U.S. Patent 4,263,153; April 21, 1981; assigned to Exxon Research & Engineering Co.* and consists in reaction of the compounds with elemental sulfur or with a sulfur compound selected from the group consisting of carbon disulfide, a sulfide of phosphorus or a phosphosulfurized olefin.

While the oxazolines of the type described are very effective dispersants for sludge and other contaminants that develop in oil compositions a problem has sometimes arisen in their use in that, under some conditions, oxidation can cause the dispersants to be somewhat corrosive to copper or its alloys. In one such instance where a bis-oxazoline dispersant of the type described was employed in an automatic transmission fluid formulation that was being tested, it was found that due to oxidation the oil cooler braze joints tended to corrode, causing engine coolant in the automotive radiator to leak into the transmission fluid cooling lines.

In this process the oxazoline is sulfurized, i.e., reacted with elemental sulfur or with carbon disulfide, a sulfide of phosphorus or a phosphosulfurized olefin, usually at a temperature of 100° to 300°C, preferably 150° to 200°C for 1 to 20 hr, usually 1 to 5 hr, in such proportions as to impart 0.3 to 7 wt %, e.g., 0.5 to 3 wt % of sulfur into the oxazoline based on the weight of the oxazoline compound. Typically after the reaction has proceeded for a period of time, e.g., one-half to one hour, a stream of inert gas such as nitrogen can be bubbled through the reaction mixture to remove unwanted volatile products or unreacted sulfur compounds while the reaction continues, or the gas bubbling can be initiated at the start of the reaction.

The phosphosulfurized olefinic material is preferably a phosphosulfurized terpene, more preferably phosphosulfurized alpha-pinene. Other suitable phosphosulfurized olefins comprise polymers of C_{2-5} olefins such as ethylene, propylene, butene, isobutene or pentene, of from 500 to 3,000 number average molecular weight that have been reacted with a sulfide or phosphorus, e.g., polyisobutene of about 1,800 molecular weight treated with P_2S_5.

The sulfurized oxazolines of this process are employed in lubricating oil compositions, e.g., mineral lubricating oil in concentrations ranging from 0.01 to 20 wt %, preferably from 0.1 to 5 wt % of the total composition, when such compositions are primarily lubricants or working fluids such as automotive crankcase lubricants or automatic transmission fluids. Ordinarily when the additives are to be employed in normally liquid hydrocarbon fuels such as fuel oil, gasoline, kerosene and the like, concentrations will range from 0.001 to 2 wt % based on the total composition.

Polyurethanes from Reaction of Diisocyanate and Diols

The objective of *R.C. Schlicht; U.S. Patent 4,235,730; November 25, 1980; assigned to Texaco Inc.* is to provide a polyurethane polymer of relatively low molecular weight and good dispersancy containing, pendant from the polymeric chain, an alkyl or alkenyl moiety of substantial chain length and a moiety within the polymer chain itself composed of an alkylene or arylene moiety having 2 to 20 carbon atoms, optionally containing inert functional groups such as tertiary amino groups. This was accomplished by the development of a polyurethane which is the reaction product of an isocyanate and a diol of the formula:

$$HO{-}R{-}(Z)_v{-}\overset{O}{\overset{\|}{C}}{-}(R^1)_x{-}\overset{O}{\overset{\|}{C}}{-}X{+}\overset{O}{\overset{\|}{C}}{-}(R^2)_y{-}\overset{O}{\overset{\|}{C}}]_z(Y)_u{-}R^3{-}OH$$

wherein

each of R, R^1, R^2 and R^3 is independently a divalent aliphatic or aromatic moiety;

u, v, x, y, and z are each independently 0 or 1 and x + y = 1 or 2;

X is a radical derived from removal of hydrogen atoms from a compound of the formula H–X–H wherein the H–X–H compound is:

(a) a diamine or polyamine bearing at least 2 secondary amino groups, optionally containing any number of tertiary amine or other functional groups inert to anhydrides, hydroxyalkylating agents and isocyanates;

(b) a monohydroxy alkyl secondary amine optionally bearing tertiary amine or other functions inert to anhydrides, hydroxyalkylating agents and isocyanates;

(c) a bishydroxyalkyl substituted mono-, di- or polytertiary amine; or

X is –O–R^4–O moiety wherein R^4 is a C_{2-30} divalent hydrocarbon radical, especially arylene or alkylene, or an oxaalkylene moiety having 1 to 15 oxygen atoms;

Y and Z are each O or NR^5 where R^5 is a C_{1-30} alkyl group which can be substituted with cyano, C_{1-32} carbalkoxy or C_{1-32} carboxamido or other substituents inert to anhydrides, hydroxyalkylating agents and isocyanates.

A fuel composition may be made which contains such a polyurethane in an amount of between 0.0002 and 0.05 wt %, preferably between 0.002 and 0.01 wt %. Fuels in which the polyurethane can be employed as dispersant include gasolines, diesel fuels, jet fuels and heating oils.

These polyurethane polymers, which are useful as dispersants for lubricants or fuels, are prepared by the reaction of isocyanates, especially diisocyanates, with bis- (or greater) hydroxyalkyl substituted alkenylsuccinamic acids, bisamides, amide-esters, polyamides (especially the low molecular weight type), polyamide-esters (especially the low molecular weight type), and esters or polyesters bearing tertiary amino groups, especially lower molecular weight type esters or polyesters. Generally, the diisocyanate is reacted with a compound of molecular weight between 300 and 3,000.

Example: *Tolylenediisocyanate reaction product of hydroxyethylated polyamide of piperazine and alkenylsuccinic acid anhydride* – (a) Preparation of polyamide: 216 g (1.0 mol) of a polybutenylsuccinic acid anhydride (prepared from ~1,250 MW polybutene) were dissolved in 2,105 g diluent oil at 60°C and 86 g (1.0 mol) of piperazine were added. The mixture was heated under a nitrogen blanket to 177°C and the reaction was continued for 5 hr, eliminating water of reaction.

The intermediate polyamide of nitrogen content 0.63 wt % (0.67 wt % calculated) had a total acid number of 6.2 and a total base number of 7.6. Thus, end group analysis indicated that the product was an oligomeric polyamide having an average of 4.3 repeating units.

(b) Hydroxyethylation of the end groups of the polyamide of (a): A catalytic amount of triethylamine (0.5 g) was added to 203.3 g (0.05 equivalence of NH and COOH groups) of the polyamide prepared in accordance with step (a) and the mixture was heated to 120°C. Ethylene oxide was added at a rate of 0.05 mol per hour for a period of two and one-half hours and the product was stripped substantially free of catalyst at 130°C at a pressure of 10 mm Hg.

Ethylene oxide uptake was 0.90 g (0.02 mol). The product had a nitrogen content of 0.63 wt % (0.65 calculated), a total acid number of 0.9 and a total base number of 2.6.

(c) Preparation of polyurethane by reaction with tylylenediisocyanate: 140 g (approximately equal to 0.03 equivalence of end-groups) of the hydroxyethylated polyamide prepared in accordance with step (b) were heated to 80°C and 2.80 g (0.015 mol) of tolylene-2,4-diisocyanate were added. The reaction was conducted at 80°C for 3 hr and for an additional 3 hr at 100°C to complete the reaction of isocyanate.

The product was filtered at 100°C and stripped of catalyst at 130°C at a pressure of 10 mm Hg. The product had a nitrogen content of 0.94 wt % (0.93 wt % calculated) and a total acid number of 0.15.

The products of steps (a), (b) and (c) were tested to determine their ability to act as an effective dispersant in a lubricating oil composition.

It was found that the final diisocyanate reaction product [step (c) product] was a very effective dispersant when tested in accordance with the Bench VC Dispersancy Test. The test data reveal that the product's precursors [the products of steps (a) and (b)] were generally much less effective as dispersants.

Hence, it is only upon converting the polyhydric alcohol to a polyurethane polymer that it becomes useful as a dispersant for a lubricant or fuel. In the case of the above example, where the precursor was initially polymeric and possessed moderate dispersancy, the further polymerization induced by reaction of the hydroxyethylated end groups with the diisocyanate resulted in improving the dispersancy to a point where the same is useful in an engine. Thus, it is only by appropriate selection of backbone polyol and conversion of the same to corresponding polyurethane that a commercially suitable dispersant for a lubricating oil or fuel is provided.

Glyoxal-Polyamine-Polybutenyl Succinic Anhydride Reaction Products

T.C. Shields; U.S. Patent 4,247,404; January 27, 1981; assigned to Ashland Oil, Inc. has developed a process for preparing a dispersant for use in lubricants and fuels comprising:

(a) Reacting an alkylamino substituted ethylene diamine or homologous polyamine of the general formula:

$$H[(NH(CH_2)_3]_mNHCH_2CH_2NH[(CH_2)_3NH]_n$$

where m + n is between 2 and 6 with a dione or mixture of diones of the structural formula:

$$R'-\overset{\overset{\displaystyle O}{\|}}{C}-\overset{\overset{\displaystyle O}{\|}}{C}-R$$

where R and R' are each either hydrogen or an alkyl radical of 2 to 4 carbon atoms; and

(b) Reacting the resulting intermediate reaction product with polybutenyl succinic anhydride.

The most preferable dione is glyoxal. Others which can be used are pyruvic aldehyde and 2,3-butanedione. Preferred alkylamino derivatives of ethylene diamine are 4,7-diazadecane-1,10-diamine and a mixture of 60% of 4,7,11-triazatetradecane-1,14-diamine and 40% 4,8,11,15-tetraazaoctadecane-1,18-diamine.

As the first step in the process, the glyoxal and the polyamine are mixed together preferably in a mol ratio of 1 mol of glyoxal to 2 mols of polyamine. An excess of either reactant can be used but is not particularly desirable. If desired, this first reaction step can be conducted in the presence of substantially inert organic liquid diluents, such as benzene, toluene, xylene, chlorobenzene, hexane, heptane, cyclohexane, mineral oil, mixtures thereof, and the like.

The reaction between the glyoxal and the amine is exothermic and the temperature is maintained between -30° and 120°C or preferably between 0° and 60°C until the evolution of heat subsides. The mixture is then refluxed for a period of time sufficient to insure reaction preferably at a temperature of about 50°C, and is then subjected to stripping preferably at a reduced pressure and a maximum temperature of 200°C to remove water of reaction and/or carrier liquid. The resulting product is the desired intermediate product.

The polyalkenyl succinic anhydride can be reacted with the intermediate product as it is upon completion of the first reaction or after it is subjected to desired purification techniques.

The intermediate reaction product from the first reaction step and the alkenyl succinic anhydride are preferably reacted at temperatures of 100° to 200°C. The time for contacting the intermediate reactive product with the alkenyl succinic anhydride will vary from 0.1 to 10.0 hr. The intermediate reaction product is reacted with the polyalkenyl succinic anhydride preferably in a ratio of between 4.5 and 3.0 mols of succinic anhydride to 2 mols of initial polyamine.

The product resulting from this second reaction between the alkenyl succinic anhydride and the intermediate reaction product is the desired dispersant. It is subjected to desired purification techniques.

When employed as sludge-dispersing additives for lubricants the oil-soluble compositions can be used in amounts such that they comprise from 0.01 to 30 wt % of the lubricant depending on the use to which the lubricant is to be put and the presence or absence of other additives, especially dispersants or detergents. Ordinarily, they will comprise at least 0.1 to 10 wt % of the lubricant although, under unusually harsh operating conditions such as are encountered in certain diesel engines, amounts of 10 to 30 wt % are beneficially employed, particularly in the absence of other detergent or dispersant additives.

In fuels, these oil-soluble compositions promote engine cleanliness by reducing or eliminating harmful deposits in the fuel system, engine and exhaust system through their dispersant capabilities. They are primarily intended for use in the normally liquid petroleum distillate fuels, that is, the petroleum distillates which boil in the range characteristic of petroleum fuels such as gasolines, fuel oil, diesel fuels, aviation fuels, kerosene and the like. When employed in fuels, they are generally employed in amounts of 0.0001 to 2 wt % and generally in amounts of 0.001 to 0.5 wt %. Other conventional fuel additives can be present in the fuel compositions, including lead scavengers, deicers, antiscreen clogging agents, neutral or basic oil-soluble alkaline earth metal sulfonates, phosphonates, or carboxylates, other ashless dispersants, antifoulants, demulsifiers, and the like.

Oxidatively-Coupled Hydroxyaromatic Compounds

J.M. Cohen; U.S. Patent 4,256,596; March 17, 1981; assigned to The Lubrizol Corporation provides additives which will impart dispersancy and oxidation-resistant properties to fuel and lubricant compositions, when used in amounts of 10 to 5,000 parts by weight per million parts of fuel or 0.1 to 10% of the total weight when used in a lubricant. These additives are oxidatively coupled products prepared from a reaction mixture comprising:

(a) At least one hydroxyaromatic compound containing no aliphatic substituent having more than 4 carbon atoms; and

(b) At least one hydroxyaromatic compound containing at least one aliphatic substituent having at least 12 carbon atoms;

with at least one position ortho to a hydroxy group in each of reagents (a) and (b) being unsubstituted. A preferred subgenus of compounds useful as reagent (a) has the formula:

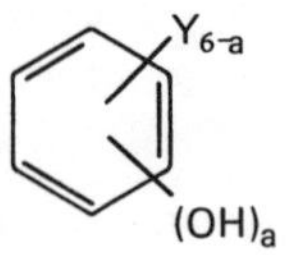

Y_{6-a}

$(OH)_a$

wherein a is 1 or 2, usually 1; each Y is individually hydrogen, halo, alkyl or hydroxyalkyl or haloalkyl of up to 4 carbon atoms, or alkoxy or alkylthio of up to 4 carbon atoms; and at least one and preferably two positions ortho to a hydroxy group are unsubstituted. Most often, up to two Y's (preferably none) are alkyl or hydroxyalkyl radicals having up to 2 carbon atoms and all others are hydrogen. It is also preferred that the positions adjacent to the previously mentioned unsubstituted ortho positions not contain bulky substituents such as tert-butyl or phenyl groups which would tend to inhibit oxidative coupling in those positions.

A preferred group of compounds useful as reagent (b) has the same general formula as that described for reagent (a) with the addition of a R group, an aliphatic radical having at least 12 carbon atoms, attached para to the hydroxy group.

The R radical preferably contains at least 40 carbon atoms, in which case it is most often derived from an olefin polymer such as polypropylene, polybutene, ethylene-propylene copolymer, butene-isoprene copolymer and the like. These polymers usually have a number average molecular weight between 500 and 15,000 and most desirably between 800 and 3,000. Especially preferred are those derived from polybutenes containing predominantly isobutene units.

These oxidatively coupled products are prepared by reacting the abovedescribed mixtures of hydroxyaromatic compounds with an oxidative coupling agent, preferably molecular oxygen in combination with copper salts and amines, such as, pyridines.

The preparation of these oxidatively coupled products is illustrated by the following examples. All parts are by weight; molecular weights are number average molecular weights and are determined by vapor phase osmometry. Polybutenyl moieties, where used, contain predominantly isobutene units.

Example 1: A mixture of 3,100 parts of o-dichlorobenzene, 48 parts of magnesium sulfate, 80 parts of pyridine and 1 part of cuprous chloride is heated to 75°C as air is bubbled beneath the surface. A solution of 300 parts of p-polybutenyl phenol (molecular weight 1,200) and 100 parts of phenol in 780 parts of o-dichlorobenzene is added and the mixture is heated at 140° to 150°C for 13 hr as air blowing is continued. Mineral oil, 500 parts, is added and the solution is filtered and stripped at 210°C under vacuum to yield an oil solution of the desired oxidative coupling product.

Example 2: A mixture of 910 parts of o-dichlorobenzene, 24 parts of magnesium sulfate, 40 parts of pyridine and 0.5 part of cuprous chloride is heated to 75°C. A solution of 100 parts of p-polybutenyl phenol (molecular weight 860) and 9.4 parts of phenol in 390 parts of o-dichlorobenzene is added and the mixture is heated at 140° to 150°C for 7 hr as air is bubbled beneath the surface. The mixture is filtered and stripped at 210°C under vacuum. Xylene, 100 parts, is added to the filtrate to yield a xylene solution of the desired oxidative coupling product.

As previously indicated, these oxidatively coupled products are useful as additives for lubricants, in which they function primarily as dispersants, oxidation inhibitors and viscosity modifiers. They can be employed in a variety of lubricants based on diverse oils of lubricating viscosity, including natural and synthetic lubricating oils.

The oxidative coupling products can also be used as dispersant and antioxidant additives in fuels. The fuel compositions of the process contain a major proportion of a normally liquid fuel, usually a hydrocarbonaceous petroleum distillate fuel such as motor gasoline, diesel fuel or fuel oil. The following are illustrative of possible fuel compositions using these additives. All parts are by weight unless otherwise indicated.

- Composition D (gasoline fuel) is gasoline containing 2 g/gal of lead as tetraethyl lead and 20 ppm of the product of Example 2.
- Composition E (diesel fuel) is diesel fuel oil containing 40 ppm of the composition of Example 1.

The patents, abstracted elsewhere in this book, which also show particularly good dispersant characteristics are U.S. Patents 4,237,022 and 4,256,596.

SEDIMENT INHIBITORS

Various types of petroleum-derived hydrocarbon oils undergo deterioration on storage or upon exposure to severe conditions. Thus, fuel oils such as gasoline, diesel fuel, jet fuel, other aviation fuel, burner oil, furnace oil, kerosene, and naphtha, for example, and other oils such as lubricating oils, cutting oils, slushing oils, etc., undergo deterioration as evidenced by such changes as, for example, formation of sediment and discoloration.

Sediment formation is undesirable for various reasons. When formed in tanks storing hydrocarbon oils, the settling of accumulated particulates requires periodic draining and cleaning of storage tanks, leading to temporary unavailability of storage capacity, substantial diversion of manpower, and waste disposal problems. Sediment formation in burner oil tends to plug strainers, burner tips, injectors, etc.

In diesel fuel such sediment tends to form sludge and varnish in the engine. If the oil is used as a heat exchange medium, as for example, with jet fuel, the sediment tends to plug exchanger coils. In gasoline the sediment may tend to deposit on sensitive parts in an internal combustion engine, such as carburetors, thereby decreasing the efficiency of combustion and causing increased fuel consumption.

It is apparent, therefore, that reduced sediment formation in hydrocarbon oils is desirable. One method of effecting such reduction would be to eliminate, to a substantial degree, those processes leading to particulate formation, such as oxidation. Another method would be to prevent agglomeration and/or settling of the formed particulate matter by effectively maintaining the fine particulates in a well dispersed state, for when the particulates are so dispersed the difficulties associated with sediment formation either do not occur or are of substantially lessened severity.

Polymeric Reaction Products of Poly(Alkoxyalkylene)Amines and Epichlorohydrin

G.W.Y. Kwong; U.S. Patents 4,239,497; December 16, 1980 and 4,252,746; February 24, 1981; both assigned to UOP Inc. has devised a method to prevent deterioration of hydrocarbon oils, as evidenced especially by sediment formation and discoloration, by the incorporation of minor amounts of a suitable additive which is prepared by reaction of poly(oxyalkylene)amine with an epihalohydrin at a temperature from 40° to 150°C, removing liberated halogen with an inorganic base, and recovering the reaction product. In a more specific embodiment the epihalohydrin is epichlorohydrin.

The poly(oxyalkylene)amines have the general formula $A-(R_1O)_n-RNH_2$ in which:

R is an alkylene moiety containing from 2 to 10 carbon atoms;
R_1 is independently selected from the group consisting of alkylene moieties containing from 2 to 10 carbon atoms;
n is an integer from 2 to 50;
A is selected from the group consisting of H_2N, alkoxy, where the carbonaceous portion contains from 1 to 40 carbon atoms; and

$$\begin{array}{l} \quad CH_2O \\ \quad | \\ B-C-CH_2O(R_1O)_yRNH_2 \\ \quad | \\ \quad CH_2O(R_1O)_zRNH_2 \end{array}$$

wherein B is selected from the group consisting of hydrogen and alkyl containing from 1 to 10 carbon atoms, and y and z are integers from 1 to 10.

The reaction products of this process may be used as additives for many kinds of hydrocarbon oils, fuel oils in particular. The additives may be employed in a concentration from 0.0001 to 1 wt %, depending upon the nature of the hydrocarbon oil, its source, its intended use, etc.

Example 1: The amine used was a methoxy poly(isopropyleneoxy)isopropyleneamine, of approximate formula: $CH_3O-[CH_2CH(CH_3)O]_9-CH_2CH(CH_3)NH_2$, supplied as Jeffamine M-600 (Jefferson Chemical Co.). Epichlorohydrin (9.2 g, 0.10 mol) was added dropwise over 1 min to a pale yellow solution of amine at 75°C in 60 g Espesol 3BC (high boiling bottoms from xylene fractionation by Charter Oil Company) and 2-propanol (20 g). The resulting solution was stirred at 75° to 78.5°C for 1 hr and 47 min, giving a solution with an acidity of 14% of the theoretical maximum.

The temperature was raised to 86°C and kept there 1 hr and 5 min, giving a solution with an acidity of 28%. Then 17% aqueous sodium hydroxide solution (25 g, 0.110 mol, 10% excess) was added in one portion. The mixture was stirred at 84°C for 1 hr and 10 min.

The layers were separated. The organic layer was azeotropically distilled with collection of the 2-propanol-water azeotrope. The solution was cooled and suction-filtered, giving a light yellow solution (114.6 g, 92%, 52.7 wt % active-ingredient).

Example 2: In this example the amine was a substituted ether of trimethylolpropane of the type:

$$C_2H_5-\underset{\displaystyle CH_2[OCH_2CH(CH_3)]_zNH_2}{\overset{\displaystyle CH_2[OCH_2CH(CH_3)]_xNH_2}{\overset{|}{\underset{|}{C}}}}-CH_2[OCH_2CH(CH_3)]_yNH_2$$

where x + y + z had an average value of about 5.3.

Epichlorohydrin (8.7 g, 0.094 mol) was added dropwise over 12 min to a stirred, pale yellow solution of the amine (44.1 g, 0.30 eq, 0.10 mol) in Espesol 3BC, (47.0 g) at 75° to 77°C. The resulting solution was stirred at 77° to 78°C for 14 min. The temperature was increased stepwise from 77.5° to 94°C over 45 min and maintained there for 13 min to an acidity of 44% of the maximum theoretical. Then 20% aqueous sodium hydroxide (21.0 g, 0.105 mol, 10% excess) was added dropwise over 13 min to the stirred solution at 94.5° to 90°C. The mixture was filtered and the layers were separated. The organic layer was filtered, giving a clear yellow pale filtrate (87.4 g, 96%, 46.1 wt % active ingredient by nitrogen-jet gun method).

Several representative products of this general type were added to fuel oil at a concentration of 17 ppm. They were compared to a commercial product, Polyflo 130 (UOP Inc.), in testing for color change and sedimentation in an accelerated storage test. The additives described herein were found to be of comparable or superior efficiency for inhibiting sedimentation and discoloration.

Polymeric Reaction Products of Alkoxyalkylamines and Epihalohydrins

A related type of sedimentation and discoloration inhibitors for hydrocarbons developed by *G. Kwong and J. Levy; U.S. Patent 4,252,745; February 24, 1981; assigned to UOP Inc.* is the polymeric reaction product of 1 molar proportion of an alkoxyalkylamine, wherein the alkoxy group contains from 1 to 25 carbon atoms (preferably 6 to 20), the alkyl group is an alkylene group containing from 2 to 10 carbon atoms (preferably propylene), and the amine is a primary amine which may contain multiple alkylene amino units, with from 0.5 to 2.0 molar proportions of an epihalohydrin (preferably epichlorohydrin), at a temperature from 40° to 150°C in the presence of an inorganic base. The amine moiety is $-NH_2$ or $-NH(CH_2)_3NH_2$.

The materials of this process have been shown to be good dispersants of particulates while being poor dispersants of water. The combination of properties is an excellent one for use of these materials as sedimentation inhibitors of hydrocarbon oils, especially fuel oils. Additionally, the materials described herein show substantial inhibition of discoloration in hydrocarbon oils. Thus, these materials are superior additives for preserving quality of hydrocarbon oils upon storage, especially at elevated temperatures or for relatively long periods of time.

Example 1: Epichlorohydrin (83.2 g, 0.90 mol) was added dropwise over 50 min to a stirred, pale yellow solution, initially at about 90°C, of tridecyloxypropylamine (280.4 g, 1.05 mols), dissolved in 222 g of Espesol 3BC. The solution was stirred about 1.5 hr at 94° to 110°C, after which a solution of 20% aqueous sodium hydroxide containing 0.99 mol of base was added over about 40 min while the reaction temperature was maintained at 86° to 94°C. The mixture was stirred about 2.5 hr at 86° to 91°C and an additional 0.09 mol of base in water was added. The mixture was cooled, layers were separated, and the organic phase was filtered to give a clear amber solution (544 g) which contained 53.2% active ingredient.

Example 2: N-tridecyloxypropyl-1,3-propylenediamine (0.740 mol) and epichlorohydrin (0.70 mol) were added concurrently over about 40 min to a mixture of Espesol 3BC (172 g) and 22% aqueous sodium hydroxide containing 0.070 mol of base at 75° to 90°C. After 8 min additional 22% aqueous sodium hydroxide (0.70 mol of base) was added over 15 min. The temperature was increased to 110°C over 1 hr and the mixture was stirred for 2 additional hours at that temperature. Layers were separated, 15 g of xylene was added to the organic phase, and water was removed by azeotropic distillation to give 423 g (97%) of a light amber solution containing 53% of active ingredient.

Alkylaryl Sulfonic Acids

W.H. Stover and S.A. Hunter; U.S. Patent 4,182,613; January 8, 1980; assigned to Exxon Research & Engineering Co. are particularly concerned with the stabilization of intermediate fuels which are blends of distillate and residual fractions from crude processing.

Various types of instability may be exhibited by residual fuel oils. Among these are: (a) separation of asphaltic or carbonaceous matter, sludge, dirt and water during storage at normal temperatures; (b) separation of black waxy material during storage at low temperatures; (c) increase in viscosity during storage at normal temperatures; and (d) incompatibility or separation of insoluble matter on mixing of fuel oils from different sources.

It has been discovered that certain alkylarylsulfonic acids will prevent or significantly reduce the amount of asphaltic sediment separating from intermediate (residuum containing) fuels made from incompatible components. Sulfonic acids with 10 to 70 total carbons in the alkyl group(s) and aromatic ring(s) are effective. Alkyl benzenes with 20 to 40 carbons in the side chain(s) are preferred. Optimally, the monoalkylbenzene with an average side chain carbon number of 28 to 32 is used.

The treatment rate required depends on the amount of sediment or precipitate that would separate from the residual fuel if it were not treated with the additive. It is generally necessary for complete dispersion to add 1.0 to 1.5 pbw of additive for 1 pbw of sediment as measured in the Sediment by Hot Filtration (SHF) Test (reported in *Industrial and Engineering Chemistry*, Vol. 10, No. 12, pp 678-680, 1938).

Of particular importance is the fact that the additive not only has the capacity to prevent sediment formation but also can resuspend sediment that has already formed in a fuel blend. Thus, there is provided here a petroleum fuel composition having a kinematic viscosity ranging from 40 Saybolt Seconds Universal (SSU) at 38°C to 300 Saybolt Seconds Furol (SSF) at 50°C comprising a residual fuel oil containing dispersed sedimentary asphaltic constituents and a minor but sediment-stabilizing proportion of an alkylarylsulfonic acid having 10 to 70 total carbons.

The Sediment by Hot Filtration Test referred to above is an analytical method developed to predict the tendency of a fuel oil to clog screens or nozzles of burners. Sediment in distillates and in residual fuels with viscosities not greater than 300 SSF at 50°C can be measured. A portion of the sample is placed in a jacketed filter and steam heated to about 95°C, and without dilution, filtered through an asbestos pad, with suction of about 250 mm Hg. The sediment remaining on the pad after washing with a nonaromatic solvent such as a high boiling naphtha is reported as weight percent to the nearest 0.01% for residual fuels (fuel containing residuum).

Especially preferred as an additive for this process is the octacosyl benzene sulfonic acid wherein the alkyl radical is derived from a nominal 28 carbon propylene oligomer.

The following experiments were conducted to illustrate that a preferred sulfonic acid, i.e., $C_{24(av)}$-alkyl substituted benzene sulfonic acid, could resuspend asphaltic material once it has precipitated as well as prevent sediment formation when added to one of the components prior to blending.

Incompatible fuel blends were prepared using 90 parts gas oil from Western Canadian crude and 10 parts pitch from South America. In one case, the additive was added to the gas oil prior to blending and homogenization at 82°C. In the other, the additive was added after blending and asphaltene separation. (The latter blend was heated to 82°C for 1 hr before spot tests were conducted.) Treats of 1.0, 1.5 and 2.0 wt % were employed.

The blotter tests showed equivalent levels of asphaltene dispersion at the same treat levels for both methods of addition.

Two sedimented incompatible blends were then treated with the additive. Changes in the level of sediment were measured using the hot filtration test. The results confirmed the effectiveness of the additive even on blends where precipitation had occurred much earlier as shown in the following table.

Resuspension of Asphaltic Sediment with Additive

Pitch Source	Fuel Sulfur (wt %)	Additive Treat (wt %)	Original SHF (wt %)*	SHF After Additive Treatment
Persian Gulf	1.0	1.0	1.62	1.20
Persian Gulf	2.5	2.0	0.18	0.09
Guanipa	0.5	1.5	1.01	0.08
Guanipa	0.5	1.0	1.01	0.40

*Sediment by hot filtration.

-4-

FLOW IMPROVERS AND POUR POINT DEPRESSANTS

Heating oils and other middle distillate petroleum fuels, e.g., diesel fuels, contain normal paraffin hydrocarbon waxes which, at low temperatures, tend to precipitate in large crystals in such a way as to set up a gel structure which causes the fuel to lose its fluidity thereby presenting difficulties in transporting the fuel through flow lines and pumps. The wax crystals that have come out of solution also tend to plug fuel lines, screens and filters. This problem has been well recognized in the past and various additives known as pour point depressants have been used to change the nature of the crystals that precipitate from the fuel oil, thereby reducing the tendency of the wax crystals to set into a gel. It is thus desirable to obtain not only fuel oils with low pour points, but also oils that will form small wax crystals so that the clogging of filters will not impair the flow of the fuel at low operating temperatures.

It is known in the prior art to employ various polymeric and copolymeric materials as pour point depressants for wax-containing petroleum fractions.

Recently, it has become known that pour point depression is not a sufficient phenomenon to alleviate some problems caused by wax crystals in various fuels, especially middle distillates. In those petroleum fractions, it has been observed that the wax crystals formed in the presence of the pour point depressant are often too large to enable the wax-cloudy fuels to pass easily through screens and orifices commonly encountered in the equipment employed either in distribution or in use of such fuels. This problem has been alleviated by the addition to the fraction of petroleum products of wax crystal modifiers which are referred to as flow and filterability improvers. This chapter will discuss various types of additives for improving flow characteristics and preventing solidification of petroleum products during low temperature storage.

FOR MIDDLE DISTILLATE FUEL OILS

Copolymer of 1-Hexene and 1-Octadecene

Although the pour point of a mineral oil is only a general indicator of its low temperature flow properties, it provides a useful function and is commonly found in an oil's specifications. Most service oils, and particularly the paraffinic base oils, require a pour point depressant as an additive in order to meet established specifications. Because of the relatively high cost of conventional pour point depressants they are generally not considered for fuel oils, rather other less expensive solutions to low temperature flow problems are utilized.

W.J. Heilman and T.J. Lynch; U.S. Patent 4,132,663; January 2, 1979; assigned to Gulf Research & Development Company have found a highly efficacious and relatively inexpensive pour point depressant which is particularly useful for lubricating oils, automatic transmission fluids, fuel oils, particularly No. 2 fuel oil, and the like.

These oils of reduced pour point according to their process are compounded with a minor amount of a copolymer synthesized from a mixture of 1-hexene and 1-octadecene. Preferably the 1-hexene comprises from about 75 to 90 mol % of the copolymer and the 1-octadecene preferably comprises from about 10 to 25 mol % of the copolymer.

The copolymers can be described as a mixture of long chain molecules formed with a series of repeating units randomly distributed in each chain, each repeating unit having the structural formula:

$$-\!\!\left[\underset{\displaystyle R}{\underset{|}{CH}}-CH_2\right]_n\!\!-$$

in which R is the group $-(CH_2)_{m-3}CH_3$ wherein m is the carbon number of the monomer from which the repeating unit is derived, that is, m is either 6 or 18, and n, the number of repeating units in an individual molecule, is at least about 50 and the maximum average number of repeating units is about 5,000. The 1-hexene and 1-octadecene are preferably copolymerized by a procedure in which they are caused to randomly join together in the polymer chain in essentially the same molar ratio that each alpha-olefin occurs in the initial reaction mixture in order to simplify the preparation of the desired product composition.

The copolymers used in forming the oil compositions of this process are preferably made by catalyzing the 1-hexene and 1-octadecene mixture using a Ziegler-Natta-type catalyst. Either continuous, semicontinuous in which only the alpha-olefins are added during reaction, or batch polymerization can be used, provided that all conditions and proportions of chemical species present in the reaction mixture are properly correlated with each other to obtain the desired copolymer product. Any Ziegler-Natta-type catalyst can be used which is useful for the polymerization of propylene. Particularly useful are titanium and vanadium salts, primarily the chlorides, in conjunction with aluminum alkyls and alkyl chlorides. It has been found that purple titanium trichloride together with aluminum triethyl constitutes an excellent catalyst. A catalyst containing from about 1 to 10 gram atoms of aluminum per gram atom of titanium is useful with a ratio of about 1.8 to 3 being preferred. The reaction is suitably carried out using about 300 to 6,000 grams of olefin per gram of catalyst and preferably from about 1,500 to 3,000 grams of olefin per gram of catalyst.

Hydrogen or another suitable molecular chain length modifier such as zinc chloride, dialkyl zinc such as diethyl zinc and the like helps to direct the reaction to the desired product. The partial pressure of hydrogen can preferably be between about 0.5 and 25 for a continuous polymerization and between about 0.5 and about 10 for a semicontinuous or batch reaction. The temperature at which the polymerization reaction is conducted can suitably be between about 100° and 250°F and preferably between about 220° and 250°F.

A suitable solvent for the reaction mixture is desirable since the resulting copolymer is highly viscous and can be substantially nonflowable at room temperature. The most preferred solvents are those light solvents which can be easily distilled from the product copolymer or mineral oils which can be incorporated in the finished oil together with the copolymer. It is found that a mineral oil which is pretreated with hydrogen in the presence of a Ziegler-Natta-type catalyst is particularly herein useful. Since a solvent is not necessary, it can be used in amounts from 0 to 75% of total charge to the reactor and when used, it is preferably used in an amount of about 25 to 60%.

Example: A copolymer synthesized from 85 mol % 1-hexene and 15 mol % 1-octadecene was tested as a pour point depressant in formulated automotive-type lubricating oils as set out in the table below.

Copolymer Added (wt %)	Pour Point, °F		
	SAE 10W	SAE 20W-20	SAE 30
0	+5	+5	+5
0.1	-5	-10	+5
0.25	-25	-20	-15
0.50	-35	-35	-15

Each oil contained 6.4 wt % of a commercially used detergent-inhibitor additive to meet SAE SE performance specifications.

Borated Mannich Bases plus a Coadditive Hydrocarbon

Heating oils and other middle distillate petroleum fuels, e.g., diesel fuels, contain normal paraffin hydrocarbon waxes which, at low temperatures, tend to precipitate in large crystals in such a way as to set up a gel structure which causes the fuel to lose its fluidity thereby presenting difficulties in transporting the fuel through flow lines and pumps. The wax crystals that have come out of solution also tend to plug fuel lines, screens and filters. This problem has been well recognized in the past and various additives known as pour point depressants have been used to change the nature of the crystals that precipitate from the fuel oil, thereby reducing the tendency of the wax crystals to set into a gel.

Recently, it has become known that pour point depression alone is not a sufficient phenomenon to alleviate some problems caused by wax crystals in various fuels, especially middle distillates. In those petroleum fractions, it has been observed that the wax crystals formed in the presence of the pour point depressant are often too large to enable the wax-cloudy fuels to pass easily through screens and orifices commonly encountered in the equipment employed either in distribution or in use of such fuels. This problem has been alleviated by the addition to this fraction of petroleum products of wax crystal modifiers which are referred to as flow and filterability improvers.

N. Feldman, W.H. Stover, and W. deWaal; U.S. Patent 4,140,492; February 20, 1979; assigned to Exxon Research & Engineering Co. have found that the borated derivative of an oil-soluble Mannich base in combination with a coadditive hydrocarbon of the class consisting of an amorphous hydrocarbon, a hydrogenated butadiene and mixtures thereof further improves the cold flow characteristics of a middle distillate petroleum fuel oil boiling within the range of about 120° to 400°C at atmospheric pressure.

In accordance with this process, a fuel composition is provided which comprises a major proportion, i.e., more than 50% by wt, of a distillate petroleum fraction preferably having an atmospheric boiling range of from about 120° to 400°C and from about 0.001 to 1.0 wt % of a flow and filterability improving combination comprising: (a) 1 to 5 parts by wt of a borated oil-soluble Mannich base of the general formula

$$R-C_6H_3(OH)-CH_2N(H)(CH_2)_xOH$$

wherein R is an alkyl group having from 8 to 40, optimally from 16 to 22 carbon atoms and x is an integer ranging from 2 to 10; and (b) 1 to 100 parts by wt of a coadditive hydrocarbon of the class consisting of an oil-soluble amorphous hydrocarbon, such as a saturated hydrocarbon fraction, having less than about 5, preferably less than about 1, wt % of normal paraffin hydrocarbons, which can be illustrated by Coray 200 petrolatum, a hydrogenated polybutadiene having from about 5 to 55, preferably 10 to 30 wt % 1-2 addition, i.e., branched units and a number average molecular weight ($\overline{M}_n$) ranging from 400 to 10,000 preferably 600 to 3,000 and mixtures thereof. It is preferred that the weight ratio of a:b is in the range of 4:1 to 1:25, optimally 2:1 to 1:8. All molecular weights herein are measured by Vapor Phase Osmometry (VPO).

Concentrates of 1 to 60 wt % of this borated additive-hydrocarbon combination in 40 to 99 wt % of mineral oil, e.g., kerosene, can be prepared for ease of handling.

The borated oil-soluble Mannich base is normally the reaction product of a C_{16-22} alkyl substituted hydroxy aromatic compound (derived from a linear alpha-olefin reacted with phenol), an aldehyde (preferably formaldehyde or paraformaldehyde), and a monoalkanol amine of 2 to 10 carbon atoms (preferably ethanolamine). The boron compound for reacting with the Mannich base may be boron oxide, a boron oxide hydrate, a boronic or boric acid or an ester of such an acid.

The amorphous hydrocarbon useful as the coadditive is fully described in U.S. Patent 3,660,058. It can be obtained by deasphalting a residual petroleum fraction and then adding a solvent such as propane, lowering the temperature of the solvent-diluted residuum, and recovering the desired solid or semisolid amorphous product by precipitation followed by filtration. The residual oil fractions from which the desired amorphous hydrocarbons are obtained will have viscosities of at least 125 SUS at 99°C. Most of these residual oils are commonly referred to as bright stocks.

Useful hydrogenated polybutadienes are described in U.S. Patent 3,600,311. Hydrogenated polybutadienes are branched polymers which have about 5 to 55, preferably 10 to 30, 1-2 addition and 45 to 95, preferably 70 to 90, 1-4 addition and usefully have a number average molecular weight ($\overline{M}_n$) of 400 to 10,000. They are readily prepared by polymerizing butadiene in a suitable solvent, e.g., hydrocarbon, in the presence of an organometallic catalyst, e.g., n-butyl lithium. The resulting polymer is then hydrogenated, e.g., with hydrogen in the presence of a nickel catalyst until saturated, e.g., <1 wt % unsaturation.

The distillate fuel oils that can be improved by additive combination include those having boiling ranges within the limits of about 120° to 400°C. The distillate fuel oil can comprise straight run or virgin gas or cracked gas oil or a blend in any proportion of straight run and thermally and/or catalytically cracked distillates.

The most common petroleum middle distillate fuels are kerosene, diesel fuels, jet fuels and heating oils. The low temperature flow problem is most usually encountered with diesel fuels and with heating oils.

The hydrogenated polybutadiene had an ($\overline{M}_n$) of about 1,250, a bromine number of 0.3 and a Fisher-Johnes melting point of 33° to 44°C. The hydrogenated polybutadiene was prepared with a n-butyl lithium catalyst and hydrogenated with hydrogen in the presence of a Raney nickel catalyst.

Testing showed that when amounts of from 0.05 to 0.10 wt % borated Mannich base and 0.05 to 0.10 wt % of hydrogenated polybutadiene or amorphous hydrocarbon were added to a middle distillate fuel, the cold flow filter plugging point of the blend was decreased 11° or 12°C below that of the nontreated fuel when measured by the Cold Filter Plugging Point Test carried out as follows. A 40 ml sample of the oil is cooled by a bath at about -34°C. Periodically (at each one degree centigrade drop in temperature starting from 2°C above the cloud point) the cooled oil is tested for its ability to flow through a fine screen in a time period. This cold property is tested with a device consisting of a pipette to whose lower end is attached an inverted funnel positioned below the surface of the oil to be tested. Stretched across the mouth of the funnel is a 350 mesh screen having an area of about 0.45 in^2.

The periodic tests are each initiated by applying a vacuum to the upper end of the pipette whereby oil is drawn through the screen up into the pipette to a mark indicating 20 ml of oil. The test is repeated with each one degree drop in temperature until the oil fails to fill the pipette within 60 seconds. The results of the test are reported as the temperature in °C at which the oils fail to fill the pipette in the prescribed time.

Amine Salts of Thiobis Lactone Acids plus Coadditive Hydrocarbons

S.J. Brois, A. Gutierrez, and N. Feldman; U.S. Patents 4,142,866; March 6, 1979; and 4,251,232; February 17, 1981; both assigned to Exxon Research and Engineering Co. report that amic acid, amic acid amine salts, bisamides and imides of thiobis(C_{12-50} alkyl lactone carboxylic acid) in combination with a coadditive hydrocarbon of the class consisting of an amorphous hydrocarbon, a hydrogenated butadiene and mixtures thereof improve the cold flow characteristics of a middle distillate petroleum fuel oil boiling within the range of about 120° to about 400°C at atmospheric pressure.

Accordingly, a fuel composition is provided which comprises a major proportion, i.e., more than 50% by wt, of a distillate petroleum fraction preferably having an atmospheric boiling range of from about 120° to 400°C and from about 0.001 to 1.0 wt % of a flow and filterability improving combination comprising: (a) 1 to 5 parts by wt of an amine derivative, i.e., an amine salt, an amic acid, amic acid amine salt, bisamide or imide of a thiobis(C_{12-50}, preferably C_{16-24}, alkyl lactone acid) to be described later; and (b) 1 to 100 parts by wt of a coadditive hydrocarbon or hydrogenated polybutadiene, both of which are thoroughly described in the previous patent.

A particularly useful amine derivative for the flow and filterability improving combination is the amine salt of a thiobis(C_{12-50}, preferably C_{18-28}, alkyl lactone acid), preferably secondary amine salts of the general formula

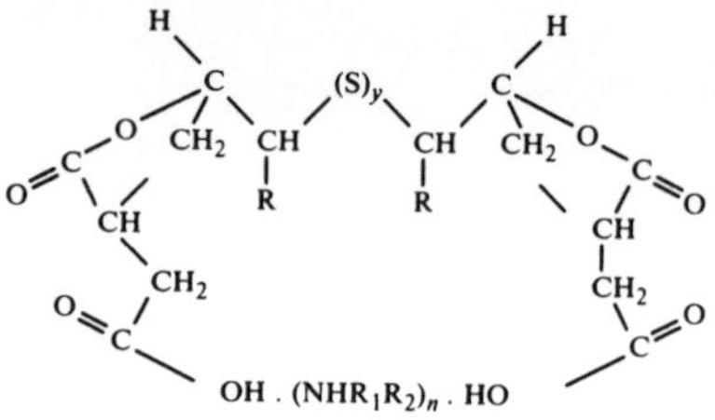

wherein R represents an alkyl group of from 8 to 40, preferably 12 to 24 carbons, y represents the number 1 or 2, n represents the number of 1 to 2, and R_1 and R_2 are selected from the class of an aliphatic hydrocarbon of 8 to 30 carbon atoms and an oxyaliphatic hydrocarbon of from 8 to 30 carbon atoms. It is preferred that the weight ratio of a:b is in the range of 4:1 to 1:25, optimally 2:1 to 1:8.

Illustrative thiobis(alkyl lactone acid) coadditive compounds include thiobis(alkyl substituted lactone carboxylic acid) wherein the alkyl substituent can be pentyl, octyl, dodecyl, tridecyl, n-tetradecyl, pentadecyl, hexadecyl, heptadecyl, nonadecyl, eicosyl, docosyl and branched analogs and mixtures thereof.

Examples of secondary amines include di-n-octyl amine, octyl lauryl amine, dodecyl hexadecyl amine, dioctadecyl amine, didocosyl amine; and useful primary amines include benzyl amine, octyl amine, dodecyl amine, hexadecyl amine, and tetracosyl amine.

Amine mixtures may also be used and many amines derived from natural materials are mixtures. Thus, coco amine derived from coconut oil is a mixture of secondary amines with straight chain alkyl groups ranging from C_{8-18}. Particularly useful is ditallow secondary amine, derived from hydrogenated tallow, which secondary amine is a mixture of C_{14-18} straight chain alkyl groups.

The amic acid and amic acid amine salts are readily prepared by mixing together neat from 1 to 2 mols of secondary amine to a molar amount of the thiobis(alkyl lactone acid) conveniently as prepared, or in an inert solvent such as xylene. The reagents are usually mixed in xylene and heated for several hours at reflux until sufficient water is azeotroped from the reaction mixture to form the amic acid and the amic acid amine salt products. Thus mild heating from ambient temperature to about 150°C may facilitate the amine acid and/or amic acid salt formation.

The bisamide products are readily produced via the condensation of thiobis(alkyl lactone acid) with 2 mols of secondary amine at higher temperatures, e.g., 210° to 240°C although temperatures of from 150° to 250°C are useful for this bisamide.

The thiobis(lactone acid) can be conveniently prepared by sulfur halide addition to the double bond in alkenyl succinic anhydride at about -60° to 100°C followed by lactonization via an internal displacement of the halide with water. Alternatively, alkenyl succinic acid can be reacted with sulfur halide at from about -60° to 100°C resulting in adduct formation which undergoes lactonization at temperatures above about 50°C since the temperature governs the displacement of the halide, e.g., chloride by a vicinal carboxylic acid group.

The preparation of alkenyl succinic anhydrides and acids is well known by either the "ene" reaction wherein an olefin is reacted with maleic anhydride or fumaric acid or a Diels-Alder reaction of a halogen substituted aliphatic material, e.g., chloropolyisobutylene with maleic anhydride or fumaric acid (see U.S. Patent 2,568,876).

The distillate oils that can be improved in this way include those boiling from 120° to about 400°C, and are more thoroughly described in the previous patent.

Example 1: *Bisamine Salt of Thiobis(Alkyl Lactone Acid)* – (This will be called the bisamine salt.) 200 g (~0.54 mol) of n-octadecenyl succinic acid were dissolved in a liter of $CHCl_3$ and 36.7 g (0.272 mol) of sulfur monochloride (S_2Cl_2) were added dropwise to the stirred solution at room temperature. The exothermic process was accompanied by vigorous HCl evolution. After refluxing the mixture for about 8 hours, the solution was cooled and solids separated. Filtration gave 19 g of solid (MP 131° to 136°C). Rotoevaporation of the supernatant gave a solid product in high yield. The proposed structure for dithiobis-(alkyl lactone acid) namely 6,6'-dithiobis(3,5-carbolactone-1-heneicosanoic acid), is as follows:

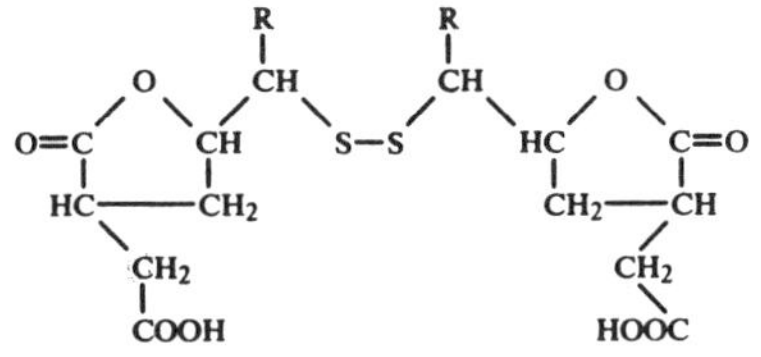

```
                R             R
                |             |
         O      CH            CH       O
       /   \   /    \      /     \   /   \
   O=C       CH      S—S          HC       C=O
     |       |                    |        |
    HC———————CH2                  CH2——————CH
      \                                   /
       CH2                             CH2
       |                                 \
       COOH                              HOOC
```

wherein R represents n-$C_{15}H_{31}$.

The bisamine salt is prepared as follows: 8 g (~0.01 mol) of the thiobis(alkyl lactone acid) as prepared above was dissolved in 25 ml of tetrahydrofuran (THF) and combined with 10.2 g (0.02 mol) of hydrogenated ditallow amine purchased as Armeen 2HT (Armak Corp.). The mixture was allowed to reflux, i.e., about 60°C for a few minutes to assure complete reaction.

While the reaction solution was still hot, acetone was added to the cloud point and a white solid precipitated out while cooling to room temperature. The solid was filtered, collected and dried at room temperature. Quantitative yield of the thiobis(alkyl lactone acid) salt was obtained. The infrared spectrum of the bisamine salt shows the characteristic bands for the lactone acid salt of 5.6 microns for the lactone carboxyl and 6.3 to 6.4 microns for the carboxylate carbonyl functionality.

Example 2: *Monoamine Salt of Thiobis(Alkyl Lactone Acid)* – The thiobis(alkyl lactone acid) salt containing only 1 equivalent of the ditallow amine was prepared in a similar manner as above, except that 0.01 mol of the thiobis(alkyl lactone acid) as shown above was treated with 0.01 mol of the ditallow amine.

Example 3: *Amic Acid and Amic Acid Amine Salt of Dithiobis(Alkyl Lactone Acid)* – (a) The amic acid amine salt of dithiobis(alkyl lactone acid) is readily obtained along with the amic acid of dithiobis(alkyl lactone acid) when one takes the procedure of preparing the bisamine salt and changes it only by heating the reactants at a higher temperature, e.g., this can be done by heating the reactants in 4 parts of xylene [1 mol of dithiobis(alkyl lactone acid) to 2 mols of the hydrogenated ditallow amine] at reflux of about 140°C for 6 hours. Water is azeotroped from the reaction mixture to form the amic acid and amic acid amine salt products. Cooling the reaction mixture to ambient temperature afforded a solid product which corresponded to the amic acid amine salt, i.e., ditallow amine salt of the ditallow monoamide of dithiobis(alkyl lactone acid).

(b) A second product crystallized from the mother liquor obtained from filtration of the crystals of the amic acid amine salt upon slight cooling and standing. This corresponds to the amic acid, i.e., the ditallow monoamide of dithiobis(alkyl lactone acid).

Example 4: *Bisamide of Thiobis(Alkyl Lactone Acid)* – The bisamide products are readily produced via the condensation of process (a) availed of for the preparation of the amic acid amine salt product except that the condensation is at a higher temperature, e.g., from 210° to 240°C, and a period of 2 hours. Infrared analysis showed the presence of a lactone amide product believed to be the bis-(ditallow amide) of dithiobis(alkyl lactone acid).

Blends of middle distillate fuels and coadditives discussed above were tested by the Cold Filter Plugging Point Test described in the previous patent. Best results for improving the flow properties were provided by a mixture of 0.02 wt % of the amic acid amine salt of Example 3 and 0.05 wt % of an amorphous hydrocarbon fraction ($\overline{M}_n$ = 775) obtained by propane precipitation from the deasphalted residuum of a Texas coastal crude oil containing no more than a trace of normal paraffin hydrocarbons; and 0.02 wt % of the same amic acid amine salt plus 0.04 wt % of the hydrogenated polybutadiene described in the previous patent.

Aliphatic Copolymer plus Derivative of a Succinic Acid plus a Nitrogen Compound

S. Ilnyckyj; U.S. Patent 4,147,520; April 3, 1979; assigned to Exxon Research and Engineering Co. has found that an oil-soluble aliphatic copolymer having the property of a nucleator for wax crystallization can be used in combination with a C_{8-28} hydrocarbyl succinamic acid mono- or disubstituted on the nitrogen atom with C_{8-28} hydrocarbyl groups or amine salts or amide derivatives which are described below to improve the cold flow characteristics of a distillate fuel oil having the property of supercooling so as to markedly improve the effectiveness of wax growth arresters in preventing the formation of large wax crystals.

Accordingly, a fuel composition is provided which comprises a major proportion, i.e., more than 50% by wt, of a distillate petroleum fraction and from about 0.001 to 0.5 wt % of a flow and filterability improving composition made up of:

(a) 1 to 20 parts by weight of an oil-soluble aliphatic copolymer functioning as a nucleator for wax crystallization in the distillate, and;

(b) 1 to 100 parts by weight of an oil-soluble succinamic acid or its derivative having the following formula:

$$\begin{array}{l} R-CH-COX \\ \quad\;\; | \\ \quad\;\; CH_2-COX^1 \end{array}$$

wherein R is a straight chain aliphatic hydrocarbon group having from 0 to 1 site of olefinic unsaturation (alkyl or alkenyl) attached at a secondary carbon atom to the succinyl group and is of at least 8 carbon atoms, and preferably in the range of 15 to 22 carbon atoms; one of X or X^1 is hydroxyl and the other is $-NYY^1$ wherein N has its normal meaning of nitrogen and Y and Y^1 are aliphatic hydrocarbyl groups of from 8 to 28 carbon atoms, more usually of from 14 to 22 carbon atoms having a total of from about 30 to 52 carbon atoms, more usually of from 32 to 48 carbon atoms, and, preferably of from 32 to 40 carbon atoms and the other of X and X^1 is of the formula:

$$OH(NHY^2Y^3)_n$$

wherein n varies from 0 to 1, Y^2 and Y^3 are selected from the class of hydrogen, an aliphatic hydrocarbon of from 1 to 30 carbon atoms and oxyaliphatic hydrocarbon of from 1 to 30 carbon atoms, and Y^2 and Y^3 may be taken together with the nitrogen to which they are attached to form a heterocyclic ring of from 5 to 7 annular members.

It is preferred that the weight ratio of a:b is in the range of 1:20 to 5:1.

It has been found that this composition prevents oil gelation and effectively controls wax crystal size in distillate hydrocarbon oils having a final boiling point in excess of about 370°C. Concentrates of 1 to 60 wt % of the additive combination in 40 to 99 wt % of mineral oil, e.g., kerosene, can be prepared for ease of handling.

Preferred among the polymeric wax nucleators are ethylene copolymers with a polymethylene backbone which is divided into segments by hydrocarbon, halogen, or oxy side chains, (usually prepared by free radical polymerization which might result in some branching) and comprises about 3 to 500, preferably 4 to 100, molar proportions of ethylene per molar proportion of an ethylenically unsaturated ester monomer (or mixture of unsaturated esters). An optimal polymer is a copolymer of ethylene with 2.0 to 12 mol % of vinyl acetate. These copolymers will generally have a molecular weight ($\overline{M}_n$) in the range of from about 500 to 50,000, preferably about 1,500 to 30,000.

Illustrative succinamic acids include N,N-dihexadecylhexadecylsuccinamic acid, N-hexadecyl-N-octadecyloctadecylsuccinamic acid, N,N-dihexadecenyl-C_{15-20}-alkenylsuccinamic acid, N-hexadecenyl-N-eicosenyloctadecylsuccinamic acid, N,N-dioctadecenyl-C_{16-18}-alkenylsuccinamic acid, etc. As indicated previously, the succinamic acid may be used as its amine salt, preferably as a mixture of acid and amine salt.

Illustrative amines which may be used to form salts are di-sec-butylamine, heptylamine, dodecylamine, octadecylamine, tert-butylamine, morpholine, diethylamine, methoxybutylamine, and methoxyhexylamine, etc.

Particularly effective is the alkenyl succinic acid composition wherein the amine employed is hydrogenated ditallowamine.

The distillate hydrocarbon oils which are treated with the additive package of this process are wax-containing distillate petroleum oils boiling in the range of 120° to 500°C, preferably middle distillates boiling from about 150° to 400°C.

The additive is particularly effective for the cold flow treatment of high end point fuels which are nonresponsive to conventional flow improvers, i.e., those fuels having a final boiling point above about 370°C (ASTM-1160).

Examples 1 through 6: The following materials were used: Polymer 1 is a copolymer of ethylene and vinyl acetate containing about 9 wt % vinyl acetate and having a number average molecular weight ($\overline{M}_n$) of 4,100 and a specific viscosity at 38°C of 0.37.

Polymer 2 is a copolymer of ethylene and vinyl acetate containing about 38 wt % vinyl acetate and having a ($\overline{M}_n$) of about 1,800 and a specific viscosity at 38°C of 0.13. All number average molecular weights reported herein are determined by Vapor Phase Osmometry (VPO).

Succinamide A is the principal ingredient of a commercial product identified as Oronite 410 (Chevron Chemical Co.) which is believed to be at least 60 wt % alkenyl succinamide and succinamic amine salt.

The properties of the middle distillate fuel tested are as follows: Cloud point, 0°C; n-paraffin carbon number, 10 to 32; distillation: IBP of 161°C, 5% at 194°C, 50% at 276°C, 95% at 398°C, and FBP of 403°C.

Blending of the polymer, polymer mixture, succinamide and succinamide-polymer mixture was accomplished by dissolution into the middle distillate fuel oil. This was done while warming, e.g., heating the oil and additive to about 90°C if the additive or additives per se were added, and stirring. In other cases, the additive was simply added with stirring to the fuel in the form of an oil concentrate which was usually about 50 wt % active ingredient dissolved in a light mineral oil.

The blends were then tested for their cold flow properties in the Cold Filter Plugging Point Test (CFPPT).

The blends prepared and the test results are summarized in the table on the following page.

Example	Active Ingredient (wt %)	Additive	CFPPT, °C
1	–	None	-1
2	0.01	Polymer 2	-3
3	0.005	Polymer 1 }	-8
	0.005	Polymer 2 }	
4	0.012	Succinamide A*	-2
5	0.006	Succinamide A* }	-3
	0.005	Polymer 2 }	
6	0.008	Succinamide A* }	-9
	0.003	Polymer 1 }	

*Introduced as Oronite 410 containing about 60 wt % Succinamide A.

Borated Derivatives of Substituted Succinic Acids or Acid Salts

N. Feldman; U.S. Patent 4,184,851; January 22, 1980; assigned to Exxon Research & Engineering Co. reports that the borated derivative of the succinic acid or derivative described in the previous patent either alone or in combination with an amorphous hydrocarbon will improve the flow characteristics of a middle distillate petroleum fuel oil boiling from 120° to 400°C.

Oronite 410 (Chevron Chemical Co.) is used as the succinamide. It is borated in the following manner: 10 g of Oronite 410 are dissolved in 160 ml of toluene and thereafter adding 8 ml of a 5 wt % solution of boric acid in ethylene glycol. The mixture was refluxed while being stirred for 6 hours, after which time the solvents were distilled off leaving 10 g of a boron-containing compound which analyzed about 0.5 wt % boron.

A combination of the borated succinamide (0.04 wt %) and 0.25 wt % of amorphous hydrocarbon (described in U.S. Patent 4,140,492) were added to a middle distillate fuel and tested by the Cold Filter Plugging Point Test (also described in U.S. Patent 4,140,492). This combination was found to give greatly enhanced results over the use of the succinamide alone, the use of the borated succinamide alone, or the use of the succinamide with the amorphous hydrocarbon.

Combination of Polyethylenes and Polyesters

M.J. Wisotsky; U.S. Patents 4,153,423; and 4,153,424; May 8, 1979; both assigned to Exxon Research & Engineering Co., describes additive combinations of polymers useful in middle distillate fuel oils having a fraction boiling above 370°C for controlling the size of the wax crystals that pour at low temperatures.

He has found that ethylene polymers or copolymers having an $\overline{M}_n$ of from 1,000 to 4,000 (U.S. Patent 4,153,424) or 5,000 to 20,000 (U.S. Patent 4,153,423) in combination with a second polymer which is a polyester, i.e., homopolymer or copolymer comprising at least 10% by wt, preferably at least 25 wt % of C_{4-16} substantially straight chain alkyl ester of an ethylenically unsaturated monocarboxylic acid, e.g., acrylic or methacrylic acid, can give synergistic results in controlling wax crystal size in distillate hydrocarbon oils.

When the polyester is a copolymer, it is limited to containing less than about 25 wt % total of one or more additional monomer moieties, i.e., in addition to the C_{4-16} alkyl ester such as alkyl ester of ethylenically unsaturated mono- or dicarboxylic acids having C_{6-44} alkyl groups extending from ester linkages.

In general, the additive combination will comprise 1 part by wt of the ethylene polymer per about 0.2 to 4 parts by wt of the polyester, i.e., polyacrylate. The distillate hydrocarbon oil compositions will contain a total preferably of 0.005 to 0.1 wt % of the additive combination. Concentrates of 1 to 60 wt % of the additive combination in 40 to 99 wt % of mineral oil, e.g., kerosene, can be prepared for ease of handling.

The ethylene polymers will have a polymethylene backbone which is divided into segments by hydrocarbon, halogen, or oxyhydrocarbon side chains. They may be simply homopolymers of ethylene, usually prepared by free radical polymerization which will result in some branching. More usually, they will comprise about 3 to 40, preferably 4 to 20, molar proportions of ethylene per molar proportion of a second ethylenically unsaturated monomer, which latter monomer can be a single monomer or a mixture of such monomers in any proportion.

The unsaturated monomers, copolymerizable with ethylene, include unsaturated mono- and diesters of the general formula:

$$\begin{array}{cc} R_1 & H \\ | & | \\ C & = C \\ | & | \\ R_2 & R_3 \end{array}$$

wherein R_1 is hydrogen or a C_{1-4} alkyl group, e.g., methyl; R_2 is a $-COOR_4$ group wherein R_4 is hydrogen or a C_{1-16}, preferably a C_{1-8}, e.g., C_{1-4}, straight or branched chain alkyl group; and R_3 is hydrogen or $-COOR_4$. The monoester, i.e., when R_3 is hydrogen, is the preferred copolymer moiety; it includes acrylic acid, its homologs such as methacrylic acid and its analogs which are characterized herein as acrylates. Thus, when R_2 is $-COOR_4$ and R_3 is hydrogen, such esters include methyl acrylate, isobutyl acrylate, 2-ethylhexyl acrylate, C_{13} oxo alcohol esters of methacrylic acid, etc. Examples of monomers where R_1 is hydrogen and R_2 and R_3 are $-COOR_4$ include mono- and diesters of unsaturated dicarboxylic acids.

Another class of monomers that can be copolymerized with ethylene include C_{3-16} alpha-monoolefins, which can be either branched or unbranched, such as propylene, isobutene, n-octene-1, etc.

The most effective polyester polymers for this purpose, from the point of view of availability and cost, are the polymerized esters of acrylic acid or alpha-methacrylic acid and monohydric, saturated, primary aliphatic alcohols containing from 4 to 16 carbon atoms in the molecule. This useful class of oil-soluble polyesters which includes the C_{4-16} alkyl esters of acrylic acid, homologs of acrylic acid and analogs of acrylic acid are here designated poly(C_{4-16} alkyl acrylates). For these purposes, an oil-soluble polymer or copolymer has a solubility in oil of at least about 0.001 % by wt at 20°C. The optimum polyesters possessing the highest solubility and stability in oils are those derived from the straight chain monohydric primary saturated aliphatic alcohols containing 8 to 16 carbon atoms such as the normal octyl, lauryl, cetyl esters. These esters need not be pure, and mixtures of two or more of these polymers may be used.

The monocarboxylic acid ester monomers described above may be copolymerized with various amounts, e.g., up to 25 wt %, of other unsaturated esters or olefins.

Examples: The following materials were used (all these materials are from U.S. Patent 4,153,424 except for Polymer 2, which is from 4,153,423):

Polymer 1 – Polymer 1 is a copolymer of ethylene and isobutyl acrylate. This copolymer was prepared by the following procedure: A 3 liter stirred autoclave was charged with 500 ml of benzene as solvent. The autoclave was then purged with nitrogen and then with ethylene. The autoclave was then heated to 90°C while ethylene was pressured into the autoclave until the pressure was raised to 3,000 psig. Then, while maintaining a temperature of 90°C and 3,000 psig pressure, 40 ml/hr of isobutyl acrylate and 70 ml/hr of a solution consisting of 11.5 wt % dilauroyl peroxide dissolved in benzene were continuously pumped into the autoclave at an even rate. A total of 100 ml of isobutyl acrylate was injected over 2.4 hr while 184 ml of the peroxide solution was injected into the reactor over a period of 2.6 hr from the start of the injection. After the last of the peroxide was injected, the batch was maintained at 90°C for an additional 10 minutes.

Then, the temperature of the reactor contents was lowered to about 60°C, the reactor was depressurized, and the contents were discharged from the autoclave. The product was then stripped of the solvent and unreacted monomers on a steam bath overnight by blowing nitrogen through the product. The final stripped product consisted of about 260 g of copolymer of ethylene and isobutyl acrylate, having a number average molecular weight of 3,545 (as measured by VPO) and an ester content of 29 wt %.

Polymer 2 – Polymer 2 was a copolymer of ethylene and isobutyl acrylate having an $\overline{M}_n$ of ~7,350. It is specifically designated Zetafax 1278 (Dow Chemical Co.). It contains about 20 wt % of isobutyl methacrylate and has a melt index of 116° to 127°C.

Polymer A – Polymer A was a polyalkyl methacrylate designated Acryloid 150. This polymer had an alkyl distribution in carbon number from C_{10-18} (largest % = C_{12}), a number average molecular weight of 82,500 and a weight average molecular weight of 798,000 (measured by Gel Permeation Chromatography).

Polymer B – Polymer B was also a polyalkyl methacrylate, Acryloid 152. The alkyl content of this polyester had a carbon number distribution from C_{12-20}. (largest % = C_{16}), with a number average molecular weight of 17,100 and a weight average molecular weight of 39,000 (determined by Gel Permeation Chromatography).

Polymer C – Polymer C is a homopolymer of n-tetradecylacrylate. The monomer was prepared as follows: To a 500 ml round bottom flask equipped with stirrer, heating mantle, condenser and Dean-Stark receiver were added 107 g n-tetradecanol, 40 g acrylic acid, 1 g hydroquinone, 3 g p-toluenesulfonic acid, and 150 ml reagent heptane. The solution was refluxed for 3 hr at which point 11 ml of water was collected in the Dean-Stark receiver. The solution was then washed with 75 ml water, 75 ml 2% sodium hydroxide solution and additional water washes till neutral. The solution was dried over magnesium sulfate, filtered and evaporated off leaving 125 g of tetradecyl acrylate.

Tetradecyl acrylate homopolymer was prepared as follows: To a round bottom microflask equipped with stirrer, condenser, heating mantle and a nitrogen inlet tube, were added 6 g of the above tetradecyl acrylate, 6 g of reagent heptane, and 0.06 g benzoyl peroxide. The solution was sparged with nitrogen then heated with stirring to about 85°C for a total of 45 minutes. Then 0.1 g hydroquinone was added and the solvent evaporated leaving 5.8 g polymer having a $\overline{M}_n$ of 6,196.

The Fuel – The distillate fuel oil tested has the following properties:

Properties	
Gravity at 16°C	0.8265
Cloud point, °C	+1
Aniline point, °C	71
Distillation, °C*	
IBP	156
20%	185
50%	261
80%	328
95%	353
FBP	355
n-Paraffin range	C_{9-30}

*Measured by ASTM D-1160

Various blends of Polymers 1 and 2 with Polymers A to C in the fuel were made by simply dissolving polymer in the fuel oil. This was done while warming, e.g., heating the oil and polymer to about 90°C if the polymer per se was added, and stirring. In other cases, the polymer was simply added with stirring to the fuel in the form of an oil concentrate which was usually about 50 wt % polymer dissolved in a light mineral oil.

The blends were then tested for their cold flow properties in the Cold Filter Plugging Point Test (CFPPT) (described in U.S. Patent 4,140,492).

The blends prepared and the test results are summarized in the following table.

Effectiveness of Polymers in the Fuel

Example	Polymer (A.I.)* (%)	CFPPT, °C
1	None	+1
2	Polymer 1 (0.01)	-2
3	Polymer A (0.02)	0
4	Polymer B (0.02)	+1
5	Polymer C (0.02)	0
6	Polymer 1 (0.005) Polymer A (0.01)	-7
7	Polymer 1 (0.005) Polymer B (0.01)	-6
8	Polymer 1 (0.005) Polymer A (0.01)	-6
9	Polymer 2 (0.005) Polymer B (0.01)	-8

*Active Ingredient.

Combination of Ethylene-Vinyl Ester Copolymers and Polyesters

In closely related work, *M.J. Wisotsky; U.S. Patent 4,153,422; May 8, 1979; assigned to Exxon Research & Engineering Co.* describes polymer combinations useful as flow improvers for middle distillates in which the ethylene polymer of the previous patents is replaced by an ethylene-vinyl ester copolymer having an $\overline{M}_n$ of under 4,000. This copolymer is then combined with a second polymer which is a polyester such as those described in the previous patent.

The ethylene copolymer is as previously described and is copolymerized with ethylenically unsaturated alcohol monoesters of C_{2-17} monocarboxylic acids, preferably C_{2-5} monocarboxylic acid, of the general formula R_1COOR_2 where R_1 is C_{1-8} straight or branched chain alkyl; and R_2 is the radical of a mono-ethylenically unsaturated C_{2-3} alcohol.

At least about 5 wt %, preferably at least about 10 wt % of the ethylene-ester copolymer will be derived from the alkyl monocarboxylic acid-ethylenically unsaturated alcohol ester monomer moiety.

An example of the preparation of such a copolymer is as follows.

Example: This polymer is a copolymer of ethylene and vinyl acetate. This copolymer was prepared by the following procedure: A 3 liter stirred autoclave was charged with 50 ml of vinyl acetate in 700 ml of benzene as solvent. The autoclave was then purged with nitrogen and then with ethylene. The autoclave was then heated to 105°C while ethylene was pressured into the autoclave until the pressure was raised to 1,400 psig. Then, while maintaining a temperature of 105°C and the 1,400 psig pressure, 20 ml/hr of vinyl acetate and a solution consisting of 5 wt % di-lauroyl peroxide dissolved in benzene, were continuously pumped into the autoclave at an even rate.

A total of 43 ml of vinyl acetate was injected over 2.1 hr while 13 g of the peroxide in the form of the solution was injected into the reactor over a period of 2.6 hr from the start of the injection. After the last of the peroxide was injected, the batch was maintained at 105°C for an additional 10 minutes. Then, the temperature of the reactor contents was lowered to about 60°C, the reactor was depressurized, and the contents were discharged from the autoclave. The empty reactor was rinsed with 1 liter of warm benzene (about 50°C) which was added to the product. The product was then stripped of the solvent and unreacted monomers on a steam bath overnight by blowing nitrogen through the product. The final stripped product consisted of about 185 g of copolymer of ethylene and vinyl acetate having a number average molecular weight of 3,047 (as measured by VPO) and an ester content of 11.6 wt %.

When this polymer was used along with such a polyester as exemplified by Polymers A, B or C in the previous patent, similar reductions in the cold filter plugging point were experienced.

Combination of Polyethylene and a Second Polymer

In yet another related patent by *M.J. Wisotsky; U.S. Patent 4,175,926; Nov. 27, 1979; assigned to Exxon Research & Engineering Co.* ethylene polymers or copolymers are combined with a second polymer having alkyl groups of 6 to 18 carbons and derived either from dicarboxylic acid esters or olefins to make useful flow improvers.

The ethylene polymer is as described in the two previous patents. The second polymer has a preferred $\overline{M}_n$ of 1,000 to 30,000 and is prepared from such dicarboxylic acid esters as fumarates or maleates which may be copolymerized with other unsaturated esters such as methacrylates or acrylates; or from olefin polymers including propylene, butene-1, hexene-1, octene-1, styrene and styrene derivatives, etc.

An example of a second polymer which, when used in amounts of 0.01 wt % with 0.005 wt % of ethylene-type polymers, resulted in a good improvement in the cold temperature flow characteristics of a middle distillate is as follows.

Example: A copolymer of a mixed C_{6-18} dialkyl fumarate and vinyl acetate was prepared as follows: A 1 liter flask equipped with a stirrer, thermometer, dropping funnel and reflux condenser with a nitrogen lead was charged with 100 g of a mixture of C_{6-18} dialkyl fumarate, 40 g of vinyl acetate, 60 g of cyclohexane as solvent and 0.5 g of tert-butyl perbenzoate. 170 g of a light mineral white oil was charged to the dropping funnel. After flushing the system with nitrogen for about 30 seconds, the above mixture was then heated to a reflux temperature of about 77°C with stirring under a nitrogen atmosphere. Heat was continued for about 19 hr over a period of 3 days after which the heat was turned off.

Then 0.5 g of p-methoxyphenol was added as a polymerization inhibitor and 170 g of additional white oil was added to the flask. The contents of the flask were removed into a beaker and washed with hexane. The material was then placed on a steam table with nitrogen blowing overnight in order to remove the solvent. A total of 285.7 g of polymer was prepared.

The aforesaid mixture of C_{6-18} fumarate was a mixture of straight chain alkyl fumarate comprising about 4.2 g of C_6 fumarate; 6.2 g of C_8 fumarate; 7.3 g of C_{10} fumarate; 43.7 g of tallow fumarate made from tallow alcohol and 38.6 g of Lorol B fumarate made from Lorol alcohol, which is a commercial mixture of coconut oil alcohols averaging about a C_{12} alcohol.

Iminodiimides of 3,3′,4,4′-Benzophenonetetracarboxylic Dianhydride

C.A. Audeh; U.S. Patent 4,194,885; March 25, 1980; assigned to Mobil Oil Corporation has developed a method for preparing the iminodiimides of 3,3',4,4'-benzophenonetetracarboxylic acid dianhydride. These compounds are valuable pour point depressants and/or UV stabilizers in various organic media, e.g., hydrocracked liquid hydrocarbons and distillate fuels.

A primary amine is reacted with benzophenonetetracarboxylic dianhydride (BTDA) under appropriate conditions to produce these compounds. The general structural formula of these compounds is as shown below:

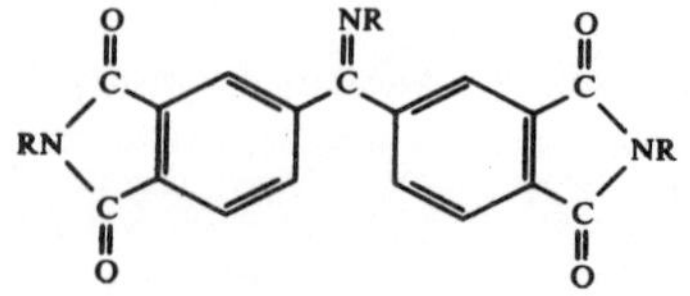

where R preferably is H or an alkyl group, branched or straight, having from 1 to about 36 carbon atoms. R may also be aryl, alkaryl or aralkyl having from 6 to 36 carbon atoms. R preferably contains 10 to 26 carbon atoms. The reaction is generally carried out at atmospheric pressure. However, pressures of up to about 200 psig and temperatures from 200° to about 400°C may be resorted to if so desired. Preferred reaction conditions are atmospheric pressure and 230° to 250°C. Further, molar ratios of the amine of the benzophenonetetracarboxylic dianhydride are from about 3 to 3.5 mols of amine to 1 mol of BTDA. Preferred molar ratio is 3 mols of amine to 1 mol of BTDA.

Any primary amine having the general structure RNH_2 can be used, where R is alkyl, aryl, alkaryl or aralkyl or any combination thereof having from 1 to about 36 carbon atoms. Some examples thereof are methylamine, tert-butyl amine, cyclohexyl amine, benzyl amine, toluidine, aniline, nonyl aniline, dodecyl aniline, and naphthyl amine.

The iminodiimides described herein are effectively used in any lubricating media in which the lubricant base is a petroleum product, such as a mineral oil or a synthetic fluid. The additives may also be used in organic media other than lubricant compositions, e.g., liquid hydrocarbons such as distillate fuel oils.

The amount of additive compound used in the organic media may vary from about 0.025 to 10% or more by wt of the base material.

Example 1 describes the general method of preparing the additive compounds and Example 2 illustrates their use as pour point depressants. The BTDA is readily obtained from commercial sources.

Example 1: 0.055 mol (17.5 g) of BTDA were mixed with 0.17 mol of $C_{16}H_{33}NH_2$ and heated until reaction was complete (about 3 hr). 0.17 mol of water (3.0 g) were collected by entrainment distillation with toluene. The resultant product was identified as having the following structure:

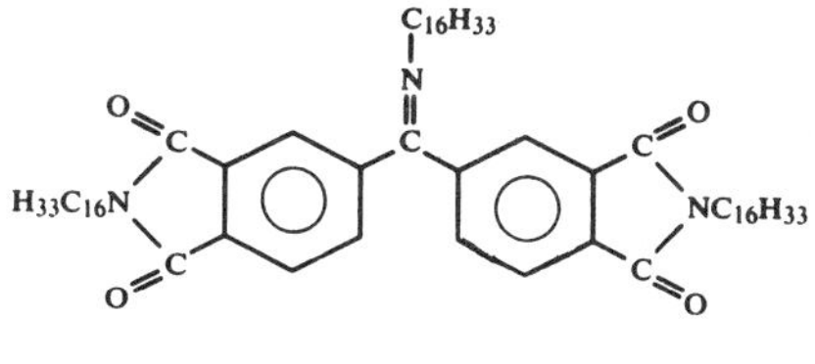

$C_{16}H_{33}$ N C O C C O $H_{33}C_{16}N$ C O C $NC_{16}H_{33}$ C O

Example 2: Three samples of the fuel oil containing 0.5% of the BTDA were subjected to the below detailed pour depressant test procedure. Each was treated as indicated prior to measuring. The tests were done with a falling ball-type viscometer in the following manner:

A sample of oil is placed in a graduated cylinder and thermostatically controlled to and maintained at the desired temperature. The oil is then cooled. A stainless steel ball is then dropped in the oil sample and its progress in the oil observed. As the oil sample cools down further, the progress of another identical stainless steel ball is observed. The temperature at which the cooled oil does not allow such a steel ball to continue falling is then recorded. A similar sample of the reference oil is treated in an identical manner. The concentration of the additive compound of Example 1 was 0.5%.

Ref. Fuel	Ref. Fuel + 0.5% Additive	Treatment*
-8°F	-14°F	A
-9°F	-15°F	B
-8°F	-18°F	B

*A–Preheat at 70°C for ½ hour. Cool to R.T., then measure.
B–Preheat at 70°C for ½ hour. Left overnight, then heat at 90°C for ½ hour. Cool to R.T., then measure.

Three-Component Additive with Ethylene Polymer plus a Nitrogen-Containing Compound

N. Feldman, J. Ryer, and M.F. Dooley; U.S. Patent 4,210,424; July 1, 1980; assigned to Exxon Research & Engineering Co. have developed a three (or more) component additive combination which is particularly useful in distillate fuel oils for improving their cold flow properties by controlling the size of wax crystals that form below the cloud point and inhibiting the agglomeration of these wax crystals.

A fuel composition is herein provided which comprises distillate fuel oil and from about 0.001 to 2.5 wt %, e.g., 0.01 to 0.5 wt %, of a flow and filterability improving, multicomponent additive composition comprising:

(A) 1 part by wt of oil-soluble ethylene backbone distillate flow improving polymer;

(B) 0.1 to 30, preferably 0.5 to 15 parts by wt of a normal paraffinic wax composed of normal hydrocarbons whose average molecular weight is within the range of from 300 to 650; preferably normal paraffins ranging from C_{23-37} inclusive; and

(C) 0.01 to 10, e.g., 0.2 to 5 parts by wt of a nitrogen compound which may be amides and/or amine salts of carboxylic acids or ammonium salts of those acids or anhydrides thereof.

If it is desired to enhance the compatibility, i.e., reduce its tendency to disperse into multiple phases, of this multicomponent additive composition, from 0.02 to 2 parts by wt, based on the weight of the oil-soluble ethylene polymer, of an auxiliary compatibility additive of the class consisting of

(1) a second oil-soluble polymer of an unsaturated carboxylic acid ester and/or olefins having straight chain alkyl groups of 6 to 30 carbons;

(2) a C_{8-18} alkanol; and

(3) mixtures thereof

are admixed with the fuel composition or concentrate containing the additive composition.

Concentrates of 20 to 90 wt % mineral oil and 80 to 10 wt % of the additive mixture of (A), (B) and (C), dissolved therein, will generally be made for ease of handling the additives.

The ethylene polymers are of the type known in the art as wax crystal modifiers. They may be simply homopolymers of ethylene as prepared by free radical polymerization so as to result in some branching. More usually, they will comprise about 3 to 40, preferably 4 to 20, molar proportions of ethylene per molar proportion of a second ethylenically unsaturated monomer, which latter monomer can be a single monomer or a mixture of such monomers in any proportion. These polymers will generally have a number average molecular weight in the range of about 500 to 50,000, preferably about 1,000 to 6,000.

The unsaturated monomers, copolymerizable with ethylene, include vinyl acetate, vinyl laurate, methyl acrylate, dilauryl fumarate; alpha-monoolefins; vinyl chloride; etc.

Also included among the ethylene polymers are the hydrogenated polybutadiene flow improvers having mainly 1,4 addition with some 1,2 addition since they can be considered as being made up of ethylene segments.

Particularly effective wax mixtures for distillate fuel oils having final boiling points in the range of 620° to 670°F are those that have normal paraffin hydrocarbons in the range of C_{24-39} inclusive. The waxes that are added include both well-defined waxes and crude waxes, such as slack wax and slop wax, as well as any of the various refinery streams wherein wax is a predominant constituent. The waxes that are used have a heat of fusion of from 40 to 55 calories per gram and are thus distinguished from petroleum resins, asphaltenes, petrolatums, and microcrystalline waxes, all of which have heats of fusion below 40 calories per gram. Suitable waxes for use in this process can be obtained by conventional dewaxing of various paraffinic petroleum refinery streams boiling within the range of about 340° to 565°C.

Particularly useful are the slack waxes obtained from the solvent pressing of oils having a boiling range of from about 345° to 500°C.

Nitrogen compounds effective in keeping the wax crystals separate from each other, i.e., by inhibiting agglomeration of wax crystals, are used as the third component of the additive mixtures. Particularly preferred are nitrogen compounds that are prepared from dicarboxylic acids, optimally the aliphatic dicarboxylic acids. Mixed amine salts/amides are most preferred, and these can be prepared by heating maleic anhydride, or alkenyl succinic anhydride with a secondary amine, preferably tallow amine, at a mild temperature, e.g., 80°C without the removal of water.

The C_{8-18}, preferably C_{13}, alkanols useful as an auxiliary compatibility additive are in general commercially available aliphatic alcohols which can be straight or branched chain. Among these alcohols are hexanol, heptanol, octanol, 2-ethyl hexanol, etc., through octadecanol with the preferred alcohol being tridecyl alcohol.

Example: *Polymer 1* – Polymer 1 was a concentrate of about 55 wt % of heavy aromatic naphtha oil and about 45 wt % of an ethylene-vinyl acetate copolymer of ethylene and about 38 wt % vinyl acetate, and had a number average molecular weight of about 1,800 (VPO).

Nitrogen Compound A – This compound is an amine salt of the monoamide of maleic anhydride prepared by reacting 1 mol of maleic anhydride with 2 mols of secondary hydrogenated tallow amine (Armeen 2HT) at a temperature such that no water is formed.

Slack Wax – This slack wax is obtained from the dewaxing of a refinery stream having a boiling range of 345° to 500°C. This slack wax contained about 6.1 wt % Solvent Neutral 150 oil, had a specific gravity (°API) of 0.8488, was a waxy solid, had an average molecular weight by Gel Permeation Chromatographic Analysis (GPC) of 435, contained about 76.1 wt % n-paraffins ranging from 19 to 37 carbons but primarily 22 to 34 carbons with an average carbon number of 27.

The utility of these additives in enhancing the cold flow properties of middle distillate fuel oils is demonstrated by the Cold Filter Plugging Point (CFPP) Test (see U.S. Patent 4,140,492) and a test determinative of the minimum screen through which the oil blend will pass at a temperature of 3°C below the cloud point of the distillate fuel oil. Also shown are the effect the additive combination has on the cold flow of bottom fractions subjected to wax settling.

Imperial Cold Filterability (ICF) Test – 200 ml of each test oil blend was cooled from 0°C at the rate of 1°C/hr to the test temperature -6°C which was 3°C below the cloud point of the fuel. At this temperature, the test blends were allowed to flow through various test screens under a vacuum of 200 mm of water to determine the finest mesh size through which at least 90% of the test oil blend would pass within 25 seconds.

Modified CFPP Test – In this modified test, the test oil blend was soaked at 6°C below its cloud point, i.e., at -9°C, for 24 hours. After this soak period, the bottom 10% fraction was removed and allowed to warm up to room temperature and remain there for at least 3 hours. These bottom fractions were then subjected to the CFPP test. This test gives an indication of the degree of deterioration in the cold flow property of a bottom fraction due to wax settling.

Oil blends 1 to 6 were made up by dissolving the additives into the Oil 1 by stirring, generally while warming the oil on a hot plate to about 60°C. The polymer additive was added in the form of an oil concentrate containing 45 wt % polymer and the nitrogen compound A plus slack wax was added to the oil directly.

The results of the tests are set forth in the following table.

Test Blend	Polymer 1 (wt %)	Nitrogen Compound A	Slack Wax (wt %)	CFPP Test (°C)	ICF Test Finest Screen for pass	Modified CFPP Test (°C)
Oil 1	0	0	0	-4	Fails 20 mesh	–
1	0.036	0	0	-8	20 mesh	+6
2	0.018	0.02	0	-6	60 mesh	+6
3	0.018	0	0.21	-11	100 mesh	+3
4	0.018	0.02	0.21	-14	325 mesh	-7
5	0	0.04	0.21	0	60 mesh	–
6	0.018	0	0.28	-11	100 mesh	+3

The data shows that the presence of n-paraffins of carbon number ranging from 23 to 37 in combination with an ethylene-vinyl acetate-type flow improver and a maleic anhydride amine reaction product in a middle distillate fuel provides it with better low temperature flow properties than would be obtained by treatment of the fuel with any of these individually or with the dual combination of any two of the three types even at higher concentration. In addition, using the three-component combination listed above has an inhibiting effect on the rate of wax settling of the flow improved fuel.

In a similar patent by *N. Feldman; U.S. Patent 4,211,534; July 8, 1980; assigned to Exxon Research & Engineering Co.* combination additives are again used to improve the cold flow properties of distillate hydrocarbon fuel oils. Component (A), the ethylene polymer or copolymer, and Component (C), the nitrogen compound are those described in the previous patent. Component (B) is, in this case a second oil-soluble polymer of monomers other than ethylene, having a molecular weight of 1,000 to 200,000 wherein at least 10% by wt of the polymer is in the form of C_{6-30} straight chain alkyls. The polymer comprises unsaturated ester, or unsaturated ester and olefin moieties, in a major weight proportion.

Example: *Polymer 1* – This was a concentrate of about 55 wt % heavy aromatic naphtha oil and 45 wt % of a mixture of 2 ethylene-vinyl acetate copolymers.

Polymer A – This was an oil concentrate of about 50 wt % of mineral lubricating oil and about 50 wt % of a copolymer of dialkyl fumarate and vinyl acetate in about equimolar proportions, having an $\overline{M}_n$ of about 15,000 (VPO) prepared in a conventional manner using a peroxide initiator. The fumarate was prepared by esterifying fumaric acid with a mixture of straight chain alcohols averaging about C_{12}.

Polymer B – Polymer B was Acryloid 157 which is a lubricating oil pour depressant for highly paraffinic oils. Dialysis indicated that Acryloid 157 consists of about 37 wt % light hydrocarbon oil and about 63 wt % of active ingredient. The active ingredient has a specific viscosity of about 0.44 at a 2% concentration in xylene at 100°C, and is a polymer comprising mainly alkyl methacrylate groups.

Nitrogen Compound A – This compound was prepared in accordance with U.S. Patent 3,982,908 and is an amine salt of the monoamide of maleic anhydride. It was prepared by reacting maleic anhydride with secondary hydrogenated tallow amine (Armeen 2HT).

The oil was a distillate fuel oil having a WAP (Wax Appearance Point as discussed in ASTM D-3117) of -1.5°C and a distillation range as follows: IBP of 162°C; 20% distillation point of 203°C; 90% distillation point of 337°C and FBP of 375°C.

Blends were made up by dissolving the additives into the fuel oil in the following amounts: Blend 1, 500 ppm Polymer 1; Blend 2, 150 ppm Polymer 1, 150 ppm Nitrogen Compound A, 250 ppm Polymer A; and Blend 3, 150 ppm Polymer 1, 75 ppm Nitrogen Compound A, and 125 ppm Polymer B.

The blends, in a 1,000 ml graduate, were cold soaked by being quietly cooled from about 20°C to -6.5°C in a cold box and then held at -6.5°C for 24 hr and 48 hr. Then the cold oil blends were visually examined. Next, the bottom 10% was drawn off and subjected to a screen test which involved using a test device, comprising a 20 ml pipette to which vacuum is applied at the upper end, while its lower end terminates in an inverted funnel across which is stretched a fine mesh screen having a diameter of about 12 mm. The test device is inserted into a 50 ml sample of the cold oil in a CFPP testing tube and vacuum of about 8" of water is applied. A "pass" result was obtained where the cloudy oil filled the pipette to the 20 ml mark without plugging the screen. Pipettes with different mesh screens were used. The smaller the size of the wax crystals that form, the finer the mesh screen which will be passed by the wax-cloudy oil.

The 3-component additive systems of Blends 2 and 3 gave a higher degree of wax dispersion, and for the bottom fraction lower wax appearance points (WAP), and smaller wax crystals as indicated by the passage of the cold oil through the 250 mesh screen, than Blend 1 containing only the ethylene polymer component.

Poly(Isomerized C_{12-50} Monoolefins) with Pour Point Depressant

Kerosene, which is a solvent for wax, had traditionally been a component of middle distillate fuel oils, e.g., diesel fuels, home heating oils, etc. With the demands for kerosene for use in jet fuels, the amount of kerosene used in distillate fuel oils has decreased over the years. This, in turn, has frequently required the addition of wax crystal modifiers, e.g., pour point depressant additives, to the fuel to make up the lack of kerosene.

H.N. Miller, M.J. Wisotsky, and R.P. Rhodes; U.S. Patent 4,255,159; March 10, 1981; assigned to Exxon Research & Engineering Co. have found that the combination of a polymeric substance featuring poly(isomerized monoolefins containing from 12 to 50 carbons) either as a polymer or a polymeric alkylation derivative of an aromatic compound and a polymeric lube oil pour point depressant having C_{6-32} pendant alkyl groups provide cold flow improvement of a middle distillate petroleum fuel oil boiling within the range of about 120° to 450°C at atmospheric pressure. In general, the (a) poly(isomerized C_{12-50} monoolefin) in combination with (b) the described lubricating oil pour depressants give synergistic results in controlling wax crystal size in distillate fuel oils.

In general, the additive combination will comprise 1 part by weight of the substance featuring poly(isomerized C_{12-50} monoolefin) per about 0.1 to 20, preferably 0.2 to 8 parts by weight of the second polymeric pour depressant. Distillate fuel oil compositions will contain a total of preferably 0.005 to 0.1 wt % of the combination. Concentrates of 1 to 60 wt % of the combination in 40 to 99 wt % of mineral oil are useful. The distillate fuel improved by the additives will have a viscosity in the range of 1.6 to 7.5 centistokes at 38°C and will have less than 3 wt %, usually less than 1 wt % of wax boiling above 350°C, i.e., wax having 20 or more carbon atoms. In general, the distillate fuel oils will boil in the range of about 120° to 450°C and will have cloud points usually from about -30° to about 5°C.

Examples: The following materials were used:

Polymeric Substance 1– Polymeric substance 1 resulted from the polymerization of isomerized alpha-olefins having an average carbon content of 28. The alpha-olefins resulted from a vacuum distillation cut of a composite of a series of ethylene polymerization runs, which composite had a boiling point such that the average molecular weight was equivalent to about a 28 carbon atom molecule. The polymer was prepared as follows: 50 g of the average C_{28} fraction was refluxed in 200 ml of heptane with 0.1 g of anhydrous $AlCl_3$. The reflux temperature of 98°C was maintained for 3.1 hr. The solvent was then stripped from the reactants by distillation. The product was washed with $NaHCO_3$ and H_2O to remove the catalyst residue and thereafter dried. The product was found to have a $\overline{M}_n$ of 1,840.

Polymeric Substance 2 – Polymeric substance 2 was a poly[isomerized C_{28} (average) alpha-olefin] alkylated toluene. The substance was prepared as follows: 50 g of the average C_{28} fraction was refluxed with 13.5 ml of toluene and 1.0 g anhydrous $AlCl_3$ in 200 ml heptane for 3.1 hr at 100.2°C. The product was recovered and purified as the Substance 1. The product was found to have a $\overline{M}_n$ of 788.

Additive A – This was prepared as a concentrate of about 50 wt % of a light mineral oil and about 50 wt % of a wax-naphthalene made from 100 parts by weight of a n-paraffin wax (having a melting point of about 73°C and chlorinated to about 12 wt % chlorine) condensed with about 8.8 parts naphthalene by a Friedel-Crafts reaction.

Additive B – This was prepared as a concentrate of about 50 wt % of a light mineral oil and about 50 wt % of a sulfone copolymer of sulfur dioxide and a mixture of C_{14-20} alpha-olefins with an olefin distribution of 0.58 mol % of C_{14}, 0.24 mol % C_{16}, 0.08 mol % of C_{18} and 0.11 mol % of C_{20}.

Polymerization was carried out on an olefin charge of about 1.0 mol of the mixture dissolved in benzene with a slight positive pressure of SO_2 on the solution and by means of a free radical initiator (tert-butyl hydroperoxide) introduced therein. About 2.8 mols of SO_2 was consumed during polymerization over a period of 65 minutes and at temperatures ranging from 10° to 24°C. The sulfone copolymer had an $\overline{M}_n$ of 12,556 with an olefin distribution of 0.575 mol tetradecene-1, 0.078 mol octadecene-1 and 0.109 mol of eicosene-1.

Various blends of Polymers 1 and 2 with Additives A and B in a conventional middle distillate fuel were made by simply dissolving the polymer or additive in the fuel oil. The blends were then tested for their cold flow properties in the Cold Filter Plugging Point Test (CFPPT).

The blends prepared and the test results are summarized as follows:

Ex. No.	Active Ingredient Polymer/Additive (%)	CFPPT (°C)
1	None	+1
2	Polymeric Substance 1 (0.04)	0
3	Additive A (0.04)	-2
4	Additive B (0.04)	+2
5	Polymeric Substance 1 (0.02) Additive A (0.02)	-1
6	Polymeric Substance 1 (0.02) Additive B (0.02)	-8
7	Polymeric Substance 2 (0.02) Additive B (0.02)	-6
8	Polymeric Substance 2 (0.02) Additive A(0.02)	-6

Three-Component Additive Including an Antiagglomerant

If the winter is particularly cold and prolonged so that bulk oil is stored for a long time during very cold weather, the bulk oil may eventually drop below its cloud point. These conditions may then result in crystallized wax settling to the bottom of the tank and in addition a bottom layer of oil forms which has an enriched wax content and a cloud point considerably higher than that of the fuel originally pumped into the tank while the upper layers of the oil are partially dewaxed and have relatively low cloud points. The crystal rich bottom layer of oil will, therefore, exhibit a greater tendency towards wax agglomeration than the upper layers and such wax agglomeration frequently leads to the plugging of screens and other flow constrictions in oil distribution systems. Since the outlets from the tanks are near the bottom, if oil is drawn off which has an abnormally high amount of wax in the form of relatively large crystallites due to the crystal agglomeration, although the agglomerates may pass through the filters on the tank, they may block protective screens or filters on the truck or clog filters or small-diameter fuel lines in the customer's storage system.

R.D. Tack, and K. Lewtas; U.S. Patent 4,261,703; April 14, 1981; assigned to Exxon Research & Engineering Co. have found that these problems may be reduced by using a three (or more) component additive combination for distillate fuel oils, comprising (A) a distillate flow improving composition; (B) a lube oil pour depressant; and (C) a polar oil-soluble compound different from (A) and (B) and of formula RX, where R is an oil-solubilizing hydrocarbon group and X is a polar group, the compound acting as an antiagglomerant for wax particles in the fuel. This combination is particularly useful in distillate fuel oils boiling in the range of 120° to 500°C, especially 160° to 400°C, for controlling the size of wax crystals that form at low temperatures.

In a preferred form, a fuel composition is provided which comprises distillate fuel oil and from 0.001 to 0.5 wt %, most preferably 0.05 to 0.1 wt % of a flow and filterability improving, multicomponent additive composition comprising: (A) 1 part by weight of a distillate flow improver composition; (B) most preferably 1 to 2 parts by weight of a lube oil pour depressant; and (C) most preferably 1 to 2 parts by weight of a polar oil-soluble compound of formula RX as already defined.

For ease of handling, the additives will generally be supplied as concentrates containing 30 to 80 wt %, a hydrocarbon diluent with 70 to 20 wt % of the additive mixture of (A), (B) and (C), dissolved therein. The process is also concerned with such concentrates.

The distillate flow improver (A) used in the additive combinations is a wax crystal growth arrestor and may also contain a nucleator for the wax crystals. It is preferably an ethylene polymer of the type for example, described in U.S. Patent 4,210,424.

The preferred ethylene copolymers are ethylene-vinyl acetate copolymers.

The lube oil pour point depressant is preferably an oil-soluble ester and/or higher olefin polymer and will generally have a number average molecular weight in the range of about 1,000 to 200,000, preferably 1,000 to 50,000. These second polymers include (a) both homopolymers and copolymers of unsaturated alkyl ester, including copolymers with other unsaturated monomers, e.g., olefins other than ethylene, nitrogen-containing monomers, etc.; and (b) homopolymers and copolymers of olefins, other than ethylene.

In the preferred lube oil pour depressant at least 10 wt % and frequently 50 wt % or more of the polymer will be in the form of straight chain C_{6-30}, e.g., C_{8-16} alkyl groups, usually of an alpha-olefin or an ester.

Preferred ester polymers, from the point of view of availability and cost, are copolymers of vinyl acetate and a dialkyl fumarate in about equimolar proportions, and polymers or copolymers of acrylic or methacrylic esters.

Compound (C) is selected from the group consisting of:

(1) $C_{24}H_{49}PhSO_4^-N^+R_4$ where R_4 is hydrogen or C_{1-30} hydrocarbyl.

(2) $C_{24}H_{49}PhSO_4^-N^+H_2(C_{12}H_{25})_2$

(3) $C_9H_{19}PhO^-N^+H_2(C_{12}H_{25})_2$

(4) $C_{17}H_{35}COO^-N^+H_2(C_{12}H_{25})_2$

(5) $CH_3(CH_2)_{15-17}NH_2$

(6) $[CH_3(CH_2)_{15-17}]_2NH$

(7) $C_{18}H_{37}OH$

(8) $C_{14}H_{29}OH$

(9) Polyisobutylene succinic anhydride where the polyisobutylene chain is about 1,000 molecular weight; and

(10) the di-n-butyl amide of (9)

where Ph represents a phenate group.

Example: Distillate fuel oil boiling at 150° to 400°C was made up into blends using as Component A an aromatic diluent of about 50 wt % of a mixture of two ethylene-vinyl acetate copolymers described in U.S. Patent 4,211,534.

The lube oil pour depressant B was an oil concentrate of about 50 wt % of mineral lubricating oil and about 50 wt % of a copolymer of dialkyl fumarate and vinyl acetate in about equimolar proportions, having a number average molecular weight (VPO) of about 15,000.

Compounds of Component C used were identified by the numbers already given above.

The behavior of the oils at sustained low temperatures was assessed by subjecting the oils to a cold soak test in which separate 500 ml samples of each test blend in an addition glass funnel were first cooled at 1° and 0.3°C/hr from room temperature of about 20° to -8°C. The test blend was thereafter held at -8°C for the indicated period. A 50 ml portion of this cooled test fuel blend was drawn off from the bottom of the funnel and transferred to another container and subjected to a modified Cold Filter Plugging Point Test (CFPPT). In this test, a sample at the cold soak temperature is sucked by 200 mm water vacuum pressure through a filter screen and the minimum mesh through which it would pass measured. The portion was then allowed to return to room temperature (about 20°C) after which it was subjected to the ASTM cloud point determination.

Visual wax settling of the fuel treated with the ethylene backbone copolymer, the lube oil pour depressant and certain of the polar compounds was observed and the following table shows the advantage of the three component mixtures in inhibiting wax settling.

Additive Concentration (ppm)				Waxy Layer (Vol %)		
A	B	C	(No.)	25 hr Soak at -8°C	37 hr Soak at -8°C	61 hr Soak at -8°C
100	–	–		15	15	14
300	–	–		15	15	13
500	–	–		15	13	12
100	100	50	of (1)	88	86	77
100	100	50	of (2)	87	86	79
100	100	50	of (3)	89	86	79
100	100	50	of (4)	89	88	85
100	100	50	of (5)	89	87	83
100	100	50	of (6)	91	89	87
100	100	50	of (7)	88	89	86

FOR HEAVY PETROLEUM OILS

1,2-Epoxy Alkanes/Cyclic Carboxylate Copolymers

A. Rossi; U.S. Patent 4,135,887; January 23, 1979; assigned to Exxon Research & Engineering Co. reports that certain copolymers of 1,2-epoxy alkanes and cyclic carboxylate compounds are useful flow improvers for petroleum fuel oils and crude oils.

He has found that about equimolar condensation polymers of 1,2-epoxy alkanes (e.g., a C_{22} 1,2-epoxy alkane) with a cyclic carboxylate compound of the class consisting of dicarboxylic acid anhydrides (preferably maleic anhydride or maleic anhydride reacted with a long chain olefin) and β-lactones (preferably hydroacrylic acid) are useful as pour depressants in residuals and crude oils. At least one of the 1,2-epoxy alkanes or cyclic carboxylate compounds or both must have a long straight chain hydrocarbon group of from 10 to 50, preferably from 20 to 40 carbons. These flow improvers will usually have number average molecular weights ($\overline{M}n$) in the range of 750 to 50,000, preferably 1,000 to 10,000.

Specific examples of suitable epoxides include ethylene oxide, propylene oxide, 1,2-tetradecene oxide; 1,2-hexadecene oxide; 1,2-octadecene oxide; 1,2-eicosene oxide; 1,2-docosene oxide; 1,2-tetracosene oxide; 1,2-octacosene oxide; 1,2-triacontene oxide; and mixtures thereof.

Preferred dicarboxylic anhydrides are maleic anhydride and alkenyl succinic acid anhydrides in which the alkenyl group contains a total of from 10 to 50, preferably 20 to 40, carbon atoms. Many of these hydrocarbyl substituted dicarboxylic acid anhydrides are commercially available, e.g., 2-octadecenyl succinic anhydride and polyisobutenyl succinic anhydride.

Representative β-lactones include hydroacrylic acid, β-docosanolactone, β-octacosanolactone, etc. These compounds are well known in the literature and generally prepared by reacting a 3-chloroalkanoic acid (e.g., 3-chloropropionic acid) and aqueous alkali. The 3-chloroalkanoic acid is readily obtained by reacting a 2,3-alkenoic acid, e.g., acrylic acid with HCl.

The oils which can be treated with these polymeric additives include straight residuum from the atmospheric distillation of crude oil or shale oil or mixtures thereof.

Some residuums, i.e., residual oils, have extremely high pour points of from 20° to 45°C, particularly those obtained from North African crudes, e.g., Libya, due to a high wax content. These oils also have low sulfur contents which make them particularly desirable because of air pollution requirements. These oils can be particularly improved by additives. Usually oils having 2 to 25 wt % wax boiling above about 345°C will give the best response to the additives of the process, while oils with lesser amounts of wax do not respond as favorably.

Example 1: *Copolymer A* – This is a condensation copolymer of about equimolar proportions of a C_{30+} alkenyl succinic anhydride and a C_{22} alkylene oxide. This copolymer was prepared by adding 10.0 g (0.030 mol) of C_{22} 1,2-epoxy alkane purchased as C_{22} α-olefin oxide (Viking Chemical Co.), 12.0 g (0.023 mol) of C_{30+} alkenyl succinic anhydride and 100 ml of hexane to a 500 ml, 4-necked flask having a stirrer, thermometer and charging funnel. The reactant mixture was heated to 70 to 75°C at which time 3 drops of triethylamine was added with stirring. The mixture was heated thereafter for about 3.5 hr at 70°C.

26.0 g of the crude product was dialyzed for 9 hr with boiling hexane solvent at 70°C in a Soxhlet extraction device, using a semipermeable rubber membrane to remove low molecular weight components, e.g., the hexane, unreacted monomers, etc. 9.4 g of residue, representing a 43% yield of the copolymer, was obtained having a $\overline{M}n$ of 1,230 by VPO.

Example 2: Blends of the above copolymer in a Racoon Blend crude oil and a waxy Brega residua oil were prepared by simply heating and stirring the oil and copolymer up to about 54° and 82°C, respectively, to dissolve the copolymer into the oil. The Brega residua oil was obtained by atmospheric distillation to a final vapor temperature of about 345°C of a crude oil from Libya which is a mixture of about 85 wt % of crude from the Zelten Field with the remainder from the Sarir and Dakar fields. This Brega residua is a very waxy black oil having about 8 to 13 wt % wax boiling above 345°C and having ASTM D-97-66 upper and lower pour point of 41°C, and an initial atmospheric boiling point of 345°C FVT (Final Vapor Temperature). The Racoon Blend crude oil is an Austin County, Texas petroleum crude having an ASTM D-97-66 upper and lower pour point of 21°C, a viscosity of 43 SUS at 38°C and an API gravity of 31.8.

The blends of oil and copolymer additive were tested for pour point depression since this is a measure of the ability of the copolymeric additive to keep the wax in suspension, thereby eliminating or reducing the amount of the wax which will deposit upon flow surfaces exposed to the oil, while improving the flow point of the oil.

The copolymer additive is as noted, blended into the petroleum oil in at least an amount sufficient to improve the flow point. Such a blending will most usually be in the amount of from about 0.0005 to 0.8, optimally about 0.03 to 0.3 wt % of copolymer additive, the wt % being based on the total weight of the oil composition.

It was found that 0.05 wt % of the copolymeric additive markedly improved the flow point of both crude and residua oils, lowering the upper pour point 58°C and the lower pour point 44°C of a crude oil, whereas a prior copolymeric additive at three times the concentration reduces upper and lower pour points 45° and 39°C, respectively.

Production of Low Sulfur Fuel Oil with Lowered Pour Point

R.F. Wilson, R.A. Peck, S. Herbstman, and L.C. Mih; U.S. Patent 4,138,227; February 6, 1979; assigned to Texaco Inc. describe a process for producing low pour point, low sulfur content fuel oils by subjecting a high sulfur, high pour point atmospheric residuum to thermal cracking and forming a residual fuel blend containing from 30 to 70 wt % of the thermally cracked product and from 70 to 30 wt % of a high wax, low sulfur, high pour point atmospheric residuum and also containing from 0.05 to 1.0 wt % pour point depressant. In a more specific process, a residual fuel oil of improved pour point and reduced sulfur content is prepared by subjecting a high sulfur, high pour point atmospheric residuum to thermal cracking, separating the thermally cracked product into a vacuum gas oil and a vacuum residuum, subjecting the vacuum gas oil to catalytic hydrogenation, blending the hydrogenated oil with the vacuum residuum and the high wax, low sulfur, high pour atmospheric residuum and adding from 0.05 to 1.0% by wt of a pour point depressant.

The pour point depressant additives suitable for use in the process preferably comprise oil-soluble ethylene-unsaturated aliphatic monocarboxylic acid ester copolymers in which the monocarboxylic acid component of the ester contains from 2 to about 6 carbon atoms, the copolymers having an average molecular weight of about 17,000 to 30,000 as determined by the membrane osmometry method, a vinyl ester content of from about 10 to 45% and a melt index of from about 7 to 476. The preferred copolymers are known as Elvax (E.I. Du Pont de Nemours and Co.), the most preferred being Elvax 250 which contains from about 27 to 29% vinyl acetate and has a melt index of 12 to 18 as determined by ASTM 1328. The pour point depressant is present in the final blend in an amount between 0.05 and 1.0% by wt, preferably between 0.1 and 0.5 wt %.

Example: In this example, an Arabian atmospheric residuum having a sulfur content of 3.1 wt %, a carbon residue of 10.2 wt %, an API gravity of 13.4°, an initial boiling point of about 650°F, an asphaltene content of 3.67 wt % and containing 6.14 wt % pentane-insoluble material is thermally treated by being passed downwardly through a reaction zone containing Berl saddles at a temperature of 825°F, a pressure of 500 psig at a liquid hourly space velocity of 0.5 and in the presence of 2,000 standard cubic feet of hydrogen per barrel of oil. Conversion to materials boiling below 650°F amounted to 12 wt % of the charge. The thermally cracked product is then separated into a vacuum gas oil having an end boiling point of about 1000°F amounting to 58 wt % basic fresh feed and a vacuum residuum having an initial boiling point of about 1000°F amounting to 42 wt % basic fresh feed and having a sulfur content of 3.7 wt %.

The vacuum gas oil which has an API gravity of 20.2°, a sulfur content of 2.6 wt %, a carbon residue of 1.29 wt % and a pour point of 90°F, is subjected to catalytic hydrogenation by being passed downwardly through a reaction zone containing a pelleted catalyst containing 3.0 wt % cobalt oxide, 15% molybdenum oxide, 3.6% silica and the balance alumina, at a temperature of 700°F, a pressure of 1,500 psig, a LHSV of 1.0 in the presence of 7,000 scf/bbl hydrogen. The hydrogenated product has an API gravity of 25.6°, a sulfur content of 0.2 wt %, a carbon residue of 0.12 wt % and a pour point of 90°F. The hydrogenated oil is blended with an Amna atmospheric residuum having an initial boiling point of about 650°F, a sulfur content of 0.28 wt %, a wax content of 22.7 wt % and a pour point of 100°F in an amount equal to the amount of

Arabian atmospheric residuum charge. The blend has a pour point of 95°F. Upon the addition of 0.1 wt % Elvax 250, the pour point is reduced to 55°F.

Ethylene-Vinyl Acetate Copolymer or A-B-A-Type Block Polymer

L.C. Parker and K.D. Miller; U.S. Patent 4,156,434; May 29, 1979; assigned to Texaco Inc. have devised a method for improving the flow properties and pour point characteristics of gas oils. They describe a fuel composition comprising a major amount of a gas oil and a minor amount of a high asphaltene residuum together with an effective pour depressant amount of, for example, an oil-soluble, high molecular weight ethylene-vinyl acetate copolymer or an oil-soluble block copolymer of the A-B-A-type wherein the A block is an acrylate ester and the B block is a copolymer of ethylene and vinyl acetate.

A particularly effective class of pour depressants already used is, e.g., an ethylene-vinyl acetate copolymer having a molecular weight ranging from about 15,000 to 60,000 and having an ethylene content of about 45 to 90%. Although the ethylene-vinyl acetate copolymers have been shown to be effective in reducing the pour points of crudes and fuel oil blends containing distillate, they have not been found to be very effective in reducing the pour points of vacuum gas oils.

These fuel oil compositions comprise a major amount of a gas oil or gas oils boiling at about 450° to about 1050°F, a minor amount (i.e., less than 15% by wt) of a high asphaltene residuum and an effective pour depressant amount of an oil-soluble copolymer selected from the group consisting of:

(a) An ethylene-vinyl acetate copolymer having a number average molecular weight between about 17,000 and 30,000 and a vinyl acetate content of about 10 to 45 wt % with the balance being ethylene;

(b) an A-B-A block copolymer wherein the A block is derived from an ester of an acrylic or methacrylic acid with a monohydric saturated aliphatic alcohol of the formula ROH, wherein R is straight chain or branched alkyl of from 12 to about 24 carbon atoms, the B block is a copolymer of ethylene and vinyl acetate, wherein the number average molecular weight of the copolymer is from 15,000 to about 35,000, the weight percent of vinyl acetate is about 8 to 25 and the weight percent of the ester of acrylic or methacrylic acid in the block copolymer is about 3 to 15 with the balance being ethylene; and

(c) a heteric polymer of ethylene, vinyl acetate and an ester of acrylic or methacrylic acid with a monohydric aliphatic alcohol of the formula ROH, wherein R is straight chain or branched alkyl of from 12 to about 24 carbon atoms, wherein the number average molecular weight of the copolymer is about 12,000 to 37,000 and the weight percent of the ethylene, vinyl acetate and ester of acrylic or methacrylic acid is the same as in copolymer B.

An especially useful group of ethylene-vinyl acetate copolymers for use as pour point depressants are the Elvax additives of Du Pont, described in the previous patent.

Block copolymers of A-B-A-type where the A block is derived from an ester of acrylic acid or methacrylic acid with a monohydric saturated aliphatic alcohol of the formula ROH wherein R is straight chain or branched alkyl having from 12 to about 24 carbon atoms and the B block is a copolymer of ethylene and vinyl acetate or a modified copolymer of ethylene and vinyl acetate which contains a small amount of copolymerized N-vinyl-2-pyrrolidone in random or heteric arrangement are also used. They can be made, for example, by redissolving an ethylene-vinyl acetate copolymer or a modified ethylene-vinyl acetate copolymer prepared by any convenient process in benzene or a hydrocarbon solvent, placing the copolymer solution in a stirred autoclave with the requisite amount of acrylate or methacrylate ester together with a peroxide-type catalyst and continuing the polymerization under the influence of heat and pressure. Temperatures of from about 250° to 400°F may be employed at autogenous pressure.

The gas oil fuels utilized in this process generally will boil between 450° and 1050°F. Typical of the gas oils which may be employed are Desulfurized Arabian Light Vacuum Gas Oil (VGO), Desulfurized Lago Medio Vacuum Gas Oil, Arabian Light Vacuum Gas Oil, Lago Medio Vacuum Gas Oil, and Amna Vacuum Gas Oil, etc.

Example: Blends are prepared by mixing Desulfurized Lago Medio Vacuum Gas Oil and Lago Medio 1050+ °F Residuum together at 220°F (with mixing) for 40 minutes and with varying amounts of an oil-soluble, block copolymer of the A-B-A-type in which the A blocks are derived from octadecyl acrylate and the B block is a copolymer of ethylene and vinyl acetate having a number average molecular weight of about 23,000 (Copolymer G). The weight percent of vinyl acetate in Copolymer G based on the total weight of the copolymer is 9.5; the weight percent of the octadecyl acrylate is 16.2 with the balance of the copolymer being ethylene.

The upper pour point of the abovedescribed blend with no added Copolymer G was determined according to the method of ASTM D-97 and found to be +80°F which is the same pour point as Desulfurized Lago Medio VGO alone whereas the upper pour point of this same blend containing 0.10 wt % Copolymer G is found to be substantially below +80°F when determined by the same ASTM method.

Alkylated Polybenzyl Polymer

In recent years the restrictive specifications limiting the sulfur content of fuel oils including those containing a major amount of gas oils have made their manufacture a much more difficult task than previously. In many localities, low-sulfur, high-pour fuels are being substituted for previously utilized high-sulfur low-pour fuels. In order to meet the stringent pour point specifications of such fuel oils, pour depressants are being used. A particularly effective class of pour depressants are the ethylene-vinyl acetate copolymers having molecular weights ranging from about 15,000 to 60,000 and having an ethylene content of about 45 to 90%. Although the ethylene-vinyl acetate copolymers have been shown to be effective in reducing the pour points of crudes and fuel oil blends containing distillate, they have not been found to be very effective in reducing the pour points of certain other hydrocarbon oils such as vacuum gas oils, etc.

It is, therefore, the main objective of *W.P. Broeckx and L.G. Dulog; U.S. Patent 4,142,865; March 6, 1979; assigned to S.A. Texaco Belgium NV, Belgium* to provide low pour hydrocarbon oil compositions based on middle distillates, gas oils, residual fuel, crude oils, etc. and containing about 0.01 to 0.50 wt % of an oil-soluble, alkylated polybenzyl polymer or copolymer with styrene, propylene, 1-hexene, etc. or mixtures thereof.

These pour depressants are prepared by polymerizing a compound of the formula:

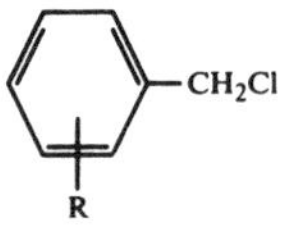

CH_2Cl

R

wherein R is alkyl of from 8 to about 26 carbon atoms as exemplified by octyl, isooctyl, tert-nonyl, isononyl, decyl, isodecyl, heptadecyl, etc. These polymers have molecular weights of about 500 to about 50,000 or more as determined by vapor pressure osmometry.

Preparation of the abovedescribed polymers can be conveniently carried out according to the process set out in U.S. Patent 3,418,259 which is incorporated herein by reference in its entirety. For example, polymerization of p-dodecylbenzyl chloride is accomplished at –55°C in the presence of a Friedel-Craft catalyst such as $AlCl_3$, etc.

Oil-soluble random copolymers of the abovedescribed benzyl compounds with alpha-olefins having from 3 to 20 carbon atoms such as propylene, 1-butene, 1-pentene, 1-hexene, 1-heptene, 1-octene, 1-nonene, 1-decene, 1-tetradecene, etc. or styrene are also useful pour depressants.

In these copolymers generally the units derived from the benzyl compounds will constitute preferably 60 to 90% by wt of the copolymer with the balance being derived from the other monomer or monomers employed. These copolymers which have average molecular weights of about 500 to 50,000 or more can be prepared in the same manner as the polybenzyl polymers.

Typical high-pour residual fuel oils useful in preparing the fuel oil compositions of this process include, for example, residual fuel oils having an API gravity of about 10.0 to about 30.00; a sulfur content of between about 0.10 to 0.96 wt %; a SUS viscosity at 122°F of about 150 to 2,000; a pour point of between about 70° and 115°F; a flash point of between about 300° and 450°F; and a wax content of between about 2 and 20%.

Example: A concentrate containing 1.5% by wt of a polybenzyl polymer having a number average molecular weight of about 2,300 and having recurring units of the formula:

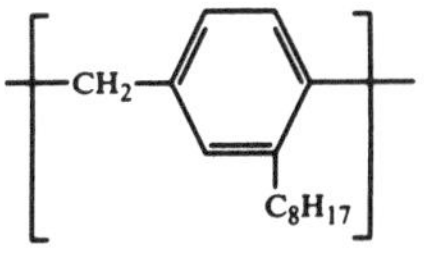

CH_2

C_8H_{17}

is prepared by adding a sufficient amount of the oil-soluble polymer to toluene at 150°F with mixing. The thus-prepared concentrate is then incorporated at 135°F with mixing in a middle distillate fuel oil between 360 to 454°F and having a pour point of 20°F to form a middle distillate composition containing 0.02% by wt of the polybenzyl polymer. After one month the pour point is again determined and found to be substantially below that exhibited initially by the pure middle distillate.

Vinyl Acetate-Ethylene-Acrylic Acid Terpolymer

The low pour point fuel oil compositions described by *W.M. Sweeney; U.S. Patents 4,155,719; May 22, 1979; and 4,160,459; July 10, 1979; both assigned to Texaco Inc.* comprise a residual fuel oil, which is preferably a waxy fuel oil, together with an effective pour depressant amount of an oil-soluble terpolymer such as a vinyl acetate-ethylene-methacrylate or acrylic ester terpolymer or a vinyl acetate-ethylene-ethoxylated or propoxylated acrylic acid terpolymer.

The terpolymers may be of two types. In terpolymer A, straight chain saturated monohydric aliphatic alcohols with from 10 to 26 carbons are used to prepare the methacrylic or acrylic esters, such as decyl, dodecyl, heptadecyl, etc. or commercial mixtures of such alcohols such as the Alfol alcohols of Continental Oil Co. may be used. In terpolymer A, the vinyl acetate units range from 20 to 45 wt %; the ethylene units from 50 to 79 wt %; and the methacrylate or acrylic ester units are the balance.

In terpolymer B ethoxylated or propoxylated acrylic acid units replace the methacrylate or acrylic acid ester units.

In preparing a type A terpolymer a mixture of the required methacrylic ester and vinyl acetate is added to an autoclave containing as a solvent a quantity of benzene, toluene, xylene, etc. following which the autoclave is purged with an inert gas such as nitrogen, argon, etc. and then with ethylene to a pressure of about 700 to 1,200 psig. Next, a free radical-type catalyst such as di-tert-butyl peroxide in, for example, benzene, is pressurized into the autoclave over a period of from about 1 to 5 hr or more during which time the temperature and pressure are usually maintained constant. Finally, the terpolymer product is recovered by stripping from the reaction mixture unreacted materials.

Terpolymers of type B are prepared in the same manner as type A polymers. The oxyalkylated acrylic acid used in preparing terpolymer B compounds are prepared by conventional methods in which ethylene oxide or propylene oxide is reacted with acrylic acid in a suitable solvent at a temperature of about 100°C in an autoclave in the presence of a basic catalyst.

The amount of the terpolymer pour depressant incorporated into the fuel compositions may vary from about 0.01 to 0.50 wt % and preferably between about 0.02 to 0.25 wt %.

Example: A mixture of 220 g of vinyl acetate, 22 g of the methacrylic ester of Alfol 1216 (a mixture of normal alcohols ranging from C_{12-16}) is metered into an autoclave containing 840 cc benzene that was purged with N_2, then with ethylene. The mixture is heated to 150°C and pressurized with ethylene to 3,000 psig. 22 g of di-tert-butyl peroxide in 66 g of benzene is also metered in over a period of 2 hours. The temperature and pressure are kept constant over this time. The terpolymer when stripped of unused reactants contains 28 wt % vinyl acetate, 3 wt % Alfol 1216 methacrylate and the balance ethylene.

A residual fuel oil composition is prepared by adding a sufficient amount of the above-prepared terpolymer to Amna 650°F residual fuel so that the concentration of the additive was 0.1 wt %. The pour point of this composition was determined by the method of ASTM D-97 and found to be substantially below that of the Amna residual fuel alone.

Crude oil compositions such as those described in the above example may be used to lower the viscosity of oil to be transported through pipelines, eliminating the necessity for the heating equipment along the pipeline which would otherwise have to be used to lower the viscosity of the crude oil to a point suitable for transporting it by pipeline.

Terpolymer or Graft Polymer

As is well known, residual fuel oils contain quantities of wax and asphaltic compounds which render them viscous and which sometimes interfere with practical use thereof. Particularly serious problems can be encountered in pumping residual fuel oils to a burner and in making them flow at low temperatures.

Various approaches to make these oils easier to handle have been made: (1) dewaxing, which is a very costly procedure; (2) cutting the heavy oils with light distillate oils, which is also expensive; and (3) incorporating additives to improve their flow characteristics.

The main objective of *W.M. Sweeney; U.S. Patent 4,178,950; December 18, 1979; assigned to Texaco Inc.* is to provide for critical blending of high pour waxy residual fuel oils with low wax, low pour residual fuel oils to give large increases in pour reduction without employing elaborate dewaxing procedures by incorporation therein of a small amount of a terpolymer of vinyl acetate, ethylene and propylene or butylene.

A further object is to provide a fuel oil blend which will be stable at different blend temperatures over prolonged storage times.

The terpolymer used in preparing the crude oil compositions of this process is a vinyl acetete-ethylene-propylene or butylene terpolymer while the graft copolymer comprises an ethylene-vinyl acetate backbone or basic chain having grafted thereto propylene or butylene.

Preferably, the residual fuel oil compositions of this process will contain about 55 to 85 volume percent of the high pour, low sulfur, waxy residual fuel; about 45 to 15 volume percent of the low wax, low pour residual fuel and about 0.01 to 0.5 wt % of the oil-soluble terpolymer or graft-type copolymer.

One type of oil-soluble terpolymers useful in this process comprises recurring units of vinyl acetate, ethylene and propylene or butylene, with an $\overline{M}n$ of from 5,000 to 80,000 and preferably 12,000 to 60,000 as determined by vapor pressure osmometry.

In the vinyl acetate-ethylene-propylene or butylene terpolymer the weight percent of the vinyl acetate units is about 10 to 45; the weight percent of propylene or butylene units is about 0.01 to 5.0 with the ethylene units being the balance.

Example 1: A terpolymer is prepared by introducing 10 parts of ethylene, 4.3 parts vinyl acetate, 0.1 part of propylene and 3 parts of benzene per hour into a stirred 2-liter autoclave maintained at a temperature of 140° to 150°C at 1,450 psig. Di-tert-butyl peroxide is employed as the catalyst and is introduced in benzene into the reactor at the rate of 0.8 lb/1,000 lb of polymer. The product is continuously removed from the reactor giving a residence time of 15 minutes. After the reaction mixture is removed from the reactor, it is stripped of solvent and unreacted materials yielding the terpolymer product. The composition of the terpolymer is about 26 wt % vinyl acetate, 0.6 wt % of propylene with the balance being ethylene. The number average molecular weight of the terpolymer as measured by vapor pressure osmometry is about 21,500.

A residual fuel oil composition is prepared by mixing at 185°F for 1 hour 65 vol % of F/18 residual fuel, about 35 vol % of Louisiana No. 6 fuel oil and a sufficient amount of the above-prepared terpolymer so that the concentration of the terpolymer is 0.15 wt %. The pour point of this composition is determined by the method of ASTM D-97 and found to be substantially below that of the same fuel oil mixture without terpolymer which exhibits a pour point of 80°F. The pour point of the F/18 residual fuel alone is 95°F while the pour point of the Louisiana No. 6 residual fuel is 30°F.

Example 2: A graft polymer is made by dissolving 1 part of di-tert-butyl peroxide and 10 parts of Elvax 410 in 90 parts of benzene in a stirred 2-liter autoclave. The reactor is flushed three times with nitrogen and then twice with the olefin to be used as the grafting agent, i.e., propylene. The reactor is heated to 300°F for 4 hr during which time the grafting reaction occurs. The product is removed from the reactor and stripped of solvent and unreacted materials. In the resulting polymer (Polymer 1) about 2 parts of propylene is grafted onto 100 parts of Elvax 410.

A residual fuel oil composition is prepared by mixing at 200°F for 1.5 hr, 75 vol % F/18 residual fuel and 25 vol % of Louisiana No. 6 fuel oil and a sufficient amount of grafted Polymer 1 made above so that the concentration of the grafted polymer is 0.18 wt %. The pour point of this composition is determined by ASTM D-97 and found to be substantially below that of the same fuel oil mixture without the grafted copolymer which exhibits a pour point of 80°F or with a similar concentration of Elvax 410 which exhibits a pour point slightly below 80°F.

W.M. Sweeney; U.S. Patent 4,178,951; December 18, 1979; assigned to Texaco Inc. demonstrates that the same terpolymer or graft polymer pour depressant described in the previous patent is effective for improving the pour point and flow properties of crude oils. This is particularly true for high-pour, high-wax, low-sulfur crude oils having an API gravity of about 30 to 40; a sulfur content of between 0.10 and 2.0% by wt; a Saybolt viscosity at 100°F of 20 to 100 SUS; a wax content of between 3 and 20% by wt; and a pour point between 40° and 100°F. A waxy, high-pour, low-sulfur crude oil which has been given particularly good results in these low-pour point oil compositions is known as Amna crude and has an API gravity of about 36.0; a Saybolt viscosity of about 69.8 SUS at 100°F; a pour point of about +70°F; a wax content of about 14.0 wt %; and a sulfur content of about 0.15 wt %; whose pour point is substantially reduced when mixed at 150°F with 0.14 wt % of the terpolymer described in Example 1 of the previous patent.

Alkenyl Succinate Diesters

The principle objective of *P.G. Pappas and A.B. Abdul-Malek; U.S. Patent 4,255,160; March 10, 1981; assigned to Standard Oil Company (Indiana)* was to economically improve the low-temperature flowability of vacuum gas oils, crude oils and residua containing a fraction of hydrocarbon having a boiling range greater than 343°C (650°F). (For the purpose of this process the term vacuum gas oil refers to the lightest fraction of hydrocarbon distillate recovered from a vacuum distillation unit.)

They have found that this objective could be obtained by forming a composition comprising a heavy petroleum product containing a fraction having a boiling range greater than 343°C (650°F), selected from the group consisting of crude oil, residuum, vacuum gas oil, and mixtures thereof and an effective amount of a linear primary monohydric alcohol diester of a linear primary alpha-olefin substituted succinic acid compound. The linear nature of the alpha-olefin substituent formed from the alcohol is critical to maintaining maximum prevention of thickening and crystal growth in the heavy petroleum product. Altering the composition of the additive by significantly isomerizing the linear substituents to branched substituents reduces the ability of the polymer to affect low temperature flowability.

Briefly, the compositions can be made by blending a heavy petroleum product with an additive prepared either by reacting a linear alpha-olefin with maleic acid, maleic anhydride or fumaric acid to produce a linear alpha-olefin substituted succinic acid or anhydride and then esterifying the substituted succinic acid or anhydride with a linear primary monohydric alcohol, or by esterifying a maleic acid or anhydride or fumaric acid with a linear primary monohydric alcohol and then reacting the ester with a linear alpha-olefin.

Examples of the useful alpha-olefins are decene, dodecene, tetradecene, pentadecene, hexadecene, octadecene, eicosene, docosene, tetracosene, hexacosene, triacontene, and mixtures thereof.

Examples of the linear primary monohydric alcohols useful for preparing these compounds are butanol, amyl alcohol, hexyl alcohol, heptyl alcohol, nonyl alcohol, decyl alcohol, dodecyl alcohol, eicosanol, heneicosanol, docosanol, triacontanol, pentacontanol, and mixtures thereof.

Example 1: To a 1,500 ml flask equipped with a reflux condenser, a Dean-Stark trap, stirring mechanism, nitrogen atmosphere, and heating mantle, were added 18.4 g (0.19 mol) of maleic anhydride and 50 g (0.19 mol) of a mixture of C_{18-20} linear alpha-olefins. The mixture was stirred and heated under a nitrogen atmosphere for 16 hr at a temperature of 193° to 205°C. The mixture was then cooled, and 171 g (0.38 mol) of a commercial grade mixture of linear primary alcohols having from about 25 to 35 carbon atoms, and 0.5 g of p-toluene sulfonic acid catalyst were added. The mixture was again heated to 200°C and maintained at the temperature until water of esterification was no longer collected. The product was then stripped at 216°C with a nitrogen atmosphere. The resultant product was cooled and collected and directly blended into petroleum products.

Example 2: Example 1 was repeated except the p-toluene sulfonic acid was added to the maleic anhydride prior to the addition of the linear alpha-olefin.

Additive in Vacuum Gas Oil

Example	Additive Concentration (wt %)	ASTM Pour Point, °C (°F)
1	0.05	4.4 (40)
	0.10	1.7 (35)
	0.20	-1.1 (30)
2	0.20	20.0 (68)
(Blank)	0.00	21.1 (70)

The table shows the improved flowability properties of the fuel containing from 500 (0.05 wt %) to 2,000 (0.20 wt %) ppm of the additive. The additive containing 2,000 ppm has an ASTM pour point reduced from 70° to 30°F, a 40° difference. The table also shows that the isomerized alpha-olefin substituted succinate diester produced by the reaction between the alpha-olefin and the succinic anhydride in the presence of an acidic catalyst has essentially no cold-flow improving properties.

Copolymer of Conjugated Diene and Alkyl Acrylonitrile as Viscosity Stabilizer

The manufacture of heavy fuel-oils and particularly No. 2 heavy fuel-oils from petroleum residues is currently performed by fluxing these residues with diluents in order to obtain suitable viscosity characteristics. These petroleum residues may particularly consist of crude oil straight-run or vacuum-distillation residues. The diluents generally used are distillates of relatively low viscosity, such as gas-oils. In view of the high cost of this type of diluents, it is desirable to make use of other products of lower cost, such, in particular, as the steam-cracking residues, which have a low viscosity, a low sulfur content and a low content of metals and which could constitute very good diluents for the treatment of viscous petroleum residues. However, mixtures of these constituents suffer, in most cases, of incompatibilities which result in a viscosity increase in the course of time and/or in the formation of sediments.

This is also the case when distillation residues (straight-run or vacuum-residues) or even residues from a visco-reduction unit are diluted with visco-reducing gas-oils. As a general rule, such phenomena are observed when a petroleum residue or a mixture of petroleum residues is to be diluted with one or more fluxing agents (such as the above-mentioned steam-cracking residues or visco-reduction gas-oils).

J.-P. Durand, F. Dawans, A. Faure and P. Maldonado; U.S. Patent 4,264,334; April 28, 1981; assigned to Institut Francais du Petrole and Elf-Union, France have found that it is possible to improve the stability of the heavy fuel-oils having the above composition, by adding thereto a sufficient proportion of certain polymeric additives.

Their compositions of heavy fuel-oils may be defined as comprising a mixture of:

(a) at least one petroleum residue selected from straight-run distillation residues, vacuum-distillation residues and visco-reduction residues;

(b) at least one diluent selected from steam-cracking residues and visco-reduction gas-oils; and

(c) a sufficient proportion for improving the stability of these mix-

> tures of at least one sequenced copolymer consisting essentially of a sequence (A) resulting from the (co)polymerization of one or more conjugated diolefins and one or more sequences (B) resulting from the polymerization of acrylonitrile or an alkylacrylonitrile in which the alkyl group contains, for example, from 1 to 20 carbon atoms.

As advantageous examples of (A), there can be mentioned those consisting of homopolymeric chains based on 1,3-butadiene, and copolymeric chains based on 1,3-butadiene and isoprene, these sequences being optionally completely or partially saturated by hydrogenation. In the case of homopolymeric chains based on 1,3-butadiene saturated by hydrogenation, the content of 1,2 units must be sufficient to make the polymers soluble in the mixture to be stabilized. For this purpose, it is preferred in this case to have a proportion of at least 50% of 1,2 units.

The average molecular weight of the sequenced copolymers is preferably from 2,000 to 10,000. Their nitrogen-containing monomer content [i.e., the proportion of (B) sequences] is preferably from 0.5 to 5% by wt.

Among the sequenced copolymers as above-defined, preferred are the bisequenced copolymers of type (A)-(B) and the trisequenced copolymers, particularly those of type (B)-(A)-(B), where (A) and (B) have each the abovementioned definition.

Example 1: In a reactor, a mixture of 57 g of 1,3-butadiene and 37 g of isoprene is added to a solution of 12.5 mmol of butyllithium in 250 ml of n-heptane. The reaction mixture is stirred for 5 hours at 50°C, which results in a complete conversion of the monomers to a butadiene-isoprene copolymer having a butadiene content of 60% by wt and an average molecular weight by number of 5,700.

To the resultant reaction mixture, there is added 4.7 g of methacrylonitrile and the polymerization is continued for 1 hour at 50°C. The copolymer is separated by precipitation in acetone and dried under reduced pressure up to constant weight. There is thus obtained a sequenced copolymer [(butadiene-isoprene)-methacrylonitrile] containing 0.5% by wt of nitrogen, i.e., 2.4% by wt of methacrylonitrile.

This product is added at 0.1% by wt to a mixture of 59% by wt of an Aramco vacuum-residue with a viscosity of 4,800 cs at 50°C and 41% by wt of a steam-cracking residue with a viscosity of 41.2 cs at 50°C, an initial distillation point of 210°C and a SG at 15°C of 1.065. The viscosity of the mixture stored at 90°C is measured in the course of time. The results are summarized in the table below showing the beneficial effect of the additive on the stability of the mixture as compared to the same mixture stored under the same conditions but without additive.

Example 2: In a reactor, there is added 94 g of 1,4-butadiene to a solution of 12.5 mmol of butyllithium and 9 g of tetrahydrofuran in 250 ml of n-heptane. The reaction mixture is stirred for 5 hours at 50°C, resulting in a complete conversion of the monomer to polybutadiene containing 60% of 1,2 units and 40% of 1,4 units and having an average molecular weight by number close to 7,000.

To the reaction mixture obtained at the end of the polymerization of 1,3-butadiene, there is added 3.5 g of methacrylonitrile and the polymerization is continued for 1 hour at 50°C, so as to obtain a butadiene methacrylonitrile sequenced copolymer.

This product, separated as in Example 1, is added in proportion of 0.1% by wt to the mixture of vacuum-residue and steam-cracking residue of Example 1.

The resulting mixture is stored at 90°C for several days and the viscosity, measured over a period of time shows the good stability of the mixture.

Example 3: Under the conditions of Example 2, the sequenced polybutadiene-methacrylonitrile copolymer is hydrogenated in the presence of a suspension resulting from the reaction of 100 mg of cobalt as octoate with 600 mg of triethylaluminum, at 130° to 140°C for 6 hours under a hydrogen pressure of 25 bars. The obtained hydrogenated copolymer is added in a proportion of 0.1% by wt to the mixture of vacuum-residue and steam-cracking residue of Example 1. The mixture including the additive is stored at 90°C for several days and its viscosity shows the good stability of the mixture.

Stability of the Vacuum-Residue and Steam-Cracking Residue Mixtures

Viscosity at 50°C (cs) After	Without Additive	.... Example Number.....		
		1	2	3
1 hour	433	388	380	375
3 hours	524	408	403	400
18 hours	595	426	420	412
8 days	650	436	430	420
32 days	760	438	432	425
40 days	1,050	440	432	425

Other flow improvers and pour point depressants will be found under the heading "Sludge Dispersants for Lubricants and Fuels" in Chapter 3.

-5-

ANTIKNOCK COMPOUNDS AND OCTANE IMPROVERS

ANTIKNOCK COMPOUNDS

Lead alkyls, most notably tetraethyl lead, have been widely used as antiknock additives for gasoline since the early 1920's. In recent years, however, steps have been taken to eventually eliminate the use of lead alkyls as fuel additives on environmental grounds. At present, resort to the use of larger quantities of aromatic hydrocarbon components is made in order to improve the octane rating of motor fuels. Large amounts of aromatics in gasoline cause undesirable exhaust emissions and increased exposure to benzene which has been linked to leukemia. In addition to this environmental penalty, it has been estimated that the removal of lead alkyls from gasoline reduces the amount of gasoline recoverable from a barrel of crude oil by about 6%. Accordingly, there exists a great need for a lead-free antiknock agent which is environmentally acceptable and does not accelerate the depletion of dwindling petroleum reserves.

A major contributing cause of engine knock is known to be due to the accumulation of carbonaceous deposits within the combustion chambers. The deposits reduce the volume of each combustion chamber resulting in an increase in engine compression ratio. The increased engine compression in turn causes a proportional increase in the octane number requirement of the engine. This increase, however, accounts for only about 10 to 20% of the octane requirement increase (ORI), the balance of the increase being due to other factors occasioned by the existence of the deposits. It has been speculated that the deposits catalyze the combustion process with a resulting tendency to cause detonation and/or preignition. It has also been thought that the deposits may exert an insulating effect which decreases the rate of heat transfer through the cylinder walls. This causes an increase in combustion temperature, a condition which makes the engine more prone to knock.

Numerous substances possessing antiknock activity have been proposed as replacements for tetraethyl lead. Methylcyclopentadienylmanganese tricarbonyl (MMT) which was developed as an antiknock additive reportedly causes the catalysts of exhaust converters to malfunction and its use has been forbidden in lead-free gasolines. Because of this fact, patents in which the cyclopentadienyl manganese compounds are essential ingredients of antiknock additives have not been abstracted in this book. For those who are interested, the following U.S. patents are included in the above group: 4,139,349; 4,140,491; 4,141,693; 4,155,718; 4,175,927; 4,191,536; and 4,266,946.

Organocerium(IV) Chelate plus p-Cresol or an Alkanediol

J.F. Deffner; U.S. Patent 4,133,648; January 9, 1979; assigned to Gulf Research & Development Company suggests a lead-free fuel composition having improved antiknock and octane characteristics which comprises a liquid hydrocarbon fuel containing an antiknock and octane-improvement amount, especially from about 0.04 to 8.0% by wt of p-cresol or an alkanediol of the formula:

$$R-\underset{\displaystyle\vert}{\overset{OH}{CH}}-(CH_2)_x-\overset{OH}{\underset{}{CH}}-R'$$

wherein R and R' can be either alike or different, selected from hydrogen or a hydrocarbon radical having from about 1 to 12 carbon atoms, preferably from about 1 to 8 carbon atoms; and x is an integer of from about 1 to 5, preferably from about 1 to 3; in combination with an organocerium(IV)chelate of the formula:

$$Ce\,[R-\underset{\underset{O}{|}}{C}=CH-\underset{\underset{O}{\|}}{C}-R']_4$$

wherein R and R' can be either alike or different members selected from the group consisting of hydrogen or hydrocarbon radicals including alkyl, aryl, aralkyl, alkaryl and cycloalkyl radicals containing from 1 to 12 carbon atoms preferably from 1 to 5 carbons, with the sum of the carbon atoms in the radicals being from 3 to 24; preferably 4 to 10.

Examples of alkanediols useful in the formula include the 1,4-, 1,3-, 1,5-, and 2,4-pentanediols; the 1,4-, 1,5-, 1,6-, and 2,5-hexanediols; the 1,4- and 2,5-heptanediols; 3,6-octanediol; 2-methyl-1,4-hexanediol; 6-methyl-1,4-heptanediol; 3-methyl-2,4-pentanediol; 2,7-dimethyl-3,6-octanediol; 3,8-dimethyl-4,7-decanediol; 2,2,8,8-tetramethyl-4,7-decanediol; 7,10-hexadecanediol, and mixtures thereof.

Preferred organocerium chelates include:

Ceric 2,4-hexanedionate
Ceric 2,4-heptanedionate
Ceric 2,4-octanedionate
Ceric 3,5-heptanedionate
Ceric 3,5-octanedionate
Ceric 4,6-nonanedionate
Ceric 5,7-undecanedionate
Ceric 7,9-pentadecanedionate
Ceric 9,11-nonadecanedionate
Ceric 9,11-eicosanedionate
Ceric 11,13-tricosanedionate
Ceric 12,14-hexacosanedionate
Ceric 2,2-dimethyl-3,5-hexanedionate
Ceric 2,2-dimethyl-3,5-nonanedionate
Ceric 2,7-dimethyl-3,5-octanedionate
Ceric 2,2,6,6-tetramethyl-3,5-heptanedionate
Ceric 3,3,7,7-tetraethyl-4,6-nonanedionate
Ceric 1-phenyl-1,3-butanedionate
Ceric 1-phenyl-1,3-undecanedionate
Ceric 1-naphthyl-1,3-butanedionate
Ceric 1-benzyl-1,3-butanedionate
Ceric 1-tolyl-1,3-butanedionate
Ceric 1-cyclohexyl-1,3-butanedionate
Ceric 1,3-dinaphthyl-1,3-propanedionate
Ceric 1,3-dibenzyl-1,3-propanedionate
Ceric 1,3-ditolyl-1,3-propanedionate
Ceric 1,3-dicyclohexyl-1,3-propanedionate

Also preferred organocerium(IV) chelates are mixtures of the above.

In order to obtain the desired result it is critical that the liquid hydrocarbon fuel contain from about 0.04 to 8.0 wt %, preferably from about 0.1 to 2.0 wt %, based on the liquid hydrocarbon fuel, of the combined p-cresol or alkanediol and organocerium(IV)chelate. In addition, the weight ratios of the p-cresol or alkanediol to the organocerium(IV)chelate should be preferably in the range of about 1.2:1 to 5:1.

Example 1: 0.90% by wt of 2,5-heptanediol and 0.15% by wt ceric 2,2,6,6-tetramethyl-3,5-heptanedionate are added to a synthetic gasoline produced from solid carbonaceous materials, having a boiling range of from about 100°F (37.8°C) to 420°F (215.6°C), a density of about 0.9 to 1.1 and a carbon to hydrogen molecular ratio in the range of about 1.3:1 to 0.66:1. The synthetic gasoline composition has enhanced octane rating and antiknock characteristics. Substantially the same results are obtained when the organocerium(IV)chelates described here are substituted for the ceric 2,2,6,6-tetramethyl-3,5-heptanedionate above and the alkanediols described herein are substituted for the 2,5-heptanediol above.

Example 2: A synthetic gasoline composition having enhanced octane rating and antiknock characteristics is prepared by mixing 0.30% by wt of p-cresol and 0.15% by wt of ceric 2,2,6,6-tetramethyl-3,5-heptanedionate with a synthetic gasoline produced from solid carbonaceous materials, having a boiling range of from about 100°F (37.8°C) to 420°F (215.6°C), a density of about 0.9 to 1.1 and a carbon to hydrogen molecular ratio in the range of about 1.3:1 to 0.66:1. The synthetic gasoline composition has enhanced octane rating and antiknock characteristics.

The organocerium(IV)chelates described herein can be substituted for the ceric 2,2,6,6-tetramethyl-3,5-heptanedionate above with substantially the same results.

Rare Earth Chelates from 2,2,7-Trimethyl-3,5-Octanedione

Numerous substances possessing antiknock activity have been proposed as replacements for tetraethyl lead. Methylcyclopentadienylmanganese tricarbonyl (MMT) which was developed as an antiknock additive reportedly causes the catalysts of exhaust converters to malfunction and its use has been forbidden in lead-free gasolines. The article "Volatile Metal Complexes" by R.E. Sievers and J.E. Sadlowski, *Science*, July 21, 1978, Volume 201, pp. 217–223, discloses a number of liquid hydrocarbon-soluble rare earth metal beta-diketonate chelates useful as fuel additives including the holmium and erbium complexes of the homologous beta-diketone, 2,2,6,6-tetramethyl-3,5-heptanedione [H(thd)]. Recently, it has been observed that rare earth chelates prepared from H(thd) promote the degradation of liquid hydrocarbon fuels to such an extent as to be virtually useless as practical fuel additives.

R.E. Sievers and T.J. Wenzel; U.S. Patent 4,251,233; February 17, 1981; assigned to University Patents, Inc. have found that liquid hydrocarbon-soluble rare earth metal chelating complexes derived from the beta-diketone 2,2,7-trimethyl-3,5-octanedione [H(tod)] intermittently or continuously present in fuels such as gasoline, diesel oil, benzene, kerosene, etc., and blends thereof, prevent formation and/or effect removal of carbonaceous deposits during the combustion of the fuels. When added to motor fuels, these rare earth metal chelates minimize or eliminate the accumulation of carbonaceous deposits in the combustion chambers of internal combustion engines in which the fuels are employed, reducing the octane requirement increase (ORI) and suppressing the tendency of the engines to knock. Unlike the rare earth metal chelates derived from H(thd), these chelates demonstrate entirely acceptable levels of stability in liquid hydrocarbon fuels. Rare earth compounds are known to be far less toxic than the lead salts produced by combustion of tetraethyl lead and unlike tetraethyl lead and MMT, are unlikely to poison noble metal catalysts such as are used in automotive exhaust converters.

The process for preparing H(tod) which possesses the structure

$$CH_3-\overset{CH_3}{\underset{CH_3}{C}}-\overset{O}{\overset{\|}{C}}-CH_2-\overset{O}{\overset{\|}{C}}-CH_2-\overset{CH_3}{\underset{CH_3}{CH}}$$

from inexpensive and readily obtainable raw materials represents another important advantage. The rare earth metal chelating complexes of H(tod) can be prepared by any suitable method, e.g., by reaction with rare earth metal halides in the presence of a strong base.

In accordance with this process, H(tod) is prepared by reacting methylisobutyl ketone (MIBK) with an ester such as ethyl pivalate,

$$CH_3-\overset{CH_3}{\underset{CH_3}{C}}-\overset{O}{\overset{\|}{C}}-OC_2H_5$$

and an alkali metal hydride of the formula MeH, in which Me is an alkali metal, to provide H(tod), a light-yellow liquid which has a boiling point of 55°C at 0.1 mm Hg. The compound exists in both the ketonic and enolic forms.

The rare earth metal (tod) chelates can be represented by the general formula:

$$\left[M \begin{matrix} O=C-CH_2-CH(CH_3)_2 \\ \quad CH \\ O=C-C(CH_3)_3 \end{matrix} \right]_x Y_a$$

in which M is a rare earth element or a mixture thereof, Y is water or an organic compound containing a donor group, a is 0 to 4 and x is 1 to 4. The rare earths include the following elements: lanthanum (La), cerium (Ce), praseodymium (Pr), neodymium (Nd), samarium (Sm), europium (Eu), gadolinium (Gd), terbium (Tb), dysprosium (Dy), holmium (Ho), erbium (Er), thulium (Tm), ytterbium (Yb), lutetium (Lu), scandium (Sc), yttrium (Y). For the sake of economy, it is preferred to employ a mixture of rare earth compounds derived from a naturally occurring rare earth ore.

The preparation of the rare earth metal (tod) chelates herein can be conducted by a procedure analogous to that disclosed in U.S. Patent 3,794,473, in which a methanolic solution of one or several rare earth metal halides is added to a methanolic solution of H(tod) and alkali metal hydroxide, ammonium hydroxide or other strong base is added under constant agitation to provide the chelate or chelate mixture which can be recovered employing standard techniques.

The rare earth metal (tod) chelates can be effectively utilized as fuel additives to obtain the beneficial effectiveness of the rare earth components in minimizing ORI, reducing surface ignition, imparting wear reduction effectiveness, improving octane ratings, alleviating spark plug fouling and other problems associated with internal combustion engines of both the spark and compression ignition types. The chelates can be dissolved and intimately mixed with liquid hydrocarbon fuels for internal combustion engines to impart the beneficial effects. Similarly the chelates herein can be added to bunker oils and furnace oils for improvement of the combustion properties of these fuels. When added to gasoline and furnace oils, for example, the chelates can generally be used at a level of from about 0.001 to 50 mmol per gallon to provide effective results.

Process for Preparing Highly Overbased Metallo-Organic Complexes

The process developed by *A. Alkaitis and P.L. Cells; U.S. Patent 4,162,986; July 31, 1979; assigned to Mooney Chemicals, Inc.* is one for preparing oil-soluble and hydrocarbon-soluble, highly overbased transition metal-organic compositions comprising a metal oxide-hydroxide-carboxylate complex. The metal content, comprising at least one metal which is a transitional metal, is in chemical combination partly with oxygen in a polynuclear metal oxide crystallite core and partly with at least two different monocarboxylic acids or a mixture of one or more monocarboxylic and monosulfonic acids containing at least two carbon atoms as hydroxyl-metal-carboxylate and hydroxyl-metal-sulfonate groups. At least one of the acids is a monocarboxylic acid containing at least seven carbon atoms, and when the second acid is also a monocarboxylic acid, the second acid contains a number of carbon atoms in its longest chain differing by at least two carbon atoms from the total number of carbon atoms in the other. At least a portion of the carboxylate and sulfonate groups are hydrogen bonded to oxygen atoms of the core, and the remainder of the carboxylate and sulfonate groups are unbonded and in equilibrium with the bonded groups. The ratio of total metal mols to the total mols of organic acid is greater than one.

The metal contained in the overbased complexes must be at least in part a transitional metal. When the complexes contain additional metals, any desirable metal can be utilized such as calcium, barium, zinc, etc. Of the transitional metals, copper and those elements found in the first transition series, namely, scandium, titanium, vanadium, chromium, manganese, iron, cobalt and nickel are preferred.

Although specific details will vary, each transition metal complex is prepared by a similar general method. When a metal oxide such as manganese oxide is used, it is slurried with an excess of water and with what, by ordinary stoichiometric considerations, would be a deficiency of organic acids at a selected mol ratio of metal-to-acid, agitated, and heated at the reflux temperature until no further reaction occurs, and the "excess" metal oxide has been completely converted to hydroxide. The batch is heated to about 120° to 149°C (300°F) or higher, until the free water is eliminated. Air is introduced into the mix at 110° to 150°C until insoluble manganous hydroxide has been solubilized by conversion into the complex. Previously bound water which is freed during air or oxygen blowing may be removed during the blowing step. Appearance and metal content analysis determination indicate when processing can be terminated.

If manganese metal is employed rather than manganese oxide, some air may be introduced to facilitate the formation of manganous oxide or hydroxide in the first step, but under some control to avoid a too large excess, which would result in formation of some insoluble higher oxides together with manganese hydroxide. A similar process is used with cobalt, nickel and iron metals. Either copper metal or cuprous oxide may be employed to form the copper complex, but in both instances, air is required for oxidation.

Various mixtures or formulations of reactant monobasic organic acids may be used to facilitate processing or for collateral reasons. Examples of useful organic carboxylic acids include propionic acid, butyric acid, 2-ethylhexoic acid, commercially available standardized nonanoic acid, neodecanoic acid, oleic acid, stearic acid, naphthenic acids, tall oil acid, and other natural and synthetic acids and acid mixtures.

The sulfonic acids include the aliphatic and the aromatic sulfonic acids. They are illustrated by petroleum sulfonic acids or the acids obtained by treating an alkylated aromatic hydrocarbon with a sulfonating agent, e.g., chlorosulfonic acid, sulfur trioxide, oleum, sulfuric acid, or sulfur dioxide and chlorine. The sulfonic acids obtained by sulfonating alkylated benzenes, naphthylenes, phenol, phenol sulfide, or diphenyl oxide are especially useful.

The major application of these complexes will be as catalysts for a variety of chemical operations in which prior art metal soaps currently are used.

The manganese, cobalt, iron, copper and nickel high metal content complexes are useful as catalysts for many chemical reactions, either directly as homogeneous catalysts or in preparation of heterogeneous catalysts. They can be used in antiknock agents for gasoline, fuel additive type combustion improvers, and smoke or toxic fume suppressants.

Coordination Compounds of Divalent Mn, Fe, Co and Ni

W.B. McCormack and C.A. Sandy; U.S. Patent 4,180,386; December 25, 1979; assigned to E.I. Du Pont de Nemours & Company have developed useful gasoline antiknock compounds which are hexacoordinated transition metal compounds having the structure

$$
\begin{array}{c}
(L_1)\text{–}M\text{–}(L_3) \\
| \\
(L_2)
\end{array}
$$

M is a divalent metal selected from the group Mn, Fe, Co and Ni;

L_1 and L_2 are the same or different chelate-forming β-diketone groups having from 5 to 20 carbon atoms, one or both of L_1 and L_2 having at least one polyfluoroalkyl group, $-CF_2X$, adjacent to the carbonyl group, where X is selected from the group H, F, Cl, phenyl, C_{1-6} alkyl and $C_nF_{2n}Y$; n is 1 to 6; and Y is H, F, or Cl;

L_3 is a ligand having the structure A–Z–B, wherein

A and B are the same or different members of the group $-NH_2$, $-NHR_5$, $-NR_5R_6$, OH, OR_5, SH, SR_5 and PR_5R_6;

R_5 and R_6 are the same or different members of the group C_{1-4} alkyl; and

Z is a divalent hydrocarbyl group having 2 to 20 carbon atoms selected from a member of the group alkylene, phenylene and cycloalkylene, each member providing 2 or 3 carbon atoms between A and B, with the proviso that when Z is phenylene, the number of carbon atoms between A and B is 2.

The preferred compounds are $Co(II)(HFAA)_2 \cdot TMED$ (HFAA = hexafluoroacetylacetone; TMED = N,N,N',N'-tetramethylethylenediamine) and $Ni(II)HFAA)_2 \cdot TMED$. These two compounds provide outstanding stability, volatility, solubility in hydrocarbons and antiknock performance.

Example 1: *Bis(hexafluoroacetylacetonato)cobalt(II)tetramethylethylenediamine, $Co(II)(HFAA)_2 \cdot TMED$* – To a 2-liter reaction flask equipped with a mechanical stirrer, a condenser, a gas inlet tube and an addition funnel, 700 ml of deionized water was added. The water was sparged for ½ hour with nitrogen at reflux and then cooled to room temperature. Hexafluoroacetylacetone was then added in an amount of 104 g over a period of 15 minutes. A temperature rise of 5°C was noted. After stirring for 15 minutes, 20.4 g of sodium hydroxide dissolved in 70 ml of water was added over 17 minutes. Cobaltous chloride solution (59.5 g of $CoCl_2 \cdot 6H_2O$ dissolved in 280 ml of water) was added over 15 minutes. Yellow precipitate formed initially and gradually turned light green. After stirring for 1 hour, 29 g of tetramethylethylenediamine was added. The color of the precipitate changed from light green to brownish-yellow. After stirring for an additional 2½ hours, the reaction mass was filtered and the solids washed with water. The filter cake (89.6 g) was heated with 450 ml of toluene and filtered. Removal of the solvent from the filtrate provided 78 g (53% yield) of rust-colored product which melted at 100° to 101.5°C.

Example 2: The compound of Example 1, $Co(II)(HFAA)_2 \cdot TMED$, was compared in antiknock efficiency with two commercial metal-containing antiknock compounds: tetraethyllead, and methylcyclopentadienylmanganesetricarbonyl, also known as MMT. The comparison was made in a commercial lead-free gasoline with a research octane number of 91.4, a motor octane number of 82.0 and an average octane number of 86.7. The knock ratings were carried out by ASTM D-909 (Research) and ASTM D-357 (Motor Methods) in duplicates, and the results are expressed in terms of octane number increase (ΔON) to the nearest tenth (0.1) of the octane numbers. The results are also expressed in terms of (R + M)/2 (average octane number) which values show good correlation with road octane improvements.

Antiknock Performance (Comparison)

Grams of Metal per Gallon	. . . . . . Octane Number Increase (ΔON)		
	Research (R)	Motor (M)	(R + M)/2
Antiknock Additive = Tetraethyllead (as Motor Mix)			
0.025	0.3	0.2	0.3
0.05	0.3	0.2	0.3

(continued)

Grams of Metal per Gallon	Octane Number Increase (ΔON) Research (R)	Motor (M)	(R + M)/2
Antiknock Additive = Tetraethyllead (as Motor Mix)			
0.10	0.8	0.5	0.7
0.15	1.0	0.6	0.8
0.20	1.5	1.0	1.3
0.25	1.7	1.5	1.6
0.30	1.9	1.6	1.8
Antiknock Additive = Methylcyclopentadienylmanganesetricarbonyl (Comparison)			
0.025	0.7	0.6	0.7
0.05	1.1	0.6	0.9
0.10	1.9	1.1	1.5
0.15	2.4	1.2	1.8
0.20	2.6	1.3	2.0
0.25	3.0	1.6	2.3
0.30	3.1	1.8	2.5

Antiknock Performance (Example 1)

Grams of Metal per Gallon	Octane Number Increase (ΔON) Research (R)	Motor (M)	(R + M)/2
Antiknock Additive = Co(II)$(HFAA)_2$·TMED			
0.025	1.7	0.9	1.3
0.05	2.2	1.4	1.8
0.10	2.8	1.9	2.4
0.15	3.3	2.1	2.7
0.20	3.7	2.4	3.1
0.25	4.1	2.6	3.3
0.30	4.1	2.6	3.4

The above results demonstrate the outstanding antiknock efficiency of the cobalt compound.

Divalent Tetracoordinated Cobalt Compounds

In further work on coordination compounds of cobalt, *C.A. Sandy; U.S. Patent 4,215,997; August 5, 1980; assigned to E.I. Du Pont de Nemours & Co.* has found that certain divalent tetracoordinated cobalt complexes, particularly Co(II) bis(β-acetyl-N-alkylvinylimines) when added to hydrocarbon fuel compositions comprising liquid hydrocarbons boiling in the range of about 20° to 400°C in an amount to provide about 0.02 to 5 g of cobalt metal per U.S. gallon of hydrocarbon fuel composition are efficient antiknock compounds. Preferred cobalt complexes are Co(II)bis(β-acetyl-N-isopropylvinylimine), and Co(II)bis(β-acetyl-N-tert-butylvinylimine).

Examples 1 and 2: The ability of the cobalt complexes to increase octane numbers was determined in a lead-free motor gasoline having a research octane number (RON) of 91.4, a motor octane number (MON) of 82.0, and an average or (RON + MON)/2 octane number of 86.7.

To separate portions of the base fuel were added cobalt(II)bis(β-acyl-N-alkylvinylimine) complexes in the amount to provide cobalt metal concentrations as indicated in the table. The knock ratings for the fuel blends were determined in duplicates on duplicate samples. The results are summarized below wherein the antiknock quality improvements are provided in terms of increase in the octane numbers (ΔON) over the research, motor and (R + M)/2 octane numbers of the base fuel.

Antiknock Response

Cobalt (g/gal)	.......Increase in Octane Number (ΔON)...... Research	Motor	(R + M)/2
Ex. 1: (Compound = Co(II)bis(β-acetyl-N-isopropylvinylimine)			
0.025	0.2	0.2	0.2
0.05	0.9	0.5	0.7
0.10	1.9	1.0	1.5
0.15	2.3	1.3	1.8
0.20	2.9	1.1	2.0
0.25	3.3	1.3	2.3
0.30	3.1	1.5	2.3
Ex. 2: (Compound = Co(II)bis(β-acetyl-N-tert-butylvinylimine)			
0.025	0.2	0.2	0.2
0.05	0.6	0.4	0.5
0.10	1.7	1.0	1.3
0.15	2.1	1.1	1.6
0.20	2.3	1.2	1.8
0.25	2.8	1.4	2.1
0.30	3.0	1.5	2.3

When cobalt(II)bis(β-acetyl-N-ethylvinylimine) was tested in the same way, increases in RONs were only on the average of about 75% of those obtained with the compound of Example 1 and about 85% of those obtained with the compound of Example 2.

Hexacoordinated Metal Compounds

There is a need for effective and stable nonlead antiknock additives to improve fuel octane ratings. *C.A. Sandy; U.S. Patent 4,189,306; February 19, 1980; assigned to E.I. Du Pont de Nemours and Company* has developed hexacoordinated compounds of Mn, Fe, Co and Ni for this purpose. These metal compounds have the formula

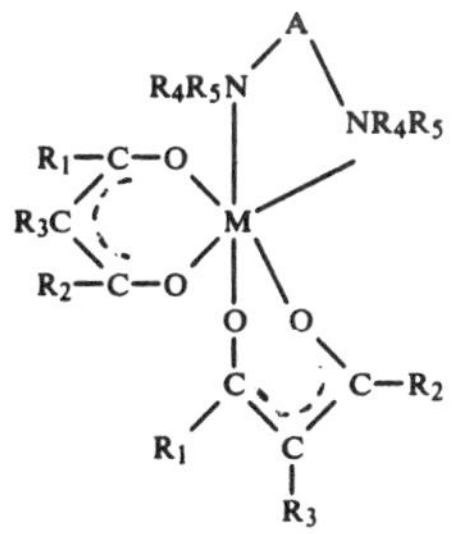

M is selected from at least one member of the group Mn, Fe, Co and Ni;

R_1 and R_2 are hydrocarbyl groups of 1 to 6 carbon atoms;

R_3 is hydrogen or an alkyl of 1 to 6 carbon atoms;

R_4 and R_5 are hydrogen or alkyl of 1 to 4 carbon atoms; and

A is a divalent hydrocarbyl group of 2 to 10 carbon atoms selected from a member of the group alkylene, phenylene and cycloalkylene, each member providing 2 or 3 carbon atoms between the nitrogen atoms with the proviso that when A is phenylene the number of carbon atoms between the nitrogen atoms is 2.

The preferred compounds are the following:

(1) Where $R_1 = R_2 = CH_3$, $R_3 = H$, $R_4 = R_5 = CH_3$ and A is $-CH_2-CH_2-$;

(2) Where $R_1 = CH_3$, $R_2 = -CH_2CH(CH_3)_2$, $R_3 = H$, $R_4 = R_5 = CH_3$ and A is $-CH_2CH_2-$;

(3) Where $R_1 = R_2 = -C(CH_3)_3$, $R_3 = H$, $R_4 = R_5 = CH_3$ and A is $-CH_2CH_2-$; and

(4) Where M is divalent cobalt or nickel, especially cobalt.

These compounds are prepared by contacting a water-soluble divalent metal salt of manganese, iron, cobalt or nickel (such as the chlorides, nitrates, acetates and the like), a β-diketone, and a diamine ligand, in the molar ratio of about 1:2:1 in the presence of aqueous alkali. Little or no heating is required. The reaction can be carried out under an inert gas atmosphere but an inert atmosphere is not required. Thus, for example, cobalt(II) dichloride, acetylacetone (AA), and tetramethylethylenediamine (TMED) are contacted in the presence of aqueous sodium hydroxide to form $Co(II)(AA)_2 \cdot TMED$ as a precipitate which can be easily isolated by filtration.

The β-diketones suitable for the preparation of these compounds are represented by the formula

$$R_1-\overset{\overset{O}{\|}}{C}-CHR_3-\overset{\overset{O}{\|}}{C}-R_2$$

where R_1, R_2 and R_3 are as defined. It is necessary that the carbon atom between the two carbonyl groups have at least one hydrogen substituent.

The diamine ligand is represented by $NR_4R_5-A-NR_4R_5$ where R_4, R_5 and A are as defined. The provision in the hydrocarbyl group, A, of 2 to 3 carbon atoms between the nitrogen atoms is important to provide increased stability by the formation of a 5 to 6 membered ring upon chelation of the diamine with the metal ion.

The amount of the compound to be incorporated into gasoline to improve its antiknock quality will depend upon the antiknock quality of the base gasoline itself and the improvements desired. In normal practice a refiner obtains the desired gasoline antiknock quality by the combination of hydrocarbon feed processing and the use of antiknock additives consistent with economy. These compounds provide antiknock improvements when used in the amount needed to provide about 0.01 g of metal per gallon and can be used in an amount to provide up to 10 g of metal per gallon. Preferably, the metal concentration will be from about 0.025 to 5 g of metal per gallon.

In addition to their utility as antiknock additives, the compounds are useful as octane requirement increase control additives when added in the fuel and/or the crankcase lubricant. Another utility is to control the formation of carbon resulting from combustion of gasolines and fuel oils. The compounds can also be employed as colorants in organic systems including hydrocarbon fuels.

The following example illustrates the formulation. Thermogravimetric analyses following the preparative details indicate the good volatility and thermal and oxidative stabilities of the compounds. The analyses were carried out on a Du Pont 990 Thermal Analyzer coupled to a Du Pont 951 Thermogravimetric Analyzer. In each characterization, approximately 20 mg of the sample was heated over a temperature range of 25° to 500°C at a heating rate of 5°C per minute with either nitrogen or air as the carrier gas at a gas flow rate of 40 ml per minute. The loss in sample weight with increasing temperature was recorded continuously. Thermal or oxidative instability is ordinarily indicated by a discontinuous loss in weight followed by little or no loss in weight since the metal residue from thermal or oxidative decomposition is not volatile. It will be noted that in each of the following analyses, employing nitrogen or air as the carrier gas, the loss in sample weight with increasing temperature is smooth and continuous.

Example: The preparation of bis(acetylacetonato)cobalt(II)tetramethylethylenediamine, Co(II)$(AA)_2$·TMED, wherein $R_1 = R_2 = CH_3$; $R_3 = H$; $R_4 = R_5 = CH_3$; and A = $-CH_2-CH_2-$, is as follows: To a 5-liter reaction flask equipped with mechanical agitation, a condenser, a gas inlet tube and an addition funnel, 3,000 ml of deionized water was added. The water was sparged for ½ hour with nitrogen. Then, 238 g of $CoCl_2·6H_2O$ were added to the reaction flask followed by addition of 200 g of 2,4-pentanedione (acetylacetone, AA) and 140 g of N,N,N',N'-tetramethylethylenediamine (TMED). Upon addition of TMED, a fine pink precipitate formed. A solution of sodium hydroxide (80 g in 700 ml of water) was then added over a period of 1½ hours during which addition, the pink-orange precipitate changed to larger blue-red precipitate. The reaction mixture was stirred for about 1½ hours following the sodium hydroxide addition and then filtered.

The precipitate collected by filtration was washed with water, then was dissolved in 1,500 ml of toluene at room temperature by stirring for 1½ to 2 hours. The toluene solution was filtered and dried over anhydrous sodium sulfate. Toluene was then removed by stripping at 90° to 95° C at 20 to 30 mm pressure. The liquid residue was poured into an enamel pan to cool. The cooled product solidified and was broken up and ground, MP 90.5° to 91.5°C. The yield of Co(II)$(AA)_2$·TMED was 315 g. The compound was stable in air, soluble in hydrocarbon fuel and monomeric in benzene as determined by cryoscopic method.

Aliphatic β-Diketone plus β-Diketone Chelate of Cerium(IV)

U.S. Patent 4,036,605 described the preparation of β-diketone chelates of cerium(IV) and their use as antiknock agents. In further research *R.J. Hartle; U.S. Patent 4,211,535; July 8, 1980; assigned to Gulf Research and Development Company* found that if sufficient aliphatic β-diketone, which is itself inactive as an antiknock agent, is added to a gasoline mixture containing the chelate of cerium(IV) with a β-diketone, the antiknock characteristics of the gasoline are further elevated.

The useful β-diketone synergists are branched compounds which are represented by the general formula $R_1-CO-CH_2-CO-R_2$ where R_1 and R_2 are alkyl radicals, with at least one being branched, having from 1 to about 8 carbon atoms and a total of at least 5 carbon atoms. Preferred compounds include 2,2-dimethyl-3,5-hexanedione; 3,3-dimethyl-4,6-octanedione; 2,2,6-trimethyl-6-ethyl-3,5-octanedione; and 2,2,6,6-tetramethyl-3,5-heptanedione (thd). The most preferred compounds are the most highly branched β-diketones.

The useful cerium(IV) chelates are prepared from β-diketones represented by the formula $R_3-CO-CH_2-CO-R_4$ where R_3 and R_4 are alkyl, aralkyl or cycloalkyl radicals each containing from 1 to 12 carbon atoms with a total of 5 to about 20 carbon atoms in the radicals.

The gasoline fuel composition desirably contains from about 0.01 to 5 g per gallon of cerium as the cerium(IV) chelate and from about 0.5 to 20 g per gallon of the β-diketone synergist for an improved antiknock rating and preferably contains from about 0.05 to 2 g per gallon of cerium as the cerium(IV) chelate and from about 2 to 10 g per gallon of the β-diketone.

Examples 1 through 8: A series of motor fuel compositions were tested for octane rating by the motor method (MON by ASTM D2700) and the research method (RON) by ASTM D2699) using a clear commercial automotive gasoline having an MON of 83.2 and an RON of 91.0. The cerium(IV) chelate that was used was tetrakis(2,2,6,6-tetramethyl-3,5-heptanedionato)cerium or $Ce(thd)_4$. The results of these experiments are set out in the following table in which each listed difference in octane numbers is based on consecutive, matched determinations with the β-diketone present and absent.

Ex.	Ce (g/gal)	β-diketone	g/gal	ΔMON	ΔRON
1	0	thd*	4.0	0	0
2	0.5	thd*	4.0	+0.6	+1.4
3	0.5	thd*	4.0	+0.2	+1.1
4	1.0	thd*	1.0	+0.3	+0.3
5	1.0	thd*	2.0	+0.1	+0.6
6	1.0	thd*	4.0	+0.3	+1.0
7	1.0	2,2-dimethyl-3,5-hexanedione	4.0	+0.4	+0.2
8	0.2	2-methyl-4,6-heptanedione	2.0	0	+0.6

*thd is 2,2,6,6-tetramethyl-3,5-heptanedione.

Halogenated Substituted Fulvenes

Many nitrogen compounds, which exhibit excellent antiknock activity, are not in commercial use because of alleged carcinogenicity or formation of nitrous oxide combustion products. One type of nonnitrogen-containing compound that has antiknock activity is substituted fulvenes. It is an object of the work described by *R.M. Parlman and L.M. Fodor; U.S. Patent 4,264,336; April 28, 1981; assigned to Phillips Petroleum Company* to improve the antiknock activity of unsubstituted fulvenes by halogenating the substituted fulvenes and utilizing the halogenated substituted fulvenes as ashless antiknock additives to increase the octane number of fuels for internal combustion engines to a greater extent than is possible with the use of only substituted fulvenes as the ashless antiknock additives. The halogenated substituted fulvenes are particularly applicable for increasing the octane number of unleaded fuels for internal combustion engines.

The halogenated substituted fulvenes useful as gasoline antiknock compounds are characterized by the formula

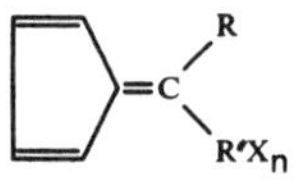

R
=C
$R'X_n$

where R can be hydrogen or any hydrocarbyl radical; R' can be any aromatic radical; X can be any halogen atom, preferably fluorine or chlorine; and n can be 1, 2, or 3. The hydrocarbyl radical will preferably have from 1 to 10 carbon atoms. The aromatic radical will preferably have from 6 to 10 carbon atoms. Some examples of halogenated substituted fulvenes include the following and mixtures of two or more thereof:

6-Methyl-6(o-fluorophenyl)fulvene
6-Methyl-6(m-chlorophenyl)fulvene
6-Methyl-6(p-bromophenyl)fulvene
6-Methyl-6(p-iodophenyl)fulvene
6-Ethyl-6(p-chlorophenyl)fulvene
6-Propyl-6(p-fluorophenyl)fulvene
6-Decyl-6(p-chlorophenyl)fulvene
6-Cyclohexyl-6(p-fluorophenyl)fulvene
6-Phenyl-6(p-chlorophenyl)fulvene
6-Methyl-6(o,p-dichlorophenyl)fulvene
6-Methyl-6(2,3,4-trifluorophenyl)fulvene
6-Methyl-6(2,4,6-trichlorophenyl)fulvene

Any method known in the art may be utilized to prepare the halogenated substituted fulvenes of the process. Preferably, the halogenated substituted fulvenes are prepared by the condensation reaction between cyclopentadiene and either a halogenated aryl ketone or a halogenated aryl aldehyde.

Suitable halogenated aryl ketones include o-, m- or p-fluoroacetophenone; o-, m- or p-fluoropropiophenone; o-, m- or p-fluorovalerophenone; and the like and the corresponding chloro, bromo, and iodo derivatives. Suitable halogenated aryl aldehydes include o-, m- or p-fluorobenzaldehyde; o-, m- or p-chlorobenzaldehyde; o-, m- or p-bromobenzaldehyde; o-, m- or p-iodochlorobenzaldehyde and the like.

Any desired amount of the halogenated substituted fulvenes may be combined with the fuel for an internal combustion engine. Generally, the amount of the halogenated substituted fulvenes employed will be in the range of about 0.01 molar (0.01 mol of halogenated substituted fulvenes in one liter of unleaded gasoline) to about 1 molar. Particularly good results are obtained when about 0.1 molar of the halogenated saturated fulvenes is added to the fuel for an internal combustion engine.

In the table are listed the results of antiknock tests employing the halogenated substituted fulvene compounds and the various comparative additives. The designation ΔRON is the difference in the RON of the base gasoline and the RON of the same base gasoline with the antiknock additive. Additive compounds in groups I and II were commercially available. The nonhalogenated substituted fulvene compounds were prepared from cyclopentadiene and the corresponding aceto- or propiophenone.

The results show that when a fluorine atom is added to toluene or aniline (Ib, IIb, IIc), both of which exhibit antiknock activity, the RON is decreased but when a fluorine atom is added to a substituted fulvene (IIIb, IVb) the RON is increased. The chlorine atom also increases RON as shown by comparing IIIa with V.

Effect of Halogen Atoms on Ashless Antiknock Compounds

	Additive Compound*	ΔRON
Ia	Toluene	0.2
Ib	p-Fluorotoluene	-0.6
IIa	Aniline	2.8
IIb	p-Fluoroaniline	2.5
IIc	o-Fluoroaniline	2.4
IIIa	6-Methyl-6-phenylfulvene	0.5
IIIb	6-Methyl-6(p-fluorophenyl)fulvene	0.9
IVa	6-Ethyl-6-phenylfulvene	0.3
IVb	6-Ethyl-6(p-fluorophenyl)fulvene	0.6
V	6-Methyl-6(p-chlorophenyl)fulvene	0.8

*0.1 molar in 1 liter of gasoline.

Aryl o-Aminoazides

R.J. Hartle and G.M. Singerman; U.S. Patent 4,266,947; May 12, 1981; assigned to Gulf Research & Development Company describe a class of metal-free antiknock agents which are capable of substantial improvement in the antiknock properties of a motor gasoline and which also provide a surprising improvement in the oxidation stability of the motor fuel.

These antiknock agents and antioxidants are aryl o-aminoazides. Surprisingly, the m- and p-aminoazides as well as the N-substituted o-aminoazides are proknocks, that is, they increase the knocking characteristics of motor gasoline. The o-aminoazides have the following general formula

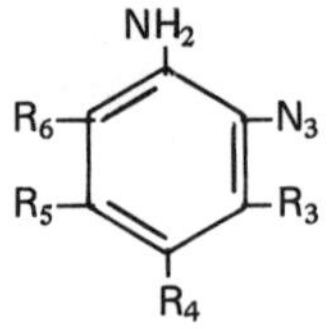

NH_2
R_6 N_3
R_5 R_3
R_4

wherein each R group is independently selected from hydrogen, alkyl having from 1 to about 4 carbon atoms, alkenyl having from 2 to about 4 carbon atoms, haloalkyl having from 1 to about 4 carbon atoms, aryl having from 6 to about 10 carbon atoms, cycloalkyl having from 5 to about 8 carbon atoms, alkoxy having from 1 to about 4 carbon atoms, carbalkoxy having from 2 to about 5 carbon atoms, carbaryloxy having from 7 to about 9 carbon atoms, alkylamino having from 1 to about 4 carbon atoms, acyl having from 2 to about 5 carbon atoms, hydroxyl, amino, cyano, nitro, halo and halophenyl, or wherein any two adjacent R groups containing a total of 4 carbon atoms are joined together in a saturated or unsaturated ring.

Compounds which are useful gasoline additives in addition to o-azidoaniline include the following:

2-Azido-3-, 4-, 5-, and 6-methylaniline;
2-Azido-3,4-, 4,5-, 5,6-, and 3,5-dimethylaniline;
2-Azido-3,4,5-, 3,4,6-, 3,5,6-, and 4,5,6-trimethylaniline;
2-Azido-3-ethyl-, 4-isopropyl-, 4-t-butyl-, 5-t-butyl- and 5-propylaniline;
2-Amino-3- and 4-azidostyrene;

2-Azido-4-, 5-, and 6-methoxyaniline;
2-Azido-4-fluoro-, chloro-, bromo- and iodoaniline;
2-Azido-4,5-difluoro-, 5,6-dichloro-, and 4,5-diiodoaniline;
2-Azido-4-chloro-5-methylaniline;
2-Azido-5-chloro-6-methylaniline;
2-Azido-3-fluoro-5-methylaniline;
2-Azido-3-fluoro-4-isobutylaniline;
2-Azido-4-iodo-5-methylaniline;
2-Azido-4-bromo-5-ethylaniline; etc.

The gasoline fuel composition desirably contains from about 0.1 to 50 g of the aryl o-aminoazide antiknock and oxidation stabilizing agent per gallon and preferably from about 0.5 to 15 g per gallon.

The aryl o-aminoazides can be prepared by the method of Smith et al, *J. Am. Chem. Soc.* 84, 485 (1962). Thus, in order to produce o-azidoaniline, o-nitroaniline is used as the initial reactant. In producing the ring-substituted o-azidoanilines, o-nitroaniline containing the desired ring substituent or substituents is used. In like manner, the polynuclear aryl o-aminoazides are prepared from the corresponding o-nitronaphthyl amines and the o-nitrotetrahydronaphthyl amines. A typical preparation is set out in the example.

Example 1: *Preparation of 2-azido-4-fluoroaniline* – A 25 g portion (0.16 mol) of 2-nitro-4-fluoroaniline was mixed with 24 g of phthalic anhydride and the mixture was heated at 180° to 210°C for 2 hours with slow stirring. After evolution of water was complete, the mixture was allowed to cool. It was ground to a powder and treated with three 100 cc portions of boiling ethanol. The residue was crude N(2-nitro-4-fluorophenyl)phthalimide (MP 238° to 240°C). A 36 g (0.126 mol) portion of this product was dissolved in 1 ℓ of acetone containing 108 cc of acetic acid and 108 cc of water. The solution was refluxed and treated with a total of 84 g of iron powder added in small portions. After 3 hours refluxing, the mixture was filtered while hot and the filtrate was neutralized with saturated sodium carbonate solution. The mixture was filtered and the filtrate poured into about 3 ℓ of ice water. N(2-amino-4-fluorophenyl)phthalimide precipitated out. This product melted at 190° to 193°C. The amine was diazotized in 1,500 cc of water containing 200 cc concentrated hydrochloric acid by treating with a solution of 12 g of sodium nitrite in 50 cc of water at 0° to 5°C. After stirring 3 hours at 0° to 5°C, the solution was filtered and the filtrate was treated dropwise with a solution of 8.2 g of sodium azide in 50 cc of water. After about 1 hour, evolution of nitrogen had stopped. The white solid was removed on a filter, washed with water, and vacuum dried. The product, N(2-azido-4-fluorophenyl)phthalimide decomposed with the evolution of nitrogen at 200° to 205°C.

This material was suspended in 325 cc of 95% ethanol and treated with 4.0 g of 95% hydrazine. The mixture was stirred for 2 hours at room temperature. The addition of 160 cc of water and 50 cc of 20% sodium hydroxide solution caused the solid to dissolve. This mixture was filtered into about 2 ℓ of ice water. The precipitate of 2-azido-4-fluoroaniline was removed on a filter, washed with water and vacuum dried (MP 44°C). It decomposes with evolution of nitrogen between 75° and 100°C.

Examples 2 through 7: A series of motor fuel compositions were tested for octane ratings by the motor method (MON by ASTM D2700) and the research method (RON by ASTM D2699) using a clear commercial automotive gasoline having an MON of 84.4 and an RON of 92.6. In these experiments various o-azidoanilines were tested using 4 g of the additive per gallon of gasoline in each test. The results of these experiments are set out in the table in which each listed difference in octane numbers is based on consecutive, matched determinations with the o-azidoaniline and the substituted o-azidoanilines present and absent.

Example	2-azidoaniline	ΔMON	ΔRON
2	Unsubstituted	+0.5	+1.2
3	4-methyl	+0.3	+0.8
4	6-methyl	0	+0.6
5	4,5-dimethyl	+0.4	+1.4
6	4-methoxy	+0.2	+1.1
7	4-fluoro	+0.3	+1.0

OCTANE IMPROVERS

When the air-fuel mixture is ignited in the cylinder of a gasoline engine, a flame front spreads out from the point of ignition and ideally this process should continue uninterruptedly until the air-fuel mixture has been completely burned. However, due to the rapid increase of pressure and temperature, which takes place in the air-fuel mixture as the flame front expands, other reactions can take place in the part of the fuel mixture not yet reached by the flame front. Such so-called delay reactions are believed to consist of temperature-pressure induced decomposition of the fuel molecules followed by a nearly instantaneous reaction with oxygen in the mixture. Such a reaction is in the nature of a detonation and results in shock waves that can impart dangerously high stresses on the engine as well as interrupting the orderly spread of the flame front. This phenomenon, commonly known as "knocking," reduces the efficiency of the engine and increases the probability of formation of fixed nitrogen (NO_x) as well as unburned hydrocarbons. Even when knocking is not readily apparent such detonations can take place locally in the air-fuel mixture but not to a degree that successively reflected shock waves reinforce each other to the point where the process becomes audible.

To counteract this phenomenon, so-called higher octane fuel is used, the octane rating being the measure of the fuel's ability to undergo compression without audible knocking during subsequent combustion, using as a reference 2,2,4-trimethylpentane (C_8H_{18}) rated at 100 octane.

```
             H
             |
      H   H-C-H   H     H      H
      |      |    |     |      |
    H-C------C----C-----C------C-H
      |      |    |     |      |
      H   H-C-H   H  H-C-H     H
             |          |
             H          H
```

Antiknock properties in the fuel are achieved by the use of certain additives, such as tetraethyl-lead, or the use of branched hydrocarbons (alkanes) which, due to the increased compactness of such molecules, are less susceptible to produce pressure-temperature induced detonation. The net effect of the above antiknock compound is that total air-fuel mixture culminates in a very high percentage burn.

The use of high octane fuel to eliminate knocking presents several drawbacks, including higher cost for such fuel over lower octane fuel and the presence of additives, such as tetraethyl-lead which is not compatible with catalytic convertors now used in most new automobiles to reduce the emission of noxious fumes.

Previous attempts to improve the efficiency of fossil fuel combustion equipment, such as the internal combustion engine, have been only marginally successful. Such measures have included water and steam injection devices, superchargers and the like. None of these, however, has been entirely satisfactory from the standpoint of substantial improvement in combustion efficiency, cost, simplicity, ease of installation and maintenance. Therefore, a simple, low-cost, efficient method and apparatus for raising the apparent octane rating of gasoline in an internal combustion engine would be of great value. Hence the interest in "octane improvers."

Organic Reactive Intermediates

R.C. DesMarais, Jr. and O.A. Sandven; U.S. Patent 4,149,853; April 17, 1979; assigned to Rigs Corporation have developed a process and associated apparatus for improving the efficiency of fossil fuel combustion equipment and raising the apparent octane rating of gasoline for use in internal combustion engines. The process comprises the steps of generating organic reactive intermediates and adding the reactive intermediates to the air-fuel mixture prior to ignition. The reactive intermediates are generated by first exposing a source compound of a volatile organic material, such as acetone, to ultraviolet radiation to produce a short-lived free radical. The short-lived free radical is then reacted with heavier hydrocarbons such as in the C_{16-20} range to produce a long-lived secondary free radical. The secondary free radicals are then injected into the air-fuel mixture prior to ignition.

Apparatus for generating the reactive intermediate compounds is also provided, comprising a container adapted to store a quantity of a suitable source compound, such as acetone, an ultraviolet radiation source communicating with the container for exposing the compound to radiation and thereby produce short-lived free radicals, a reaction stage including a source of heavy hydrocarbons, preferably in the C_{16-20} range, for converting the short-lived free radicals into long-lived free radicals, and means for delivering the long-lived free radicals to the air-fuel combustion mixture prior to ignition.

In accordance with the process, organic reactive intermediates are generated from a source comprised of a volatile, organic compound, such as acetone, by first exposing the compound to ultraviolet radiation and, thereby, producing short-lived free radicals. The action of the ultraviolet radiation upon the acetone in the gas phase breaks down some of the acetone molecules into free radicals.

(1)
$$\begin{matrix} & H & O & H & \\ & | & \| & | & \\ H- & C- & C- & C & -H \\ & | & & | & \\ & H & & H & \end{matrix} \rightarrow \begin{matrix} & H \\ & | \\ H- & C^{\circ} \\ & | \\ & H \end{matrix} + \begin{matrix} & H & O \\ & | & \| \\ H- & C- & C^{\circ} \\ & | & \\ & H & \end{matrix}$$

The acetone source compound has the properties of being sufficiently volatile and capable of forming suitable reactive intermediates of a nature beneficial to the combustion process. In practice, the desired reaction is achieved by ultraviolet radiation with a wavelength whose maximum peak is 24 nm (nanometers).

The above relatively unstable, short-lived (10^{-3} to 10^{1} seconds) free radicals are then reacted with alkanes with carbon chains in the range of C_{16-20} to convert the short-lived, UV-induced, free radicals to relatively stable, longer-lived (10 seconds to ±24 hours) radicals which are formed by the short-lived free radicals combining with the heavier hydrocarbon molecules.

These secondary radicals are then combined with the air-fuel mixture as in the intake system of an internal combustion engine, or into the firebox of a furnace, for example. A relatively small and almost trace amount of the reactive intermediates is required to produce satisfactory results when combined with the air-fuel mixture.

While acetone has been found to be a very satisfactory source compound because it is readily available and low in cost, other compounds may also be used to advantage, such as ethene, ethanol, 1,3-butadiene, 1,3,5-hexatriene and others.

Referring to Figure 5.1, there is illustrated a block diagram of a system embodied in an internal combustion engine which typically may be installed on a motor vehicle.

Figure 5.1: Block Diagram for Use of Reactive Intermediates in an Internal Combustion Engine

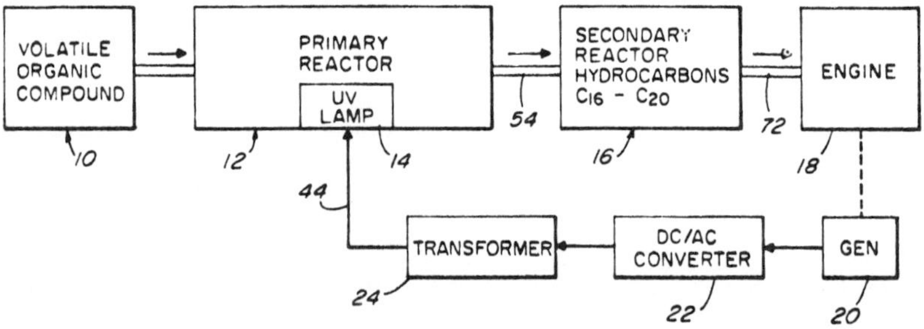

Source: U.S. Patent 4,149,853

As shown, this system includes a source **10** of a volatile organic compound, such as set forth above, and typically may be acetone, for example. The vapors from the acetone are fed from the source to a primary reactor **12** where they are exposed to radiation from an ultraviolet lamp **14** whose maximum peak preferably is on the order of 280 μ. The action of the ultraviolet radiation upon the acetone in the gas phase breaks down some of the acetone molecules into primary, short-lived, free radicals as shown in (1) above. The above free radicals are then carried to a secondary reactor **16** where the short-lived free radicals produced in the primary reactor are combined with hydrocarbons typically in the C_{16-20} range to form secondary, long-lived free radicals. The secondary reactor **16** typically includes alkanes with carbon chains in the C_{16-20} range to provide the hydrocarbon molecules which produce the more stable, free radicals. The secondary radicals produced in the secondary reactor **16** are then delivered to an engine **18**, typically to the carburetor portion or its equivalent, to mix with the air-fuel mixture prior to ignition. The engine **18** is provided with a generator **20**, the output of which may be used, at least in part, to drive a DC to AC convertor **22** which, in turn, operates a transformer **24** as the power supply for the UV lamp **14**.

The patent contains detailed drawings and descriptions of the generator for making reactive intermediates and views of the system in use with a furnace.

Preparation of Methyl tert-Butyl Ether

In recent years, there has arisen the problem of environmental pollution with lead which is caused by the exhaust gases of internal combustion engines. In order to produce unleaded gasoline while maintaining octane value as in the past and without varying the compounding ratio of basic gasoline, an octane value improving agent must be added to gasoline.

There are known many kinds of compounds as the octane value improving agents. Particularly, ethers having branched alkyl groups are well-known. For example, it is well-known that the octane values of methyl tert-butyl ether (hereinafter referred to as MTBE), ethyl tert-butyl ether and isopropyl tert-butyl ether are very high.

With regard to the method for producing MTBE, it is known that MTBE is prepared by the reaction between methyl alcohol and isobutylene in the presence of a catalyst. Proposed in some methods is the use of strongly acidic cation-exchange resin as a catalyst. In this method, however, an acid substance is extruded from the strongly acidic cation-exchange resin during the reaction. In addition, during the heating necessary for distillation the yield of MTBE is disadvantageously reduced.

Therefore, the objective of *T. Takezono and Y. Fujiwara; U.S. Patents 4,182,913; January 8, 1980; and 4,256,465; March 17, 1981; both assigned to Nippon Oil Company Ltd., Japan* is to eliminate the abovedescribed disadvantages in the conventional art and to provide an industrially serviceable method for producing MTBE which can be mixed into a fuel composition for internal combustion engines.

According to their process, a mixture of isobutylene-containing hydrocarbon mixture and methyl alcohol at a molar ratio (isobutylene:methyl alcohol) of 1:0.6 to 1:1.4 is used. The mixture is passed through a fixed bed which is filled with the particles of strongly acidic cation-exchange resin at a temperature in the range of 0° to 100°C, a liquid space velocity of 0.1 to 50 (1/hr) and a pressure of 1 to 50 atm. The cation-exchange resin has a mean grain diameter of 0.2 to 10 mm. The reaction mixture obtained through the resin bed is then passed through a fixed bed filled with a water-insoluble solid particulate acid-neutralizing agent having a mean grain diameter of 0.1 to 10 mm at a temperature of 0° to 100°C. The reaction mixture is further introduced into a flashing tower so as to flash-remove unchanged hydrocarbon, thereby obtaining a mixture containing MTBE from the bottom of the flashing tower.

The above mixture containing MTBE may be further subjected to distillation in a multistage distillation column, thereby continuously obtaining MTBE from the bottom of the distillation column and an azeotropic mixture of MTBE and methyl alcohol from the top of the column, which azeotropic mixture is then recycled to the initial reaction vessel of the cation-exchange resin tower. A fuel composition for internal combustion engines is then prepared by mixing 2 to 30% by volume of the above obtained MTBE into the internal combustion fuel.

The strongly acidic cation-exchange resins used in this method are cation-exchange resins which may be the styrene sulfonic acid type or the phenol sulfonic acid type.

The water-insoluble, solid particulate acid-neutralizing agent is a member selected from the group consisting of hydrotalcite, magnesium oxide, alumina, silica, silica-alumina, double oxides containing magnesium, aluminum and/or silicon, and their hydrates. Preferably, the neutralizing agent is a natural hydrotalcite having a molar ratio (magnesium to aluminum) of 3 to 1 or a synthesized hydrotalcite in which the molar ratio (magnesium to aluminum) is 1 to 10.

Figure 5.2 is a flow diagram of the process. The characteristic features of the process will be further made clear with some examples in which the bold numbers refer to Figure 5.2.

Figure 5.2: Process for Preparing Methyl tert-Butyl Ether

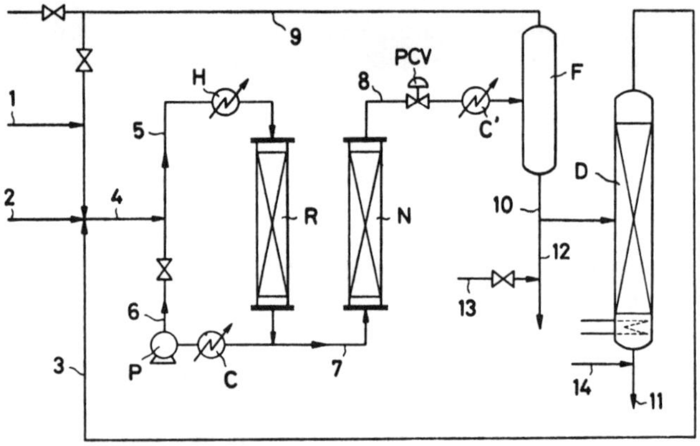

Source: U.S. Patent 4,182,913

Example 1: A reaction tower **R** was filled with 30 ℓ of catalyst, styrene type ion exchange resin (Amberlyst 15). A neutralizing tower **N** was filled with 20 ℓ of hydrotalcite ($6MgO{\cdot}Al_2O_3{\cdot}CO_2{\cdot}12H_2O$ having a mean grain diameter of 0.7 mm). Isobutylene (purity: 99%) was fed through a pipeline **1** at a flow rate of 105.0 kg/hr and methyl alcohol (purity: 99%) was fed through a pipeline **2** at a flow rate of 58.5 kg/hr (1.83 kg-mol/hr). Unreacted isobutylene was recycled for reuse through a pipeline **9** at a flow rate of 28.3 kg/hr. Thus the flow rate of isobutylene as a whole in the pipeline **4** was 133.3 kg/hr (2.38 kg-mol/hr).

The liquid space velocity of the raw materials was 10 (1/hr). The pressure in the reaction system was maintained at 15 kg/cm^2 gauge with a pressure controlling valve **PCV**. The mixed raw materials of isobutylene and methyl alcohol were combined with the fluid from the recycled pipeline **6** and the mixture was then fed into the reaction tower **R** through a pipeline **5**.

The temperature of the inlet of reaction tower **R** was set to 50°C by means of a heat exchanger **H**. The flow rate of the fluid in the pipeline **6** was controlled to 7 times as much as the flow rate of raw material feed through the pipeline **4**, by means of a circulation pump **P**. The flow rate of the fluid from the reaction tower **R** through the pipeline **7** was 191.8 kg/hr, the fluid contained 85.2% by wt of MTBE, 14.0% by wt of unreacted isobutylene, 0.8% by wt of isobutylene dimer and 0% of unreacted methyl alcohol, and the concentration of acid in this fluid was 3.5×10^{-4} eq/ℓ.

The fluid was passed through the pipeline **7** into the neutralizing tower **N** and further through the pipeline **8** into the flashing tower **F** at normal pressure. Unreacted isobutylene was flashed off from the top of the flashing tower **F** and it was supplied into the raw material isobutylene line through the pipeline **9**. From the bottom of the flashing tower **F**, MTBE-containing mixture of 99.1% in purity was obtained at a flow rate of 160 kg/hr. The impurity in the product was isobutylene dimer and the acid concentration was very low at 1.4×10^{-7} eq/ℓ.

Example 2: To the MTBE-containing mixture obtained in Example 1 was added 1,926 ℓ/hr of regular gasoline through the pipeline **13**, thereby obtaining a fuel composition for internal combustion engines which contained 10% by volume of MTBE-containing mixture. The research method octane value of the fuel composition was 95.

Process for Making a Hydrocarbon Fraction Containing tert-Amyl Methyl Ether

It is well-known in the art of preparing gasolines for internal combustion engines that various dialkyl ethers can be used to improve the octane ratings of the gasolines. The two major sources of branched chain olefins to make ethers for inclusion in blended gasolines (gasoline pools) are the light catalytically cracked gasoline fraction (LCCG) from gas oil cracking operations and the partially hydrogenated pyrolysis gasoline fraction (HPGB) or "dripolene" from steam cracking of naphtha or heavier distillate fractions. In the prior art of converting these branched chain olefins (specifically tertiary olefins) or ethers for inclusion in gasoline pools, it has been the general practice to etherify the maximum possible proportion of the tertiary olefins with the minimum amount of processing, for example, by etherifying in the presence of excess primary alcohol and/or by etherifying the tertiary olefins in admixture with one another as well as with all the other hydrocarbons occurring in olefinic LCCG or HPGB gasoline fractions. The results have been less than satisfactory, primarily because of the problems involved in separating, from the etherifying reaction effluent, any material which does not beneficially go directly into a gasoline pool or blending tank.

J.D. Chase, H.J. Woods and B.W. Kennedy; U.S. Patent 4,193,770; March 18, 1980; assigned to Gulf Canada Limited, Canada have found that the 4 and 5 carbon tertiary olefin content of effluents from hydrocarbon cracking operations are more beneficially etherified for octane improvement of blended gasoline by fractionally distilling the cracked effluent to separate therefrom two particular individual streams, one portion containing predominantly hydrocarbons of 4 carbon atoms (C_4 stream) and the other predominantly hydrocarbons of 5 carbon atoms (C_5 stream) and etherifying only the tertiary olefins in the two portions containing predominantly 4 and 5 carbon atom hydrocarbons respectively, each of the two portions being etherified in a manner preferred for the predominant tertiary olefin in the portion.

By this procedure, a cracked hydrocarbon stream containing a range of tertiary olefins which can be etherified is fractionated to provide a first portion containing predominantly C_4 hydrocarbons, a second portion containing predominantly C_5 hydrocarbons, and a remainder of which an appropriate part can be passed directly to a gasoline pool. The foregoing C_4 hydrocarbon portion containing isobutylene is then etherified with methanol in any of the known ways of etherifying such isobutylene-rich fractions, the conventional processes generally providing excellent conversions of isobutylene and yields of methyl tertiary butyl ether (MTBE). This ether has a high octane blending number and is a highly advantageous ingredient for addition to a gasoline pool. The entire etherified C_4 portion containing the ether can be blended directly into a gasoline pool, or more advantageously, the MTBE can be separated from the stream and added to a gasoline pool while the unreacted ingredients of the C_4 stream are diverted to other applications, for example an alkylation process for alkylate gasoline preparation.

According to this process, the predominantly C_5 hydrocarbon portion, separated from a cracked hydrocarbon effluent stream, boiling at atmospheric pressure in the range from 80° to 122°F (27° to 50°C), and containing the isomers 2-methyl butene-1 and 2-methyl butene-2, is etherified with methanol separately from the etherification of the C_4 fraction in a processing sequence by which optimum conversions of the foregoing isomers and optimum yields of tert-amyl methyl ether (TAME) are obtained. It may be noted in passing that the 3-methyl butene-1 isomer also is present in the C_5 hydrocarbon fraction but does not etherify to form TAME and is substantially inert under these reaction conditions. It can also be noted that, although several alcohols may be used to etherify 2-methyl butenes, methanol is the most practicable, particularly from an economic standpoint, and therefore is used in this process.

The process for preparation of gasoline containing tert-amyl methyl ether (TAME) comprises:

> Separating, from lower and higher boiling compounds, an olefinic hydrocarbon portion boiling at atmospheric pressure in the range from 80° to 122°F (27° to 50°C) and containing a mixture of hydrocarbons having predominantly 5 carbon atoms each, of which hydrocarbons at least 10% are 2-methyl butenes;
>
> Passing the olefinic hydrocarbon portion, together with methanol in a proportion of from 0.5 to 3.0 mols of methanol per mol 2-methyl butenes present, into contact with a bed of solid acidic etherifying catalyst in a reactor at temperature in the range from 150° to 240°F (66° to 116°C) under pressure sufficient to maintain the passing material in the liquid phase, contact being of sufficient duration to etherify from 15 to 60% of the 2-methyl butenes in the passing material during the contact;

Passing a proportion of the effluent stream from the reactor to a distillation column wherein it is fractionally distilled under reflux and wherein a bottom fraction containing substantially all the ethers entering the column is withdrawn from the bottom of the column and a distillate fraction is withdrawn from the top of the column;

Blending the ether containing the bottom fraction and any remainder of the effluent stream passing directly from the reactor into a gasoline product;

Recycling a proportion of the distillate fraction from the top of the column to the reactor for additional contact with the catalyst; and

Passing a proportion of the distillate fraction into the gasoline product when the entire effluent stream from the reactor is passed to the distillation column.

In preferred embodiments of the process, a proportion of from 10 to 85% of the effluent stream from the reactor is passed to the distillation column for fractional distillation, although this proportion can be as high as 100% of the effluent. When a proportion of 100% of the reactor effluent is passed to the distillation column, it becomes necessary to bleed off a proportion, for example from 20 to 30%, of the distillate from the column into the gasoline product in order to prevent accumulation of volatile material in the system; when a significant proportion of the ether containing reactor effluent is withdrawn and blended directly into the gasoline product, it is not necessary to bleed off any proportion of the distillate and the entire distillate is recycled to the reactor. The size of the preferred proportion of the effluent stream from the reactor to be passed to the distillation column depends largely on the extent to which it is desired to increase the conversion of 2-methyl butenes into TAME at the cost involved in recycling unreacted distillate through the reactor.

(Throughout this patent the percentages and proportions referred to are percentages and proportions by weight unless otherwise specifically indicated.)

In preferred embodiments, the mol proportion of methanol per mol of 2-methyl butenes in the hydrocarbon feed stream is in the range from 0.7 to 1.5 and most preferably it is 0.9 to 1.1. More particularly, it is preferred that the proportion of methanol to 2-methyl butenes in the feed to the reactor be kept low enough so that the proportion of methanol in the reactor effluent also is appropriately low, for example from 2 to 9%, and/or correspondingly the proportion of methanol in the distillation column bottom fraction, containing substantially all the ethers, is appropriately low, for example below 5%. The efficiency of the separation in the distillation column obviously affects the proportion of free methanol in the distillation bottom fraction.

The liquid hourly space velocity (LHSV) of the reactor feed in contact with catalyst in the process (and accordingly the contact time) is preferably adjusted to achieve etherification of from 25 to 50% of the 2-methyl butenes in the streams being fed to the reactor, most preferably from 32 to 40%. Space velocities (LHSV) in the range from 1.5 to 4 are preferred. Pressure during the reaction must be sufficient to maintain the reactants in the liquid phase, and thus may vary with reaction temperature; generally from 5 to 10 atm is adequate, and the desired LHSV is readily achieved. The preferred reaction temperature for etherification is in the range 165° to 225°F (74° to 107°C), most preferably 175° to 200°F (79° to 94°C).

Solid acidic etherifying catalysts are well-known in the art, and include the sulfonated, crosslinked, polystyrene ion-exchange resins in their acid form, for example Dowex 50, Amberlyst 15, Ionac C-252, Rexyn 101(H), and Nalcite HCR. The foregoing commercial products are the preferred catalysts.

Preparation of Phenyl tert-Butyl Ether

The objective of the process described by *W.M. Sweeney and F.S. Bove; U.S. Patent 4,244,704; January 13, 1981; assigned to Texaco Inc.* is to provide a gasoline composition having an improved octane number by the addition of a small amount of phenyl tert-butyl ether.

The ether in question may be prepared by (1) reaction of Grignard reagent phenyl magnesium bromide, in ethyl ether, with tert-butyl perbenzoate, or (2) by Williamson synthesis reaction of sodium phenate and tert-butyl chloride, etc. Phenyl tert-butyl ether is a liquid.

The ether may be added to the gasoline in an amount of typically 1 to 10%, say 2 to 5%, by volume of the gasoline.

Example 1: A Grignard reagent, phenyl magnesium bromide, is formed by mixing 100 g magnesium turnings with 607.7 g bromobenzene and 6,170 g anhydrous ethyl ether. After formation of the Grignard reagent, tert-butyl perbenzoate (449 g) in ether is added at 15° to 25°C under nitrogen with stirring. Reaction product is shaken with cold, dilute hydrochloric acid, then washed with dilute aqueous sodium hydroxide, and then washed with water. It is then dried and distilled to yield 269.2 g of product phenyl tert-butyl ether which is analyzed to give 79.1% C (calc 80.0%) and 9.2% H (calc 9.3%).

Examples 2 through 7: In this series of examples, the phenyl tert-butyl ether is tested for octane appreciation in a lead-free gasoline. Other octane appreciators were also tested. In each case the ether appreciator was added in an amount of 10 volume percent and the increase in research octane number (RON) and motor octane number (MON) was measured. The charge gasoline was a nominal 91 octane, lead-free gasoline.

Example	Additive (Ether)	ΔRON	ΔMON
2	Phenyl tert-butyl	3.5	2.8
3-4	Methyl and ethyl tert-butyl	2.9	1.8
5	Methyl tert-amyl	2.6	1.5
6	Diisopropyl	1.5	2.0
7	Methyl tert-hexyl	0.2	1.0

It will be apparent from inspection of the table that a significantly greater increase in MON and RON is achieved with phenyl tert-butyl ether than with any of the other materials.

Preparation of Methyl tert-Butyl and Methyl tert-Amyl Ethers

S. Herbstman; U.S. Patent 4,252,541; February 24, 1981; assigned to Texaco Inc. has developed a process for preparing ethers, particularly unsymmetrical ethers, in high yield and purity from hydrocarbon streams produced during petroleum refining. The process is discussed at length and consists of the steps of isomerization, separation, dehydrogenation, etherification, washing and separation.

The product ether prepared by this process may be a pure product, e.g., methyl tert-butyl ether, methyl tert-amyl ether, etc. In one preferred embodiment however, it is a mixture containing 40 to 70 parts, say 58 parts of methyl tert-butyl ether and 50 to 90 parts, say 84 parts of methyl tert-amyl ether.

This preferred mixture of product ethers permits attainment of desired product gasolines. Specifically it is found that the use of pure methyl tert-butyl ether in gasoline may contribute to undesirable increase in gasoline relative vapor pressure (RVP). Use of this combination of ethers permits attainment of increased octane numbers in gasolines without any undesirable increase in RVP. This may be particularly useful in preparation of gasolines suitable for use in summer–in which a high RVP is to be avoided.

The preferred ether mixtures of this process may include 30 to 60%, say 41 wt % of methyl tert-butyl ether and 50 to 70%, say 59 wt % of methyl tert-amyl ether. This may correspond to a weight ratio of the butyl ether to the amyl ether of 0.6 to 0.86, preferably 0.65 to 0.75, say 0.69.

Preferred product gasoline compositions may include 1,000 to 2,200 parts, say 1,000 parts of gasoline, 40 to 70 parts, say 41 parts of methyl tert-butyl ether, and 50 to 90 parts, say 59 parts, of methyl tert-amyl ether. The most preferred product ether mix contains 57 parts (41 wt %) of methyl tert-butyl ether and 83 parts (59 wt %) of methyl tert-amyl ether.

It is a particular feature of this preferred blend of ethers that it unexpectedly permits maximum improvement in properties of blended gasolines.

It is found that improvements are attained in research octane number (RON) and motor octane number (MON) when using a gasoline containing either the methyl tert-butyl ether (MTBE) or methyl tert-amyl ether (MTAE) or mixtures of e.g., 41 wt % of the former and 59 wt % of the latter. In comparative tests wherein each of MTBE, MTAE, and the mixture of both were present in a gasoline in an amount of 10% volume of the gasoline, the clear octane numbers and the RVP (psi at 77°F) are as noted in the following table.

Composition	RVP	RON	Change RON	MON	Change MON
No ether	8.4	92.1	–	83.7	–
MTBE	9.1	94.9	2.8	85.5	1.8
MTAE	8.4	94.2	2.1	85.3	1.6
Mixture	8.4	94.7	2.6	85.4	1.7

From the above table, the following will be noted:

Use of MTBE alone gives an increase in RON of 2.8 units and in MON of 1.8 units–however it undesirably increases the RVP to 9.1.

Use of MTAE alone permits maintenance of desired RVP (of 8.4 psi) but yields a lower RON and MON than may be achieved with MTBE alone.

Use of the preferred mixture of MTAE and MTBE permits attainment of the following desirable and unexpected results:

Attainment of RON which is greater than that attained with MTAE and about as good as that attained with MTBE alone;

Attainment of MON which is greater than that attained with MTAE and almost equal to that attained with MTBE alone; and

Attainment of product gasoline mixture unexpectedly characterized by RVP of 8.4 which is substantially equal to that of gasoline containing no ether.

It is unexpected to find that it is thus possible to achieve such increases in octane number coupled with the ability to maintain the RVP constant. This is particularly important for example in preparing gasolines to be used for summer driving—which gasolines require lower RVP. The ability to prepare gasolines characterized by improved octane and constant volatility represents a desideratum which is attained to maximum advantage by this technique.

Another additive which improves the octane rating of fuels in which it is used is described in Chapter 7 (see U.S. Patent 4,244,703). Compounds which prevent the ORI in nonleaded fuels are described under a special heading in Chapter 1.

-6-

CORROSION AND OXIDATION INHIBITORS

ANTICORROSION ADDITIVES FOR FUELS AND LUBRICATING COMPOUNDS

Certain Alkenylsuccinic Ester Acids and Their Amine Salts

H.J. Andress, Jr.; U.S. Patent 4,148,605; April 10, 1979; assigned to Mobil Oil Corporation has found that certain alkenylsuccinic ester acids and amine (e.g., tolyltriazole) salts thereof, are effective rust or corrosion inhibitors, and organic compositions containing them substantially prevent and/or reduce the formation of rust on metallic surfaces with which the compositions are in contact. These can be used for preventing and/or inhibiting the formation of rust on metal surfaces in contact with organic media such as petroleum distillate hydrocarbon fuel oils, gasoline, naphthas, various burning oils, diesel fuel oil, furnace oils, oils of lubricating viscosity and greases derived from both synthetic and mineral oil base stocks.

These antirust additives are compounds having the following general formula:

$$\begin{array}{l} R{-}CH{-}COOH \\ \quad\;\; | \\ \quad\;\; CH_2{-}COO{-}\underset{\displaystyle R^1}{\underset{|}{CH}}{-}COOH \end{array}$$

where R is alkenyl having from about 4 to 24 carbon atoms and R^1 is selected from the group consisting of H, C_1 to C_{16} alkyl, aryl, alkaryl, carboxy, C_1 to C_{16} alkyl carboxy and combinations thereof having up to about 16 carbon atoms. R preferably contains from about 8 to 16 carbon atoms and R^1 is preferably alkyl and contains from 1 to about 4 carbon atoms.

Generally, the compounds are prepared by condensing an appropriate alkenylsuccinic anhydride with a suitable aliphatic hydroxy acid, e.g., lactic acid, under ambient pressure (although the reaction may be carried out under pressure if desired) at a temperature of from about 105° to 130°C. Usually no solvent is used but any suitable hydrocarbon diluent, e.g., xylene may be advantageously used. The amine salts thereof are then prepared by reacting the ester acid under similar temperature and pressure conditions with an amine, e.g., tolyltriazole, in a mol ratio of amine to ester acid of from about 1:1 to 2:1.

Lactic acid is the preferred aliphatic hydroxy acid for use in this formulation and benzotriazole and tolyltriazole are the preferred amines. The preferred succinic anhydride is dodecenylsuccinic anhydride.

The amount of additive added to a particular composition will vary dependent on the intended use of such composition and the specific nature of the organic media. Usually, however, the compositions will contain from about 0.1 to 100.0 lb of the additive per 1,000 bbl of the total composition.

Example 1: A mixture of 266 g (1 mol) of dodecenylsuccinic anhydride and 90 g (1 mol) lactic acid was gradually heated to 120° to 125°C with stirring and held there for about 4 hours to form the ester acid.

Example 2: A mixture of 178 g (0.5 mol) of the product from Example 1 and 67 g (0.5 mol) of tolyltriazole was stirred and gradually heated to 110°C and held there for about 2 hours and then further heated to 125°C and held there for about 2 hours to form the final product–a monoamine salt.

Example 3: A mixture of 178 g (0.5 mol) of the product from Example 1 and 133 g (1.0 mol) of tolyltriazole was stirred with gradual heating to 115°C for about 3 hours to form the final product–a diamine salt.

The compounds of Examples 1 through 3 were then tested in a gasoline blend in accordance with ASTM Rust Test D-665 with results as given in the table below.

The inhibitors were blended in a gasoline comprising 40% catalytically cracked component, 40% catalytically reformed component, and 20% alkylate–approximately 90° to 410°C boiling range.

ASTM Rust Test D-665

Fuel	Additive, PTB	Rust Rating
Base fuel	0	Heavy rust
Base fuel plus		
Example 1	1	No rust
Example 2	1	No rust
Example 3	1	No rust

The results clearly demonstrate that the compounds in accordance with the process completely prevented the formation of rust while the gasoline blend without the additive resulted in the formation of heavy rust.

Oxazoline Reaction Products

J. Ryer, H.N. Miller, J. Zielinski and S.J. Brois; U.S. Patent 4,157,243; June 5, 1979; assigned to Exxon Research & Engineering Co. describe oxazoline reaction products of hydrocarbon substituted dicarboxylic acid, ester, or anhydride, for example, octadecenylsuccinic anhydride, with 2,2-disubstituted-2-

amino-1-alkanols, such as tris(hydroxymethyl)aminomethane (THAM), and their derivatives which are useful antirust agents for gasoline.

Such compounds are also useful as sludge dispersants and are described for use in that role in U.S. Patent 4,195,976 (which see). It is also of note that these compounds may be stabilized against oxidation with sulfur, a process described in U.S. Patent 4,263,153 (which see).

Example: 584 g (1.5 mols) of 90% active 2-octadecenylsuccinic anhydride and 4.0 g of zinc acetate dihydrate were slurried in 200 ml of tetrahydrofuran. To this was added 363 g (3.0 mols) of 2-amino-2-(hydroxymethyl)-1,3-propanediol. The temperature was gradually raised to 160°C effectively removing most of the solvent. Following this the mixture was heated to 208°C over a 3-hour period during which water distilled. The contents were found to have a prominent absorption band at about 60 μ. Recrystallization of the crude reaction product gave a white solid which melted at 104° to 106°C. The recrystallized product showed the following analysis: 66.80% by weight carbon (calculated 66.83% by weight); 9.86% by weight hydrogen (calculated 10.10% by weight); 4.59% by weight nitrogen (calculated 5.20% by weight); and 18.40% by weight oxygen (calculated 17.82% by weight). Calculated values were based on $C_{30}H_{54}N_2O_6$.

This product was tested for its effectiveness as a gasoline antirust agent. It was first dissolved in xylene, and the solution was added to the gasoline to incorporate the additive at a treat rate of 10 lb of oxazoline additive per 1,000 bbl of gasoline, i.e., about 0.024% by weight. The gasoline so treated was then tested for rust according to ASTM D-665M Rust Test. In brief, this test is carried out by observing the amount of rust that forms on a steel spindle after rotating for 1 hour in a water-gasoline mixture. In this case, the oxazoline treated gasoline gave no rust indicating that it was very effective as an antirust additive since the untreated gasoline will give rust over the entire surface of the spindle.

The oxazoline reaction products of the process which are primarily useful as an antirust additive and/or detergent for gasoline will generally have hydrocarbyl substituents having from about 12 to 49 carbons (preferred is about 18 as exemplified by 2-octadecenyl); whereas, for applications as a dispersant or detergent in lubricants it is preferred that the hydrocarbyl substituents have from about 30 to 49 carbons, e.g., 35 to 45 carbons.

Magnesium-Containing Dispersions from Magnesium Carboxylates

W.J. Cheng and D.B. Guthrie; U.S. Patents 4,163,728; August 7, 1979; 4,179,383; December 18, 1979; and 4,229,309; October 21, 1980; all assigned to Petrolite Corporation describe a process for preparing stable, fluid magnesium-containing dispersions by the high-temperature decomposition of magnesium salts (or magnesium hydroxide) to magnesium oxide.

In U.S. Patent 4,229,309, the process comprises heating $Mg(OH)_2$ above its dehydration temperature in the presence of a fluid of low volatility containing a dispersing agent soluble in that fluid.

This process, in essence, comprises an almost explosive dehydration of magnesium hydroxide to magnesia according to the equation

$$Mg(OH)_2 \rightarrow MgO + H_2O$$

During this dehydration, $Mg(OH)_2$ is disintegrated into minute particles of MgO which are immediately suspended and become stabilized in the fluid by the presence of a dispersing agent.

In U.S. Patent 4,179,383, magnesium salts of carboxylic acids (magnesium carboxylates) in a dispersant-containing fluid are also explosively decomposed to magnesia. During this decomposition, the magnesium carboxylate is disintegrated into minute particles of MgO which are immediately suspended and stabilized in the fluid by the presence of a dispersing agent.

However, it is to be noted that a stoichiometric amount of carboxylic acids, based on $Mg(OH)_2$, or equivalent, is employed in forming these magnesium carboxylates.

U.S. Patent 4,163,728 uses a process essentially the same as that of U.S. Patent 4,179,383, except that less than stoichiometric amounts of the carboxylic acid are used. Since a low stoichiometric amount of carboxylate is employed, the reactant is in essence a mixture of magnesium carboxylate and magnesium hydroxide.

Any suitable carboxylic acid at low stoichiometry can be employed. These include mono- and polycarboxylic acids including aliphatic, aromatic, cycloaliphatic, etc., carboxylic acids. Representative examples include: formic acid, acetic acid, propionic acid, butyric acid, acrylic acid, maleic acid, etc. In view of the high volatility of aliphatic carboxylic acids, those of the formula $H(CH_2)_nCOOH$, where n is from 0 to 5, especially 1, are preferred.

The fluid dispersions of MgO from magnesium carboxylate-magnesium hydroxide mixtures are achieved at the same low temperature range as the dispersions of MgO derived only from stoichiometric magnesium carboxylate such as magnesium acetate. The temperature required to effect such decomposition is significantly lower than the temperature range required to form dispersions of MgO by the dehydration of $Mg(OH)_2$. The preferred decomposition range of magnesium carboxylate-magnesium hydroxide mixtures of this process is about 280° to 330°C. This is about 30° to 50°C lower than the temperature range required for dehydration of $Mg(OH)_2$ which is about 310° to 380°C.

The conversion of Mg carboxylate-$Mg(OH)_2$ mixtures to dispersed MgO is practically quantitative whereas the conversion of commercial technical grade $Mg(OH)_2$ in the absence of small amounts of magnesium carboxylate to dispersed MgO is only about 50%. Therefore, in this process, commercial technical grade $Mg(OH)_2$ is first transformed to magnesium acetate in situ in low stoichiometric amounts by the addition of acetic acid and then quantitatively converted to dispersed MgO at the process temperature in a dispersant-containing fluid.

Any suitable nonvolatile process fluid capable of being heated to the decomposition temperature of the magnesium carboxylate-magnesium hydroxide mixture can be employed. The process fluid should be relatively stable and relatively nonvolatile at the decomposition temperature. However, any volatility encountered is readily controlled by refluxing and condensing apparatus. Examples of such fluids are as follows: hydrocarbons (such as mineral oil, paraffin oil, or aromatic oil), diphenyl oxide fluids, silicone oils, polyglycol ethers or vegetable oils, etc., solely the dispersant, or any combinations thereof.

The nonvolatile process fluid should contain a dispersant(s) capable of retaining the magnesium compound formed by decomposition in stable suspension. Any suitable dispersant which is relatively stable under these decomposition conditions can be employed, such as saturated and unsaturated fatty acids (such as stearic acid and oleic acid) and derivatives thereof (such as sorbitan monooleate), sulfonic acids (such as mahogany or petroleum-derived sulfonic acids and synthetic sulfonic acids), naphthenic acids, oxyalkylated fatty amines, alkylphenols, sulfurized alkylphenols, oxyalkylated alkylphenols, etc.

The reaction is carried out as follows. Since the decomposition temperature of magnesium acetate is about 323°C (613°F), the reactant mixture containing both magnesium acetate and magnesium hydroxide is heated at about this temperature. The decomposition products from such mixture are removed from the reaction by their volatility. In practice, temperatures of about 230° to 400°C or higher are employed, but preferably from about 280° to 350°C.

The particle size of the resulting MgO so formed in general should be of a size which is stable and fluid. In practice, the particle size is no greater than about 5 μ, such as no greater than about 2 μ, but preferably no greater than about 1 μ.

The concentration of the magnesium compound so formed in the nonvolatile process fluid should be no greater than that concentration which maintains suitable fluidity. In general, the final concentration based on nonvolatile fluid and other materials is from about 1 to 32% when calculated as percent magnesium, but preferably from about 4 to 23%.

These MgO dispersions can be further reacted to form dispersions of the corresponding derivatives. For example, after decomposition in accord with this process, the MgO dispersions can be further reacted with CO_2 to form $MgCO_3$ dispersions, reacted with H_2O to form $Mg(OH)_2$ dispersions, etc.

These compositions have a wide variety of uses. The following are illustrative:

(1) as a combination anticorrosion and acidic neutralization additive for lubricating oils and greases;

(2) as a combination anticorrosion and acidic neutralization additive during the combustion of fuels such as residual fuel, pulverized sulfur-containing coal, or mixtures thereof; and

(3) as corrosion inhibitors, particularly in fuels containing vanadium.

Example 1: *Reaction without acetic acid* – To a 500-ml reaction flask equipped with agitator and thermometer are charged 150 g high-boiling hydrocarbon oil and 112.8 g (about 0.4 equivalent) naphthenic acids. The contents are heated to 85° to 95°C and 93.6 g (1.6 mols) commercial technical grade magnesium hydroxide are added.

When all the magnesium hydroxide has been added, the contents of the reactor are heated to 140°C at which temperature water begins to come off by distillation. The heating is continued to 400°C and until no further liberation of water is indicated. The net weight is 285.6 g. A 50.0 g quantity of reaction mass is diluted with 50 g kerosene and the muddy-looking slurry is centrifuged. The weight of sediment collected, upon suitable extraction and drying is about 4.4 g.

The conversion of commercial technical grade magnesium hydroxide to fluid dispersion of MgO is calculated at only 55.6%.

Example 2: *Reaction with about 5 mol percent acetic acid* – To the reactor of Example 1 are charged 112.8 g naphthenic acids and 140.0 g high-boiling hydrocarbon oil. The contents are agitated and heated to 80°C. Commercial magnesium hydroxide 94.0 g (about 1.6 mols) is added through a powder funnel. The contents are heated to 210°C and the water formed by the reaction of naphthenic acid with magnesium hydroxide is returned to the flask by a reflux condenser. The contents are cooled to 90°C and 10 ml (0.17 mol) of glacial acetic acid and 10 ml H_2O are added. The contents are then heated to about 125°C under reflux. After 2 hours, the contents are heated such that free water is removed. Then heating is continued to 280° to 350°C to remove the necessary decomposition products during the formation of dispersed MgO.

A sample of the reaction product is centrifuged; based on the amount of sediment and by the appropriate calculation, the conversion of commercial technical grade magnesium hydroxide to dispersed MgO is indicated to be 92%.

Example 3: *Reaction with 5% acetic acid and colloidal milling* – To a suitable vessel equipped with agitators are charged 390 g hydrocarbon oil, 293.3 g naphthenic acids, 244.4 g (about 4.19 mols) commercial technical grade magnesium hydroxide, and 24.4 g (0.41 mol) glacial acetic acid. The contents are stirred and charged to a colloid mill at a setting of 0.001" clearance. The product of the first pass is passed twice more. A 366.4 g sample is then heated as in Example 2 to remove volatile matter and until no further decomposition products are removed during the formation of dispersed MgO.

The product is centrifuged to remove undispersed material. Calculations based on the amount of sediment collected indicate that the conversion of commercial technical grade magnesium hydroxide to dispersed MgO is about 96%.

Examples 4 through 8: The process of Example 2 is repeated with the same high yields at the following mol percents of acetic acid–Example 4, at 10 mol percent acetic acid; Example 5, at 20 mol percent acetic acid; Example 6, at 30 mol percent acetic acid; Example 7, at 40 mol percent acetic acid; and Example 8, at 50 mol percent acetic acid.

In addition, magnesium carboxylate-magnesium hydroxide mixtures can also be prepared by reacting other magnesium salts with carboxylic acids, for example, by reaction of $MgCO_3$, MgO, basic magnesium carbonate, etc., with the low stoichiometric amount of carboxylic acid necessary. The resulting magnesium carboxylate-magnesium hydroxide mixture can then be decomposed, in accord with this process.

The products of this process at a metal ratio of about 15 to 16/1 can be employed as hyperbasic additives for lubricating oils.

Magnesium-Containing Dispersions from Magnesium Carbonate

In still another patent on the preparation of magnesium-containing dispersions, *W.J. Cheng and D.B. Guthrie; U.S. Patent 4,226,739; October 7, 1980; assigned to Petrolite Corporation* heat magnesium carbonate in the presence of a fluid of low volatility containing a dispersing agent soluble in the fluid to effect its decomposition into MgO and CO_2 at temperatures substantially lower than required when $MgCO_3$ in the dry state is decomposed into the aforementioned products.

During this decomposition, $MgCO_3$ is disintegrated into minute particles of MgO which are immediately suspended and become stabilized in the fluid by the presence of a dispersing agent.

Any suitable magnesium carbonate capable of being subdivided upon decomposition into submicron particles of magnesia can be employed.

The nonvolatile process fluids and dispersants for use in this process are the same as those described in the previous patent.

Whereas the temperature for the decomposition of dry $MgCO_3$ which is not wet with any very high-boiling fluids evolves CO_2 primarily at 450° to 600°C, the CO_2 liberated by this process is removed from the reaction by its volatility at substantially lower temperatures.

In this process, temperatures of about 250° to 450°C are employed, with temperatures from about 320° to 380°C being preferred.

Example: To a 1-liter glass reaction flask were charged 210 g of a high-boiling hydrocarbon oil, 39.5 g of naphthenic acid (about 0.103 equivalent), and 3.0 g of $Mg(OH)_2$ (0.103 equivalent). The contents were stirred and heated to about 225°C under reflux to form magnesium naphthenate and by-product water. The mass was cooled to 150°C and 35.5 g of magnesium carbonate (0.842 equivalent) were added. The contents were stirred and heated to about 320°C by which temperature the water of neutralization had been removed and decomposition of the magnesium carbonate commenced as observed by the explosive sputterings and entrainment of hydrocarbon oil as carbon dioxide was evolved. The heating of the contents was continued up to 380°C and was held there until no further decomposition was observed. The reaction mass was cooled, and weighed 249.2 g. The contents were centrifuged for 2 hours and the amount of undispersed solids was only 0.1% by volume.

The calculation for both oil-soluble and oil-dispersed magnesium content is 4.61% as Mg. The ratio of equivalents of total metal as Mg to equivalents of acid dispersant is 9.17/1; the overbased value is 8.17/1 or 817%.

These MgO dispersions can be further reacted to form dispersions of the corresponding derivatives. For example, after decomposition according to this process, the MgO dispersions can be further reacted with CO_2 to form $MgCO_3$ dispersions, reacted with H_2O to form $Mg(OH)_2$ dispersions, etc. These compositions have a wide variety of uses. Of interest here is their use as a combination anticorrosion and acidic neutralization additive for lubricating oils and greases.

Dispersions Containing Calcium Carbonate

When processes analogous to those described in U.S. Patents 4,163,728 and 4,179,383 for the high-temperature decomposition of magnesium acetate to magnesium oxide in a dispersant-containing fluid are run using calcium acetate in place of magnesium acetate, analogous products are not obtained. *W.J. Cheng and D.B. Guthrie; U.S. Patent 4,164,472; August 14, 1979; assigned to Petrolite Corporation* have found, for example, that when calcium acetate is decomposed in a dispersant-containing fluid, the product of decomposition is a dispersion of calcium carbonate rather than calcium oxide.

They have also found in decomposing calcium acetate to calcium carbonate in a dispersant-containing fluid, that in order to achieve a superior highly overbased calcium-containing material, it is necessary to use stoichiometric amounts of acetate in relation to basic calcium. In their process, the decomposition to calcium carbonate is virtually quantitative such that no clarification is required for the resulting product. The process also results in a 1000% overbased calcium-containing product.

Any suitable calcium carboxylate capable of being subdivided upon decomposition into submicron particles of calcium carbonate in a dispersant-containing fluid can be employed. Calcium acetate is the preferred starting molecule.

The carboxylic acids suitable for use in preparing stoichiometric amounts of calcium carboxylate in this process include aliphatic carboxylic acids, for example, formic, acetic, propionic, acrylic, butyric. Carboxylic acids containing other functional groups are useful. These include, for example, hydroxycarboxylic acids such as lactic. This group also includes ketocarboxylic acids such as pyruvic, acetoacetic. Dicarboxylic acids are also used; examples of such acids are malonic, maleic, succinic. The aromatic and substituted aromatic carboxylic acids are also useful materials. Some examples are benzoic and salicylic.

Process fluids and dispersants used are the same as those described in the previous patents.

Example: To a reactor are charged 12.0 g naphthenic acids at 365 equivalent weight, 250 g hydrocarbon oil and 1.21 g calcium hydroxide. The contents are stirred and heated to about 250°C at which temperature the water of calcium naphthenate formation is removed. The reactor contents are then heated to 355°C. An aqueous solution of calcium acetate containing 28.9 g calcium acetate monohydrate is added slowly while maintaining the reactor temperature at 355° to 370°C. During the addition some of the hydrocarbon oil distills out but is returned as an upper layer from the water-oil separator.

After all of the aqueous calcium acetate has been added, the reactor contents are heated to 385°C at which temperature no more water is removed. The weight of the reaction product is 274.3 g. The amount of insoluble matter is 0.05% as determined by centrifugation for 2 hours. The calcium analysis of the product is 2.78% which calculates to a 1060% overbasing value.

The products of this process have a wide variety of uses, such as a combination anticorrosion and acidic neutralization additive for lubricating oils and greases, and as a combination anticorrosion and acidic neutralization additive during the combustion of fuels such as residual fuel, pulverized sulfur-containing coal, or mixtures thereof.

Zinc-Containing Dispersions from Zinc Carbonate

W.J. Cheng and D.B. Guthrie; U.S. Patent 4,193,769; March 18, 1980; assigned to Petrolite Corporation have developed a process for preparing zinc-containing dispersions which is analogous to those processes of their development for preparing magnesium- and calcium-containing dispersions.

This process comprises heating $ZnCO_3$ in the presence of a fluid of low volatility containing a dispersing agent soluble in the fluid to effect its decomposition into ZnO and CO_2 at temperatures substantially lower than required when $ZnCO_3$ in the dry state is decomposed into the aforementioned products.

The process, in essence, comprises an almost explosive decomposition of zinc carbonate to zinc oxide according to the equation

$$ZnCO_3 \rightarrow ZnO + CO_2$$

During this decomposition, $ZnCO_3$ is disintegrated into minute particles of ZnO which are immediately suspended and become stabilized in the fluid by the presence of a dispersing agent.

Any suitable zinc carbonate capable of being subdivided upon decomposition into submicron particles of zinc oxide can be employed.

The same process fluids (e.g., paraffin oil) and dispersion fluids (e.g., oleic acid) as those described in the previous processes are used here; the temperature for the decomposition is preferably from 225° to 350°C, and the particle size of the resulting ZnO is again preferably no greater than 1 μ.

The compositions are useful as corrosion inhibitors.

Example 1: To a 1-liter glass reaction flask with agitator are charged 500 g hydrocarbon oil and 128 g (about 0.4 equivalent) naphthenic acids. The contents are heated to 200°C and 23.3 g (about 0.4 equivalent) basic zinc carbonate (at 70.0% ZnO content) are added. The contents are stirred at 190° to 200°C for 1½ hours. The mass is then heated to 260° to 310°C and placed under slightly reduced pressure to facilitate the removal of by-product H_2O. Upon cooling and centrifugation, there is only a trace of white solids at the bottom of the centrifuge tube.

Example 2: To a 500-ml glass reaction flask with agitator are charged 162.8 g of zinc naphthenate solution from Example 1 and 30.0 g basic zinc carbonate (at 70.0% ZnO) content. The contents are heated to 260°C. There is elimination of some vapors including hydrocarbon oil. Heating is continued to 330°C at which temperature there is no further decomposition taking place. Upon cooling, the weight of the mass is 162.5 g; a centrifugation test indicates only 0.075% insolubles. A portion of the clear, bright supernatant layer upon ashing gives 15.8% as ZnO or 12.74% when calculated as Zn.

Example 3: To the glass reactor of Example 2 are charged 125.0 g hydrocarbon oil, 32.0 g naphthenic acid and 4.0 g ZnO. A solution of zinc naphthenate is made similar to that of Example 1. To the solution of zinc naphthenate is added 90.0 g basic zinc carbonate and the contents heated as in Example 2.

Upon cooling, the weight of the reactor contents is 207.2 g. Upon centrifugation, the insoluble layer measures 1.5%. The clear, bright supernatant layer upon ashing gives 32.5% as ZnO or 26.1% when calculated as Zn.

Metal Salts of Organosulfonic Acids

Salts of high molecular weight sulfonic acids of organic compounds have found use as rust inhibitors in motor fuels and lubricating oils, and as rubber plasticizers. See, for example, U.S. Patent 2,764,548. Particularly valuable salts are alkali metal, alkaline earth metal, lead and zinc salts of such organosulfonic acids as dinonylnaphthalene mono- and disulfonic acid. Special mention is made of such salts, and especially the sodium, potassium, lithium, calcium, magnesium, barium and zinc salts of dinonylnaphthalene disulfonic acid. The latter family of salts is discussed in U.S. Patent 2,764,548. The barium, calcium and lithium salts, particularly, form products having exceptional rust-inhibiting properties.

Commercially, such salts are often prepared by using oxides or hydroxides of the corresponding metal to neutralize the sulfonic acid. However, this is disadvantageous because the oxides and hydroxides are highly caustic and, in some cases, toxic. Moreover, end-point control is difficult and requires accurately stopping the flow of neutralizing agent. Exact neutrality can be very important because, for example, overneutralized metal salt sulfonates tend to be difficult to filter. Also, the salts of sulfonic acids are frequently used in combination with ester lubricants or other additives including amines and weak acids which cannot tolerate free acidity or basicity.

L.V. Gallacher and R.G. King; U.S. Patent 4,164,474; August 14, 1979; assigned to King Industries, Inc. have devised a process for preparing metal salts of such organosulfonic acids which may be produced in completely neutral solution. The process consists of the following steps:

(1) Providing a mixture comprising the organosulfonic acid and a small, effective amount of water;

(2) Adding to the mixture a compound of at least one alkali metal salt, alkaline earth metal salt, lead or zinc in the form of a carbonate, in an amount sufficient to provide a molar excess of the compound of at least about 1%;

(3) Reacting the mixture until the carbonate/acid equilibrium point is reached; and

(4) Adding a small amount of a compound which has a base strength greater than that of bicarbonate ion, sufficient to effect complete neutralization of the sulfonic acid.

When used herein, the term small, effective amount of water in step (1) means at least 2 mols of water per equivalent of the sulfonic acid.

The term until the carbonate/acid equilibrium point is reached in step (3) means that point in time in the reaction cycle where the hydrogen ion concentration measured on a substantially carbonate-free sample of the reaction mixture becomes stabilized at a fixed value. Conveniently, and preferably, the equilibrium point is determined by following the pH of a carbonate-free sample or samples. It will always become fixed at a value in the range of from about 6.0 to 6.38 (carbonate-free basis). When used herein, carbonate-free also means bicarbonate-free.

The term small amount of a metal compound in step (4) means up to about 1% by weight, based on the weight of the carbonate.

In the following examples all percentages are by weight, except where otherwise indicated.

Example 1: 9,071 pounds of a solution of approximately 35.5% dinonylnaphthalene sulfonic acid, 45% heptane, 15% water and 5% unsulfonated nonylnaphthalenes, produced by the process described in U.S. Patent 2,765,548, is added to a steam-jacketed 1,600 gallon reactor at 50°C. A slurry of 759 pounds of barium carbonate (7.5% molar excess based on the sulfonic acid) in about 200 gallons of mineral oil diluent is prepared in a separate vessel. The slurry is added to the acid mixture with strong agitation over a period of 1.5 hours.

Carbon dioxide is evolved with foaming and heptane vapor is condensed and returned to the reactor-neutralizer. Then 100 gallons more of mineral oil is added. The mixture is heated to refluxing at 78°C and refluxed for 3 hours with agitation to produce an equilibrium with the carbonate. After settling the excess carbonate, the pH of a carbonate-free sample [1:10 wt/wt in a 50:50 mixture of heptane (88% isopropanol-12% water)] is 6.17. The acid number is 0.03. Then ½ pound of calcium hydroxide is added as the strong base to the mixture at about 71°C and the mixture is agitated without supplying more heat for about 1 hour.

A 10 g sample, diluted as above, has a pH of 7.5. The base number is 0.03 (ml of 0.1 N HCl per 1 g of sample). The mixture is cooled to 65° to 70°C, allowed to settle and the lower water layer is drawn off. The remaining water and heptane are removed by distillation under vacuum to 138°C and the product is filtered to remove unreacted carbonate. There is produced a clear, neutral solution of the barium salt of dinonylnaphthalene sulfonic acid in mineral oil.

The procedure is repeated on a smaller scale with a 2 and a 4% molar excess of barium carbonate, respectively, and using barium octahydrate as the strong base. Substantially the same results are obtained.

Example 2: 200 g of dinonylnaphthalene sulfonic acid solution of the composition used in Example 1 is placed in a 1,000 ml flask fitted with a stirrer, thermometer and condenser. The flask is heated and 8.55 g of powdered calcium carbonate (10% excess based on the sulfonic acid) is added with agitation. After a few minutes, carbon dioxide evolution ceases. The mixture is refluxed for 30 minutes, then the solids are allowed to settle. A 10 ml sample diluted with a mixture of heptane and isopropanol as described in Example 1, and then filtered to remove the excess carbonate, has a pH of about 6.0, indicating that the carbonate/acid equilibrium has been reached.

Then 2.0 g of calcium hydroxide in 98 g of mineral oil is prepared and 2 ml of the mixture is added (0.02 g of strong base). Finally, 60 g of mineral oil is added, and the heptane and water are distilled off. The remaining fluid is filtered hot, at 130° to 140°C, through a pressure filter. There is produced a completely neutral solution of calcium dinonylnaphthalene sulfonate in mineral oil.

Overbased Manganese Salts of Organic Acids

Overbased manganese salts of organic acids, which are compounds in which manganese is present in excess of the stoichiometric amount required to react with the acidic groups of the organic acids, are widely used as additives in liquid hydrocarbon fuels and in lubricants for internal combustion engines. The basicity of these additives counteracts the corrosive acidic compounds that are formed during the operation of the engines and inhibits the formation of deposits of soot, lacquer, and sludge in the engines. In addition, these additives act as smoke suppressants in the fuels and improve the detergency of the lubricants.

The overbased salts are commonly produced by a process in which a basic manganese compound, such as manganous oxide, is suspended in an inert solvent containing an organic acid, a promoter, and a copromoter, and an acidic gas, which is usually carbon dioxide, is passed through the suspension to reduce its basicity. This process produces a product in which the manganese compound is complexed or dispersed in the solvent.

A. Ali, R. Caruso, A. Fischer, A.J. Deinet, P.P. Minieri and H. Sidi; U.S. Patent 4,179,385; December 18, 1979; assigned to Tenneco Chemicals, Inc. have developed a process for the production of overbased manganese salts that quickly and efficiently converts manganous oxide to highly overbased manganese salts of organic acids.

This process involves contacting with carbon dioxide at a pressure in the range of 1 to 10 atm a dispersion that contains (1) manganous oxide, (2) at least 1 organic carboxylic acid or organic sulfonic acid, (3) a promoter selected from the group consisting of ammonium halides, ammonium nitrate, mono-, di-, and trialkylamine hydrohalides, ammonium sulfide, and ammonium peroxydisulfate, (4) a copromoter that is a metal halide, (5) water, and (6) a solvent system that comprises an alcohol and an inert hydrocarbon or halogenated hydrocarbon. The products of this carbonation reaction are overbased manganese salt solutions that contain from 2 to 10% by weight of manganese and that can be concentrated to clear, fluid products containing high levels of manganese, usually from 20 to 28% by weight of manganese, that are useful as additives for fuel oils and lubricants.

In addition to providing high yields of overbased manganese salt solutions that contain at least 20% by weight of manganese, the process has the advantages of yielding these products in reaction times that are far shorter than those required by the processes of the prior art and of consuming much less carbon dioxide than do those processes.

The organic acids that are used in the process are organic carboxylic acids and organic sulfonic acids that are oil-soluble and that form manganese salts that are oil-soluble. They are preferably aliphatic and cycloaliphatic monocarboxylic acids having 4 to 10 carbon atoms, aromatic monocarboxylic acids having 7 to 12 carbon atoms, and mixtures thereof. Examples of these preferred acids include butyric acid, valeric acid, hexanoic acid, heptanoic acid, n-octanoic acid, 2-ethylhexanoic acid, n-nonanoic acid, isononanoic acid, neononanoic acid, n-decanoic acid, neodecanoic acid, naphthenic acids, benzoic acid, toluic acid, tert-butylbenzoic acid, hydroxybenzoic acids, chlorobenzoic acids, chlorotoluic acids, and the like.

The promoters that are used include ammonium halides, ammonium nitrate, mono-, di- and trialkylamine hydrochlorides, ammonium sulfide and ammonium peroxydisulfate. Among the useful copromoters are alkaline earth metal halides, aluminum chloride, and ferric chloride. Excellent results have been obtained using 3 to 5% of ammonium chloride as the promoter and 7 to 9% of calcium chloride or barium chloride as the copromoter, based on the weight of manganous oxide in the reaction mixture.

Best results have been obtained using a solvent system that contained 30 to 35% by weight of 2-methoxyethanol and 65 to 70% by weight of naphtha and having a pH in the range of 1.5 to 2.5.

The amount of water that is in the reaction mixture is that which will catalyze the carbonation reaction and cause this reaction to take place quickly and to give a nearly quantitative yield of highly overbased manganese salt. Optimum rate of reaction and yield result when the reaction mixture contains at the start of the carbonation reaction from 0.5 to 0.6% by weight of water, based on the weight of the solvent system.

The process is further illustrated by the following examples. In these examples, all parts are parts by weight, and all percentages are percentages by weight.

Example 1: To a 1 liter flask equipped with a heating mantle, thermometer, gas inlet tube, stirrer, and reflux condenser were charged 43.5 g (0.333 mol) of heptanoic acid, 50 g (0.698 mol) of manganous oxide (99% MnO), 2 g of ammonium chloride, 4 g of anhydrous calcium chloride, 431 g of naphtha (boiling range, 160° to 167°C), and 215 g of 2-methoxyethanol.

The reaction mixture was stirred at 90° to 95°C under atmospheric pressure and sparged with carbon dioxide for 1 hour. Then 5 g of water was added, and the reaction mixture at 90° to 95°C was sparged with carbon dioxide at about 90 ml/min for 5 hours. After the addition of 2 g of filter aid, the reaction product was cooled to 25°C and filtered. The filter cake was washed with 70 g of naphtha, and the washings were combined with the filtrate to give 685 g of an overbased manganese heptanoate solution that contained 4.5% of manganese.

The yield of overbased manganese heptanoate was 80.4%, based on the weight of manganous oxide charged.

Example 2: To a 1 liter, stainless steel autoclave were charged 43.5 g (0.333 mol) of heptanoic acid, 50 g (0.698 mol) of manganous oxide (99% MnO), 2 g of ammonium chloride, 4 g of anhydrous calcium chloride, 431 g of naphtha (boiling range, 160° to 167°C), and 215 g of 2-methoxyethanol.

The autoclave was sealed and sparged with carbon dioxide to remove air from it. The reaction mixture was then stirred and heated at 90° to 95°C for 1 hour during which time the pressure in the autoclave was maintained at 2 atm by the addition of carbon dioxide as needed. Then 5 g of water was added, and the reaction mixture was stirred at 90° to 95°C under a pressure of 2 atm for 5 hours.

2 g of filter aid was added to the reaction product, which was then cooled to 25°C and filtered. The filter cake was washed with 70 g of naphtha, and the washings were combined with the filtrate.

The resulting overbased manganese heptanoate solution, which contained 5.0%

of manganese, was obtained in a yield of 99.4%, based on the weight of manganous oxide charged.

Certain Amide and Polyamine Derivatives of Lactam Carboxylic Acids

R.C. Schlicht and W.M. Cummings; U.S. Patent 4,185,965; January 29, 1980; assigned to Texaco Inc. has found that certain 3-amido, and 3-tetrahydropyrimidyl and, particularly, 3-dihydroimidazolinyl derivatives of hydrocarbyl substituted butyrolactams are possessed of efficacious detergent and acid-corrosion-inhibiting properties when incorporated in gasoline formulations used in internal combustion engines.

Illustrative of the preferred products of this kind are 3-polybutenyl-3-dihydroimidazolin-2'-yl-4-phenyl-5-methyl butyrolactam; 3-dodecenyl-3-dihydroimidazolin-2'-yl-4-phenyl-5-methyl butyrolactam; 2-dodecenyl-3-dihydroimidazolin-2'-yl-4-phenyl-5-methyl butyrolactam; 3-octadecyl-3-tetrahydropyrimidin-2'-yl-4-phenyl-5-methyl butyrolactam; ethylene-bis(2-polybutenyl-3-tetrahydropyrimidin-2'-yl-4-phenyl)butyrolactam; 3-dodecenyl-3-tetrahydropyrimidin-2'-yl-4-phenyl-5-methyl butyrolactam; and isomeric mixtures thereof.

The prescribed compounds or isomeric mixtures thereof are employed as gasoline motor fuel compositions in a concentration to provide both effective carburetor detergency and corrosion-inhibiting properties. In general, an effective concentration of the additive ranges from about 0.001 to 0.1% by weight with a preferred concentration ranging from about 0.01 to 0.075% by weight. The limits of the preferred range correspond respectively to about 25 to 200 PTB (pounds of additive per 1,000 barrels of gasoline).

Example 1: To 1,930 g (equivalent to 0.96 mol) of 2(3)-polybutenyl-3-carboxy-4-phenyl-5-methyl butyrolactam was added 181 g (0.96 mol) of the tetraethylene pentamine in 400 ml of xylene. After refluxing for 19 hours at 30° to 177°C with removal of the water of reaction, the product was diluted with 1,500 ml of n-heptane, filtered and stripped to 150°C at 15 mm Hg pressure.

A yield of product amounting to 2,081 g was secured. This product was redissolved in 6,000 ml of heptane and then extracted with two separate 500 ml portions of methanol. The heptane raffinate was then stripped to 150°C at 15 mm pressure. The product recovered included predominantly the isomeric mixture, 2(3)-polybutenyl-3-[1'-(3,6,9-triazanonyl)-imidazolin-2'-yl]-4-phenyl-5-methyl butyrolactam. Recovery in the extraction was 90% by weight.

Example 2: 2(3)-polybutenyl-3-carboxy-4-phenyl-5-methyl butyrolactam (wherein the polybutenyl group has an average molecular weight of about 1,290 and is composed of 85 to 98% by weight of high molecular weight monoolefins, the balance, isoparaffins, was reacted in an amount of 1,350 g (equivalent of 0.645 mol basis theory) with 66.5 g (0.645 mol) of diethylene triamine in 600 ml of xylene as solvent. After 16 hours of refluxing at 155°C, 22.6 ml of water phase was collected (as against a theoretical recovery of 23.2 ml for the dihydroimidazoline).

The reaction mixture was cooled, diluted, filtered and stripped in the manner described in Example 1 and the product including predominantly (i.e., from at least 60% by weight to about 80% by weight) the isomeric mixture, 2(3)-

polybutenyl-3-(1'-aminoethyl dihydroimidazolin-2'-yl)-4-phenyl-5-methyl butyrolactam, was recovered in an amount of 139 g.

Example 3: The additive compositions prepared as described in Examples 1 and 2 were tested for their corrosion-inhibiting properties using the Colonial Pipeline Rust Test, the procedure for which follows.

A steel spindle, 3³⁄₁₆" long and ½" wide, made from ASTM D-665-60 steel polished with Crystal Bay fine emery paper, is placed in a 400 cc beaker with 300 cc of fuel sample, which is maintained at 100°F for ½ hour. Then 30 cc of distilled water is added. The beaker and contents are kept at 100°F for 3½ hours. The spindle is thereafter visually inspected and the percentage of rusted surface area is estimated.

The Base Fuel, designated as Base Fuel A, employed in the following examples was a premium grade gasoline having a Research Octane Number of about 100 and contained 3 cc of tetraethyllead per gallon. This gasoline consisted of about 25% aromatic hydrocarbons, 10% olefinic hydrocarbons and 65% paraffinic hydrocarbons and boiled in the range from about 90° to 380°F. The results secured in performing the foregoing test procedure with the products of Examples 1 and 2 are shown in the table as follows:

Colonial Pipeline Rust Test

Additive Composition	Concentration	% Rust
Example 1	Base Fuel A + 25 PTB additive	1-5*
Example 2	Base Fuel A + 25 PTB additive	1-5*

*An effective corrosion inhibition was found to exist as a result at the levels tested.

The additives of Examples 1 and 2 were also tested in the Chevrolet Carburetor Detergency Test and the additives demonstrated outstanding carburetor detergency properties when used in amounts of 75 PTB.

Ether Diamine Salts of N-Acylsarcosines

A constant concern of manufacturers and distributors of liquid hydrocarbons, such as fuel oil, is the corrosion of metal used for storage and transport. Corrosion of metal in storage tanks and pipelines can, of course, lead to loss of the liquid hydrocarbon, replacement and repair of tanks, lines, and fittings, and undesirable contamination of the environment. Such corrosion is primarily caused by the presence of water contamination. However, in certain liquid hydrocarbons, such as gasoline, which contain many ingredients, corrosion may even be promoted by the presence of such additives.

Many solutions have been suggested to prevent or inhibit the corrosion of construction materials, in particular mild steel, used for containment or transportation of liquid hydrocarbons. However, many such solutions entail the addition of additives to the system in typically large quantities. Furthermore, such additives often promote the formation of water-in-oil emulsions, which is undesirable. Also, certain physical properties of some suggested corrosion inhibitors, such as viscosity, or ease of handling in the field, have been found to be undesirable.

The process developed by *T.C. Newman; U.S. Patent 4,195,977; April 1, 1980; assigned to Akzona Incorporated* provides a corrosion inhibitor for use in liquid hydrocarbons which is effective at low dosage levels and does not unduly promote the formation of water-in-oil emulsions.

Ether diamine salts of N-acylsarcosines having the formula:

$$[R_1O(CH_2)_3NH(CH_2)_3NH_2]\cdot[HO_2CCH_2(CH_3)NOCR_2]$$

wherein R_1 is an aliphatic group containing from about 6 to 18 carbon atoms and R_2 is an aliphatic group containing from about 8 to 18 carbon atoms, have been found to be effective corrosion inhibitors for use in liquid hydrocarbons in amounts of 1 to 10 ppm by weight.

Typical of the ether diamines which are useful in the process are Duomeen EA-13, Duomeen EA-25, Duomeen EA-26 and Duomeen EA-80 ether diamines (Armak Co.).

The N-acylsarcosines which are useful in the process are, for example, N-lauroyl and N-stearoyl sarcosine (Ciba-Geigy).

The ether diamine salts of the N-acylsarcosines may be prepared utilizing any recognized neutralization technique. For example, simply stirring the desired reactants together for a sufficient length of time, such as one hour, produces the desired salt.

It has been found that the ether diamine salts are most desirable corrosion inhibitors due to the fact that, in addition to their excellent corrosion inhibition properties, they have low viscosities at room temperature, are less prone to form stable water-in-oil emulsions, and do not discolor the fuel oil to which they may be added, as does sarcosine alone.

Example: *60% Sarcosine salt of N-tetramethylnonyloxypropyl-1,3-propane diamine* – 51 g (0.3 equivalent) of N-tetramethylnonyloxypropyl-1,3-propane diamine (Duomeen EA-13) having a neutralization equivalent of 168 was weighed into a 200 ml beaker and 49 g (0.18 equivalent) of N-lauroyl sarcosine (Sarkosyl L, Ciba-Geigy) having an average molecular weight of 270 was subsequently added thereto. The resultant mixture was stirred for approximately one hour at room temperature to produce a clear amber liquid, which exhibited, upon testing under quite severe conditions, great effectiveness as a corrosion inhibitor when added to fuel oil in amounts of 10 ppm or less.

Methylol Polyester Derivatives of C_{12} to C_{22}-Substituted Succinic Anhydrides

E.D. Winans, J. Ryer, A. Gutierrez and H. Shaub; U.S. Patent 4,209,411; June 24, 1980; assigned to Exxon Research & Engineering Co. have found that hydrocarbon-soluble methylol polyester derivatives can be formed from the equimolar reaction of a C_{12} to C_{22} (preferably C_{18}) hydrocarbyl substituted succinic anhydride or acid and a cyclic polymethylol compound of the class consisting of 2,2,6,6-tetramethylol-cyclohexanol (TMC), tetrahydro-3,3,5,5-tetrakis-(hydroxymethyl)-4-pyranol (AEH) and tetrahydro-3,3,5-tris-(hydroxymethyl)-5-methyl-4-pyranol (tris-AEH).

The aliphatic hydrocarbon substituent of these polyesters can be branched and can possess unsaturation. For applications of the additive compounds in fuels such as gasoline, the carbon chain length is from 12 to 20, preferably 18, carbon atoms. One operational embodiment thus is a composition comprising a major proportion of a liquid hydrocarbon of the class consisting of fuels and lubricating oils and a minor amount of a hydrocarbon-soluble methylol polyester, the polyester preferably being from 0.001 to 20% by weight of the lubricating oil composition and from 2 to 10 ppm for the fuel composition.

Example 1: *Preparation of TMC* – 98 g (1 mol) of cyclohexanone was combined with 166 g (5 mols) of paraformaldehyde and 900 ml of water in a 2-liter 4-neck flask. This mix was cooled by an ice bath to 5°C and 35 g (0.61 mol) of calcium oxide was added over a 20-minute period. The temperature was allowed to rise slowly. After 1 hour approximately 3 ml of formic acid was added to neutralize the mix. Stirring was continued overnight. The mixture was evaporated on a rotafilm evaporator to remove all the water; the residue was dissolved in 750 cc hot absolute methanol and filtered through a steam suction filter to remove calcium formate. The filtrate yielded 134 g of white solids on cooling, a 61% yield of product, TMC.

Example 2: *Preparation of AEH (Anhydroenneaheptitol)* – 74 g (1 mol) of calcium hydroxide was added with stirring to a mixture of 116 g (2 mols) of acetone and 485 g (16.2 mols) paraformaldehyde in 1 liter of water. External heat was applied to 40°C to initiate the reaction which is exothermic. The reaction was not allowed to exceed 55°C and was kept at this temperature for 2 hours. The almost clear solution was neutralized with approximately 80 g of concentrated H_2SO_4, followed by the addition of 1 mol of oxalic acid.

The white solid was filtered and the filtrate stripped under vacuum. The residue was dissolved in methanol and filtered. This filtrate was vacuum evaporated to yield 375 g of crude product. The yield was 85%.

Example 3: 1.34 mols (470 g) of octadecenyl succinic anhydride and 1.34 mols (295 g) of TMC were combined with 741 g of a diluent oil and heated at 180°C for 3 hours. A nitrogen sparge was used to strip off the water.

Example 4: *Preparation of the methylol polyester of 2-octadecenyl succinic anhydride and AEH* – 0.5 mol (175 g) of octadecenyl succinic anhydride was heated to 100°C and 0.5 mol (111 g) of AEH added. The mix was heated to 180°C for 2 hours and 15 ml of water was collected from the reaction. The product was dissolved in hexane, filtered, evaporated and diluted to 50% by weight with solvent oil.

The products of Examples 3 and 4 were tested for their effectiveness as gasoline antirust agents. Each product was first dissolved in xylene and the solutions added to the gasoline to incorporate the additive at a treat rate of 1.5 and 3 pounds of polyester additive per thousand barrels of gasoline. The gasoline so treated was then tested for rusting according to ASTM D-665M rust test. In brief, this test is carried out by observing the amount of rust that forms on a steel spindle after rotating for 1 hour in a water-gasoline mixture. In each case, the polyester-treated gasoline gave a value of 1.0, i.e., no rust, indicating that each product was very effective as an antirust additive since the untreated gasoline will form rust over the entire surface of the spindle.

When gasoline treated with 1.25 PTB of the product of Example 3 was subjected to the National Association of Corrosion Engineers Rust Test [also known as the Colonial Pipeline Rust Test (see U.S. Patent 4,185,965 above)], it gave a reading of B+ whereas the untreated gasoline resulted in a reading of E thus further showing the antirust activity of these polyester products in fuels.

Bridged Phenol Metal Salt/Halocarboxylic Acid Condensate Additives

The ability of a lubricant or normally liquid fuel to inhibit corrosion of metals with which it comes in contact is becoming an increasingly sought-after property. Such antirust and/or anticorrosion properties are usually enhanced in lubricants or normally liquid fuels through use of additive organic compounds.

T.F. Steckel; U.S. Patent 4,216,099; August 5, 1980; assigned to The Lubrizol Corporation has provided such an additive by reacting (1) at least one phenoxide metal salt of a bridged phenol having (a) at least 2 and up to about 20 phenolic moieties or thiophenol analogs thereof and (b) at least 1 to about 19 bridging linkages independently selected from the group consisting of covalent carbon-to-carbon bonds, ether linkages, sulfide linkages, polysulfide linkages of 2 to 6 sulfur atoms, sulfinyl linkages, sulfonyl linkages, methylene linkages, alkylene linkages, di(lower alkyl)methylene linkages, lower alkylene ether linkages, lower alkylene sulfide linkages, lower alkylene polysulfide linkages of 2 to 6 sulfur atoms, amino linkages, polyamino linkages and mixtures of such divalent bridging linkages with (2) at least one carboxylic acid reagent having 1 to about 3 carboxyl-based groups and a halogen-substituted hydrocarbon-based aliphatic or alicyclic group containing a halogen atom.

Valuable posttreated compositions may be made by further reaction of these compositions with mono- or polyhydric alcohols, basically reacting metal compounds, amino compounds such as monoamines, hydrazines, hydroxyamines, polyamines, hydroxypolyamines and small ring heterocycles.

Preferably, the phenoxide metal salts contain one or more phenolic moieties substituted with 1 to 3, preferably 1, hydrocarbon-based aliphatic or alicyclic groups of from 1 to about 300, preferably about 30 to 250, carbon atoms.

The sources of the hydrocarbon-based groups include principally the high molecular weight substantially saturated petroleum fractions and substantially saturated olefin oligomers and polymers, particularly oligomers and polymers of monoolefins having from 2 to about 30 carbon atoms.

A preferred class of bridged phenols that can be used is represented by the general formula:

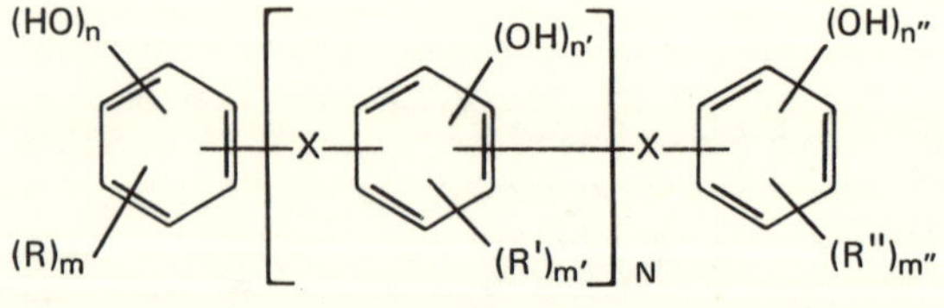

$(HO)_n$, $(OH)_{n'}$, $(OH)_{n''}$, X, X, $(R)_m$, $(R')_{m'}$, N, $(R'')_{m''}$

wherein n, n' and n'' are each independently integers of 1 to 3 but preferably 1 each; R, R' and R'' are each independently aliphatic hydrocarbon-based groups, generally, alkyl or alkenyl groups, of 1 to about 300 carbon atoms, preferably about 6 to 200 carbon atoms each, and usually about 30 to 250 carbon atoms each; m, m' and m'' are each independently integers of 0 to 3, but generally 1 or 2 each; N is an integer of 0 to 20 but usually 0 to 5; and X is a divalent bridging linkage.

Preferred carboxylic acid reagents have the formula, $A-(Cox)_{1\text{-}3}$, wherein A represents a halogenated, hydrocarbon-based aliphatic or alicyclic group of 1 to about 20 carbon atoms having a number of unsatisfied valences corresponding to the number of Cox groups present, and each Cox independently represents a carboxyl-based group. Exemplary of such A groups are chloro- and bromomethyl; 1- and 2-chloro- and bromoethyl; 1, 2- and 3-chloro- and bromopropyl, etc.; 1-, 2-, 3-, etc., chloro- and bromocyclopentyl and cyclohexyl groups.

The carboxylic acid ester reagents useful in making these compositions can be made from relatively higher mono- and polyhydric hydrocarbon-based alcohols. The term higher alcohols refers here to alcohols which contain either 8 or more carbon atoms and/or 3 or more (up to 8) hydroxyl groups. Preferably they contain 8 to about 30 carbon atoms and 1 to about 6 hydroxyl groups. The higher alcohols can be aliphatic, alicyclic, mixed aliphatic alicyclic (e.g., pentylcyclohexyl), aromatic (e.g., the naphthols, benzohydroquinones, and phenylphenols), mixed aliphatic-aromatic (e.g., β-phenylethanol, 3-phenylpropanol, etc., as well as the ethyl-, n-decyl-, n-pentadecyl-phenols, etc.) and alicyclic-aromatic (e.g., the cyclohexylphenols, phenylcyclohexanols, etc.) monohydric alcohols.

The carboxylic acid reagents used can be a nitrogen-containing carboxylic acid reagent made from one or more amino compounds. Generally, these amino compounds result in the carboxyl-based group Cox being a carboxamide, although more complex groups such as carboximide or carboxamidine groups can be present.

Polyamines are also useful for preparing the nitrogen-containing carboxylic acid reagents.

Polyamines useful in this process are exemplified specifically by: ethylenediamine, triethylenetetramine, tetraethylenepentamine, pentaethylenehexamine, etc., with ethylene-polyamine being the most useful.

Among the particularly preferred carboxylic acid reagents used to produce these compositions are α-halocarboxyl acid reagents having 2 to 20 carbon atoms and being aliphatic- or alicyclic-based, especially those in which Cox is an ester group of a lower alkanol. Exemplary of such carboxylic acid reagents are: ethyl chloroacetate; methyl-7-bromostearate; sodium 1,1-dichloropropionate; calcium bromosuccinate; 3-chloroglutaric acid; 8-chloromethyl stearate; 3-bromobutyramide; chlorosuccinic anhydride; etc.

In the following examples all percentages and parts are by weight and the molecular weights are number average molecular weights ($\overline{M}n$) as determined by gel permeation chromatography (GPC) or vapor phase osmometry (VPO).

Example 1: A mixture of 1,120 parts of a poly(isobutene)-substituted phenol ($\overline{M}n$ 960 VPO), 200 parts of xylene, 300 parts of mineral oil and 14 parts of

50% aqueous sodium hydroxide solution is heated at 75°C. At 78°C, 29.7 parts of paraformaldehyde is added, and the reaction mixture is heated at reflux for 2.5 hours. An additional 66 parts of 50% aqueous sodium hydroxide solution is added, and the mixture dried by azeotropic distillation. At 73°C, 110 parts of ethyl chloroacetate is added and the mixture is held at 130° to 147°C for 2.5 hours. At 60°C, 100 parts of a commercial mixture of alcohols containing approximately 61% isobutyl alcohol and 39% amyl alcohol, 10 parts of aqueous hydrochloric acid and 30 parts of water is added. The mixture is refluxed at 105° to 107°C for 1 hour, then azeotropically dried. The mixture is stripped to 170°C under vacuum and filtered. The filtrate contains the desired product and 19.9% mineral oil.

Example 2: A mixture of 782 parts (0.5 equivalent) of the product solution of Example 1, 276 parts of mineral oil and 30.7 parts (0.75 equivalent) of a commercial ethylene polyamine mixture, wherein the amines have an average of 3 to 10 nitrogen atoms per molecule, containing about 34% nitrogen is heated at 160° to 165°C for 5 hours. The mixture is stripped to 170°C under vacuum and filtered to yield an oil solution containing 0.97% nitrogen.

Example 3: A mixture of 2,366 parts of a poly(isobutene)-substituted phenol ($\overline{M}n$ 1,369 VPO), 350 parts of toluene and 41.5 parts of paraformaldehyde is heated at 79°C. At 85°C, 56 parts of 50% aqueous sodium hydroxide solution is added and the reaction mixture is heated at reflux for 3 hours. The mixture is then dried by azeotropic distillation. At 142°C, 86 parts of ethyl chloroacetate is added and the reaction mixture is cooled to 80°C, whereupon 250 parts of a commercial mixture of alcohols described in Example 1, 25 parts of aqueous hydrochloric acid and 60 parts of water are added. The reaction mixture is refluxed at 89° to 91°C for 4 hours, then azeotropically dried. The mixture is stripped to 170°C under vacuum and 1,000 parts of mineral oil is added. The product solution, which contains 29.1% mineral oil, is obtained by filtration.

Example 4: A mixture of 5,600 parts of a poly(isobutene)-substituted phenol ($\overline{M}n$ 885 VPO), 1,600 parts of xylene and 80 parts of 50% aqueous sodium hydroxide solution is heated at 65°C. At 67°C, 148.5 parts of paraformaldehyde is added and the reaction mixture is heated at reflux for 4 hours. A charge of 171 parts of flake sodium hydroxide and 60 parts of water is made to the reaction mixture at 82°C and then dried by azeotropic distillation. At 70°C, 750 parts of a commercial mixture of alcohols described in Example 1 and then 524 parts of sodium chloroacetate are added and the mixture is held at reflux (124° to 126°C) for 10.5 hours. The mixture is stripped to 168°C and 1,000 parts of mineral oil is added. At 98°C, 500 parts of aqueous hydrochloric acid and 100 parts of water are added to the reaction mixture, which is refluxed for 2.5 hours followed by azeotropic distillation. The mixture is stripped to 170°C under vacuum and 1,200 parts of mineral oil is added. The desired product solution (containing 27.1% mineral oil) is obtained upon filtration.

Example 5: A mixture of 912 parts of the oil-containing product solution of Example 4, 150 parts of toluene, 150 parts of mineral oil and 33.2 parts (0.8 equivalent) of a commercial ethylene polyamine mixture corresponding in empirical formula to pentaethylenehexamine is heated at 134° to 149°C for 6 hours, then stripped to 161°C under vacuum and filtered to yield the desired product solution. The product solution contains 36.5% mineral oil and 0.99% nitrogen.

Example 6: A mixture is prepared by the addition of 24 parts of aqueous hydrochloric acid to 798 parts of tetrapropenyl-substituted phenol, 300 parts of xylene and 45 parts of paraformaldehyde at 65°C. The reaction mixture is held at 100° to 107°C for 3 hours, then dried by azeotropic distillation. The reaction mixture is cooled to 35°C and 256 parts of 50% aqueous sodium hydroxide solution is added. The mixture is heated to reflux and azeotropically dried. At 62°C, 130 parts of toluene, 360 parts of a commercial mixture of alcohols described in Example 1 and 349.5 parts (3.0 equivalents) of sodium chloroacetate are added. The mixture is refluxed at 114°C for 7.5 hours, stripped to 162°C and then 550 parts of mineral oil is added.

After cooling to 120°C, 400 parts of toluene, 500 parts of aqueous hydrochloric acid and 250 parts of water are added. The mixture is stirred at 65°C for 3 hours, stripped to 101°C, azeotropically dried and then stripped to 165°C under vacuum. Filtration yields the desired oil-containing product solution.

Example 7: A mixture of 385 parts of the oil-containing product solution of Example 6, 50 parts of mineral oil, 75 parts of toluene and 43 parts (1.05 equivalents) of the ethylene polyamine mixture used in Example 2 is reacted according to the procedure described in Example 5. The product solution contains 40% mineral oil and 3.06% nitrogen.

These compositions are useful in and of themselves as antirust and anticorrosion agents for fuels and lubricants, particularly when they are free acids, esters of the abovedescribed higher alcohols, carboxamides or ammonium carboxylates of the abovedescribed polyamines. These esters, carboxamides and carboxylates can also function in fuels and lubricants as detergents and dispersants for sludge and varnish formed in internal combustion engines.

Trisubstituted, Hydrocarbon-Soluble Chromium Compounds

The advantages of the use of chromium compounds as fuel additives in such things as gas turbine engines to prevent corrosion has been well recognized in the prior art. However, because of the numerous process steps required to make conventional reactants needed to produce these valuable complexes, the expense of conventional synthesis methods has been somewhat prohibitive.

Accordingly, what has been lacking in the prior art is a relatively simple method of producing chromium compounds useful as anticorrosion fuel-soluble additives which utilize readily available and inexpensive chromium-containing starting materials.

It has been found by *R.A. Pike; U.S. Patent 4,218,385; August 19, 1980; assigned to United Technologies Corporation* that by utilizing a two-step synthesis, relatively available chromic acid anhydride (CrO_3) can be used as the chromium source in the production of such compounds. It has also been found that such compounds can be formed by such method with no sacrifice in product quality over conventional fuel-soluble additives in spite of the relatively simple synthesis scheme.

The first stage in the synthesis comprises reducing the relatively toxic hexavalent chromium source to trivalent chromium by means of reducing agents such as alcohols.

The second phase of the process comprises completing the trisubstituted soap formation by reacting the required stoichiometric amount of carboxylic acid with the complex in the presence of a hydrocarbon carrier and a base to give the trisubstituted soap compound. While any carboxylic acid which is hydrocarbon-soluble can be used, aliphatic carboxylic acids having at least 7 carbon atoms are preferred and 8 to 12 carbon atoms most preferred. Preferred hydrocarbons are kerosene, mineral spirits, No. 2 home heating oil and high flash point oil, with kerosene being most preferred. While any standard base such as sodium hydroxide or other alkali metal salt may be used, ammonium hydroxide is preferred because it produced no contaminating metal ions which must be removed for anticorrosive fuel additive purposes.

After washing with a 1% by weight solution of ammonium chloride in water, the oil (hydrocarbon) layer can be used as produced for such purposes as a fuel additive in gas turbine engines to produce corrosion resistance.

An essential part of the reaction scheme is the use of at least a 7.5 times the amount of alcohol necessary to reduce the Cr^{6+} to Cr^{3+} in the initial complex formation reduction step. With less than this amount the intermediate Werner complex formed insolubilizes, adversely affecting the remainder of the reaction and attainable yields.

Example: The following describes a run carried out on a 1-liter scale. To a 1-liter 3-necked round-bottom flask, having a drain plug on the bottom, equipped with a mechanical stirrer, thermometer, reflux condenser, and dropping funnel was charged 112.5 g methyl alcohol, 5.5 g xylene, 22 g neodecanoic acid and 4 g of 2,4-pentanedione. Through the dropping funnel was added dropwise with stirring a solution of 10 g of water, 16 g CrO_3 and 14 g of concentrated hydrochloric acid over a 20-minute period. After 10 minutes the temperature had risen to 50°C and in 15 minutes to 60°C. At the completion of the addition, the UV transmission at 440 mμ was 81%. The solution was heated an additional 40 minutes (UV transmission at 440 mμ = 93%) at 60° to 65°C.

To the dark green solution, with stirring, was then added in consecutive order 55.5 g (70 cc) of kerosene, 100 g water and 52 g of 2-ethyl hexanoic acid. During these additions the temperature dropped to 50°C. Through the dropping funnel was then added 31 g of concentrated ammonium hydroxide over a 10-minute period, the temperature rose to 62°C. The resulting mixture was stirred at 60° to 61°C for ¾ hour. The stirrer was stopped and the layers allowed to separate over a 15-minute period. The bottom aqueous layer which was drawn off had a pH of 7.5 and amounted to 325 cc.

To the purple oil layer was then added 325 cc of 1% NH_4Cl solution with stirring. The temperature dropped to 40°C. The mixture was heated to 50°C and stirred ½ hour. After phase separation for 15 minutes the wash water was drawn off and the product collected. The weight of the product was 123 g, viscosity 120 cs and Cr content 7.2%.

Tartarimides

D.E. Barrer; U.S. Patent 4,237,022; December 2, 1980; assigned to The Lubrizol Corporation has developed a class of additives for motor fuels and lubricants which are effective rust inhibitors, dispersants and demulsifiers in fuel compositions and which provide reduced friction in automatic transmission and power steering fluids as well as giving improvement in fuel economy. These additives are tartarimides of the general formula

```
          O
          ‖
      H   C
HO—C ⁄     \
   |        N—R
HO—C \     /
      H   C
          ‖
          O
```

wherein R is a hydrocarbon-based radical of about 5 to 150 carbon atoms or R'OR'' in which R' is a divalent alkylene radical of 2 to 6 carbon atoms, and R'' is a hydrocarbyl radical of about 5 to 150 carbon atoms, or R.

The tartarimides are prepared conveniently by reacting equimolar amounts of tartaric acid with the one or more of the corresponding primary amines. The tartaric acid used for preparing the tartarimides of this process is the commercially available type (obtained from Sargent Welch), and is likely to exist in one or more isomeric forms such as d-tartaric acid, l-tartaric acid or mesotartaric acid, often depending on the source (natural) or method of synthesis (from maleic acid). The reaction is carried out at temperatures sufficiently high to form the imide, most preferably from about 130° to 165°C.

The reaction is considered complete when 2 mols of water are removed per mol of tartaric acid and mol of the primary amine. The formation of the imide can be checked by known chemical techniques such as infrared or NMR spectra. The foregoing reaction is carried out conveniently in a solvent, such as toluene, mineral oil, xylene, ethylbenzene, or even benzene.

Primary amines particularly suitable for use in this preparation are those of formula RNH_2, where R represents a long hydrocarbyl radical, preferably of from 8 to 18 carbons. Representative of these are the aliphatic primary fatty amines called Armeen (Armour Chemicals). Other commercially available amines are Surfam P14AB, Surfam P16A and Surfam P17AB (Worth Chemical Co.).

The fuel compositions comprising these tartarimides contain a major proportion of a normally liquid fuel such as aviation or motor gasoline or diesel fuel or fuel oil in an amount sufficient to impart rust-inhibiting and demulsifying properties to the fuel, preferably 10 to 100 pbw of the reaction product per million pbw of fuel. The preferred gasoline-based fuel compositions generally exhibit excellent antirust properties, particularly under acidic and neutral conditions, and alkaline conditions up to pH 10. In addition, they exhibit excellent demulsifying properties thus not allowing the formation of harmful emulsions.

Example 1: 150 parts of tartaric acid (obtained from Sargent Welch) is added to 173 parts of toluene and the mixture is heated with stirring to 100°C. 281 parts of Armeen O (essentially oleylamine, Armak Chemicals) is added slowly and in small portions while the system is kept under nitrogen purge. After the addition of the Armeen O is completed the contents are heated to 130°C at which temperature the water formed is removed by azeotroping, and collected. The temperature is then raised to 160°C, after no more water is collected and kept at about 160°C for 1 hour under nitrogen purge to remove the toluene.

The liquid residue is then filtered through diatomaceous earth to yield the desired N-oleyltartarimide. Analysis shows nitrogen content of 3.42% and hydroxyl content of 8.33%. The tartarimide can be expressed as

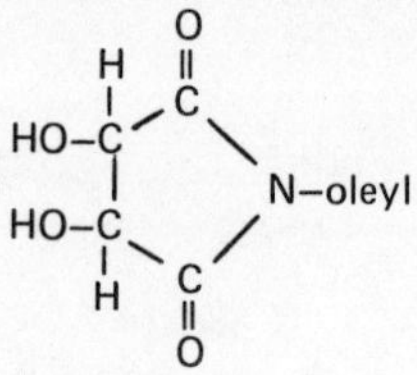

```
           O
       H   ‖
       |  C
  HO–C ⁄    \
       |      N–oleyl
  HO–C \    /
       |  C
       H   ‖
           O
```

Example 2: Following the procedure described in Example 1, 150 parts of tartaric acid are added to 173 parts of toluene and the mixture heated to 80°C. 258 parts of Surfam P17AB (branched C_{17} etheramine, Worth Chemical) is added to the toluene-tartaric acid mixture and the entire contents are heated to 130°C under nitrogen purge. The temperature is increased to 140°C until the water is removed (32 parts collected). The toluene is stripped at 160°C and the final imide product is filtered and collected.

Example 3: To a normally liquid hydrocarbon composition suitable for use as a fuel in an internal combustion engine is added such a tartarimide as that of Examples 1 or 2 from a concentrate in a proportion of 9 PTB. The treated gasoline is found to exhibit no rust under acidic, neutral and alkaline pH (up to pH 10) conditions. Moreover, the tartarimide provides desirable demulsifying properties in minimizing the formation of emulsions which give rise to harmful carburetor and engine deposits.

Oxazolonium Hydroxides

Certain types of fuels contain increased amounts of cracked stocks resulting in a higher olefin content and an increased susceptibility to the formation of gum. These fuels generally contain minor amounts of impurities which can promote corrosion during the time when the fuel is transported and stored and even in the fuel tank, fuel lines and carburetors of the engine. Consequently a motor fuel must contain a corrosion inhibitor.

For obvious reasons, it would be advantageous to use a multipurpose additive which provides both detergency and corrosion-inhibiting properties to the fuel.

R.L. Sung and P. Dorn; U.S. Patent 4,257,780; March 24, 1981; assigned to Texaco Inc. provide oxazolonium hydroxides having utility as such fuel additives which are represented by the following structure:

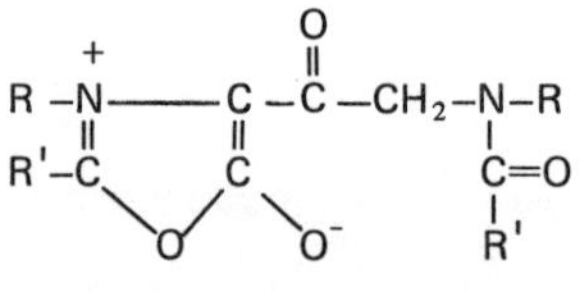

```
                      O
     +                ‖
  R –N———— C – C –CH₂–N–R
     ‖        ‖           |
  R'–C        C           C=O
      \    /    \         |
        O         O⁻      R'
```

wherein R stands for a hydrocarbyl radical having from 1 to 5 carbon atoms and R' is a hydrocarbyl radical having from 10 to 20 carbon atoms.

Preferred compounds are those where the substituent R group has from 1 to 3 carbon atoms and R' has from 12 to 13 carbon atoms in the chain.

Both the R and R' radicals can be straight chain or branched and may be substituted with one or more typical noninterfering substituents such as halogen, cyano, trifluoromethyl, nitro or alkoxy.

A motor fuel composition comprising a mixture of hydrocarbons in the gasoline boiling range may be made which contains a minor, detergent and corrosion inhibiting amount of at least one of the above compounds. Preferably, the amount ranges from 20 to 200 parts per 1,000 barrels of fuel to provide both detergency and corrosion-inhibiting effects. As little as 2.5 parts per 1,000 barrels will provide corrosion inhibition.

The substituted oxazolonium hydroxides are prepared by dissolving the corresponding glycine in an inert solvent, preferably nitromethane, and adding the resulting solution to dicyclohexyl carbodiimide. The mixture is heated to about 43°C and filtered to give the solid compound.

Example 1: *Preparation of anhydro-2-lauroyl-3-methyl-4-(N,N'-methyl-lauroyl-glycyl)-5-hydroxy-1,3-oxazolonium hydroxide (Compound A)* – 45 parts of N,N'-methyl-lauroyl-glycine were dissolved in 125 parts nitromethane. To the mixture, 35 parts of dicyclohexyl carbodiimide in 125 parts nitromethane was added. The mixture was heated at 43°C for 2 hours, then filtered. The filtrate was stripped to yield the desired product. The structure was confirmed by IR and NMR.

Example 2: The procedure of Example 1 was repeated using N,N'-methyl-cocoyl-glycine yielding anhydro-2-cocoyl-3-methyl-4-(N,N'-methyl-cocoyl-glycyl)-5-hydroxy-1,3-oxazolonium hydroxides.

Results of the Chevrolet Carburetor Detergency Test show that on a weight-to-weight basis these compounds are more effective than commercial additives as carburetor detergents.

These additives also have anticorrosion properties as shown by their performance in the National Association of Corrosion Engineers (NACE) Rusting Test. In this test a determination is made of the ability of motor gasolines to inhibit the rusting of ferrous parts when water becomes mixed with gasoline. Briefly, the test is carried out by stirring a mixture of 300 ml of the test gasoline and 30 ml of water at 37.8°C with a polished steel specimen completely immersed therein for a test period of 3½ hours. The percentage of the specimen that has rusted is determined by comparison with photographic standards. Further details of the procedure appear in NACE Standard TM-01-72 and ASTM D6651 1P-135 (Procedure A).

The table below shows the results of this test for the additive of Example 1 at different concentrations in and against an unleaded base fuel (UBF).

Concentration (PTB)	Additive	Rust Rating (%)
Unleaded base fuel	–	50-100
7.5	Compound A	1-5; Trace
5.0	Compound A	Trace
2.5	Compound A	Trace

Containing a Hydrocarbon-Substituted Lactono-Imidazoline Reaction Product

M.E. Davis and K.L. Dille; U.S. Patent 4,263,014; April 21, 1981; assigned to Texaco Inc. have developed a rust-inhibiting motor fuel composition comprising a mixture of hydrocarbons in the gasoline boiling range containing minor rust-inhibiting amounts of a polymeric dimer or trimer acid and a hydrocarbon-substituted lactono-imidazoline reaction product that are combined to allow interaction of acidic and basic entities prior to being added to gasoline.

More specifically, this motor fuel composition comprises a mixture of hydrocarbons in the motor fuel or gasoline boiling range containing in critical combination about 0.000003 to 0.0003% by weight of a dimer or trimer of a dienoic or trienoic acid containing from 16 to 18 carbon atoms and from about 0.001 to 0.02% by weight of a hydrocarbon-substituted lactono-imidazoline reaction product. This reaction product is obtained by reacting diethylenetriamine with a lactone reaction product under amidation conditions at a temperature in the range of 80° to 170°C employing about 1 to 2 mols of the amine per mol of lactone reaction product. The lactone reaction product is obtained by treating a hydrocarbon succinic acid, represented by the formula:

$$\begin{array}{l} X-CH—COOH \\ \quad\;\; | \\ \quad\;\; CH_2-COOH \end{array}$$

in which X is a hydrocarbon radical having an average molecular weight ranging from about 300 to 1,500, under substantially anhydrous esterification conditions at a temperature ranging from about 50° to 100°C in the presence of a protonating agent.

The polymer acid component of the additive combination comprises a dimer of a dienoic or trienoic acid containing from about 16 to 18 carbon atoms, specifically linoleic, linolenic, 9,11-octadecadienoic and eleostearic acids. Effective polymeric acids can be prepared from naturally occurring materials, such as linseed fatty acids, soybean fatty acids and other natural unsaturated fatty acids. Suitable polymeric acids are also available commercially, such as Empol 1022 Dimer Acid, a dimer of linoleic acid, which is well-known as a rust inhibitor for a motor fuel composition and will provide a gasoline qualifying under the NACE Rust Test when employed in a concentration of 6 PTB, i.e., about 0.002% by weight.

Example 1: *Polyisobutenylsuccinic anhydride reaction product* – To a solution of 126 g (0.05 mol) of crude polyisobutenylsuccinic acid (prepared from polyisobutene of about 1,300 molecular weight and maleic anhydride by thermal alkenylation with about 50% unreacted polyisobutene) in a 50% by weight mineral oil solution was added 1.25 g (0.0125 mol) of concentrated sulfuric acid.

The mixture, containing about 0.0125 mol of sulfuric acid or about 0.025 mol of available protons was held at 90°C for 3 hours. IR analysis of the product from the foregoing reaction showed a high conversion to 5- and 6-membered lactones, with the yield estimated to be greater than 85 mol percent.

Example 2: *Polyisobutenyl (1290) lactono-aminoethyl imidazoline product* – To 105.4 lb of polyisobutenyl (1290) lactono-carboxylic acid in 132 lb of xylene is added 8.3 lb of diethylenetriamine. The molar ratio of amine to acid is 1.46. The mixture is heated to 293°F and held there for 9 hours. About 3.3 lb of water is collected. The reaction mixture is allowed to cool to room temperature to permit settling of sludge (amine salt of sulfuric acid). The reaction mixture is filtered, and the xylene removed from the filtrate under reduced pressure.

The IR absorptions at 5.65 μ (lactono) and 6.25 μ (imidazolino) show that the above additive is formed.

The lactono-imidazoline component of the additive combination is employed in a concentration range of about 0.001 to 0.02% by weight, which corresponds to about 3 and 50 PTB. The preferred concentration range is from 0.002 to 0.01% by weight. Additive A was a 60% active solution of 2-polyisobutenyl (335 molecular weight) lactono-1-aminoethylimidazoline in unreacted polyisobutylene.

The base fuel employed in the following examples was a lead-free gasoline having a Research Octane Number of about 91. This gasoline consisted of about 10% olefinic hydrocarbon and 60% paraffinic hydrocarbons and boiled in the range from about 90° to 370°F.

The antirust properties of the fuel composition of the described formulation and of a comparison fuel composition in the NACE Rust Test described in the previous patent is shown in the following table. The amount of additive added to the Base Fuel is given in pounds per thousand barrels (PTB).

NACE Rust Test

Run	Dimer Acid* (PTB)	Additive A (PTB)	NACE Rust Rating (%)**
1	0.25	–	100
2	–	10	78
3	0.25	10	18

*~85% dimeric acids and 12% trimeric acids of C_{16}-C_{18} dienoic or trienoic acid, i.e., Emery Empol 1022.

**Fuel with more than a trace of rust.

The fuel composition of Run 1, wherein Dimer Acid was employed failed the NACE Rust Test 100% of the time. The fuel composition of Run 2 failed the NACE Rust Test 78% of the time. The fuel composition of Run 3, which is representative of the described formulation of this process was outstandingly effective with a NACE Rust Test pass-fail ratio of 82 to 18%.

Thixotropic Magnesium-Containing Complexes

J.W. Forsberg; U.S. Patents 4,253,976; March 3, 1981; and 4,260,500; April 7, 1981; both assigned to The Lubrizol Corporation describes thixotropic magnesium-containing compositions with a number of uses, e.g., as protective coatings for metal surfaces and as additives for lubricants and fuels. The complexes are prepared by heating, at a temperature above about 30°C, a mixture comprising:

(A) At least one of magnesium hydroxide, magnesium oxide, hydrated magnesium oxide and a magnesium alkoxide;

(B) At least one oleophilic organic reagent comprising a sulfonic acid, a pentavalent phosphorus acid, a mixture of a major amount of either of the above with a minor amount of a carboxylic acid, or an ester or alkali metal or alkaline earth metal salt of any of these;

(C) Water, if necessary, to convert a substantial proportion of Component (A) to magnesium hydroxide or hydrated magnesium oxide; and

(D) At least one organic solubilizing agent for Component (B);

the ratio of equivalents of magnesium to the acid portion of Component (B) being at least about 5:1, and the amount of water present, if any, being sufficient to hydrate a substantial proportion of Component (A) calculated as magnesium oxide.

Component (B) is preferably an alkylbenzene sulfonic acid. Component (D) may be a wax, a hydrocarbon resin, such as polyethylene, a natural resin such as copal, an addition polymer resin such as styrene-butadiene, a polymer resin, or a solid plasticizer such as triethylene glycol dibenzoate. Mixtures of materials all of which are liquid at normal ambient temperatures (e.g., about 20° to 30°C), such as mineral oil-toluene, Stoddard solvent-toluene, mineral oil-alkylbenzene, Stoddard solvent-alkylbenzene; of materials all of which are solid at normal ambient temperatures, such as paraffin wax-polyethylene wax, paraffin wax-polyethylene wax-C_{20} to C_{40} alcohol wax; or of materials which are both liquid and solid at normal ambient temperatures, such as mixtures of the abovementioned normally liquid diluents and a resin or hydrocarbon wax (e.g., paraffin wax-toluene, polypropylene-toluene, polypropylene-mineral oil) are also useful.

Following are examples of magnesium complexes which are useful as corrosion inhibitors, vanadium scavengers and smoke suppressants in fuels. In the examples, all parts are by weight.

Example 1: 600 parts magnesium oxide is added to a solution in 478 parts of Stoddard solvent and 244 parts of mineral oil of 308 parts of an alkylbenzene sulfonic acid having an equivalent weight of about 430 and containing about 22% unsulfonated alkylbenzene. Water, 381 parts, is added and the mixture is heated under reflux for 15 minutes. It is then cooled to room temperature, yielding the desired magnesium oxide-sulfonate complex in the form of a gel.

Example 2: A mixture of 125 parts of toluene, 225 parts of the alkylbenzene sulfonic acid of Example 1, 680 parts of mineral oil, 550 parts of magnesium oxide and 200 parts of water is heated slowly to reflux temperature (about 100°C) and excess water (about 43 parts) is removed by azeotropic distillation. The residue is stripped under vacuum at 170°C as 117 parts of volatiles are removed. The residue from the stripping is cooled to yield the desired magnesium oxide-sulfonate complex in the form of a gel.

Example 3: A mixture of 2,050 parts of water, 30 parts of magnesium oxide, 294 parts of the alkylbenzene sulfonic acid of Example 1, and 520 parts of oil is heated to 35° to 40°C, and an additional 715 parts of magnesium oxide is added slowly. An exothermic reaction takes place and the magnesium oxide addition is regulated so as to cause a temperature increase of about 10°C per hour to a maximum temperature of about 65°C. Heating is continued until the temperature reaches 85°C, whereupon a gel is obtained containing a clear water layer on top. The excess water (about 1,552 parts) is decanted. To the residue are added 375 parts of mineral oil and 165 parts of toluene, and an additional portion of water (259 parts) is removed by stripping under nitrogen (180°C) and the residue is screened to yield the desired magnesium oxide-sulfonate gel.

Example 4: A mixture of 1,280 parts of water, 18 parts of magnesium oxide, 180 parts of the alkylbenzene sulfonic acid of Example 1, and 215 parts of mineral seal oil is heated to 45°C and an additional 534 parts of magnesium oxide is added, with stirring. The mixture is heated to 80° to 85°C over 2 hours whereupon partial coagulation takes place and a water layer separates. The water layer (about 1,050 parts) is decanted and an additional 290 parts of mineral seal oil is added. The mixture is purged with nitrogen at 140° to 145°C for 2½ hours to remove excess water and is then passed through a 20-mesh screen to yield the desired magnesium oxide-sulfonate complex.

These complexes are homogeneously incorporated into fuel oils, bunker fuels or the like in an amount preferably from 4 to 1,000 ppm, normally in combination with auxiliary detergents of the ashless type, other corrosion- and oxidation-inhibiting agents, etc.

These magnesium complexes can be added directly to the lubricant or fuel. Preferably, however, they are diluted with a substantially inert, normally liquid organic diluent such as mineral oil, naphtha, benzene, toluene or xylene, to form an additive concentrate. These concentrates generally contain about 20 to 90% by weight of the magnesium complex and may contain in addition one or more of the other additives described hereinabove.

Containing Acyl Glycine Oxazolines

R.L. Sung; U.S. Patent 4,266,944; May 12, 1981; assigned to Texaco Inc. has provided a motor fuel composition containing an amount varying from 20 to 200 parts per 1,000 barrels of fuel of certain detergent and corrosion-inhibiting additives. These compounds have the following structure:

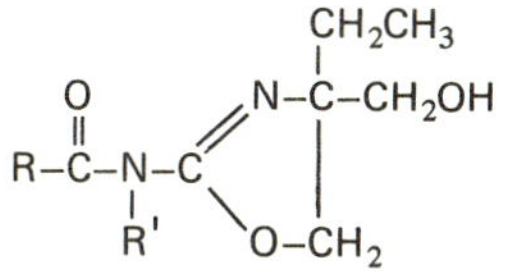

```
                       CH2CH3
                        |
       O             ,N-C-CH2OH
       ||           //  |
     R-C-N-C          |
         |    \       |
         R'    O-CH2
```

where R is lauryl, $C_{11}H_{23}$, oleyl or stearyl; R' is hydrogen or (lower) alkyl. Preferably R and R' taken together contain from 13 to 21 carbon atoms.

Preferably, both the R and R' radicals are straight chain; however they also can be branched and may be substituted with one or more noninterfering substituents such as halogen, cyano, trifluoromethyl, nitro or alkoxy.

These additives preferably are synthesized by reacting a 2-amino-2-(lower)-alkyl-1,3-propanediol with an N-acyl sarcosine in an inert solvent preferably xylene, refluxing the reaction mixture for about 8 hours to azeotrope the xylene and the water of reaction, filtering and stripping the filtrate under vacuum to isolate the product.

N-acyl sarcosines suitable as reactants include lauroyl sarcosine, cocoyl sarcosine, oleoyl sarcosine, stearoyl sarcosine and other fatty sarcosines containing from 8 to 22 carbon atoms. The preferred propanediol is 2-amino-2-ethyl-1,3-propanediol.

N-acyl sarcosines also known as Sarkosyls previously suggested as corrosion inhibitors for fuels were found to be completely extracted into caustic water bottoms so that the fuels lost their corrosion inhibiting properties. However, it was found that the oxazolines of these compounds were completely unextractable by acidic or basic water bottoms.

Example: *Synthesis of oxazoline of Sarkosyl 0* – A mixture of 0.7 mol of oleyl sarcosine and 0.7 mol of 2-amino-2-ethyl-1,3-propanediol in 600 parts of xylene was refluxed and water of reaction was azeotroped over. After 8 hours of reflux, the reaction mixture was cooled and filtered, then stripped under vacuum. The residue was analyzed by IR and elemental analysis.

That these additives have anticorrosion properties was shown by their performance in the NACE Rusting Test, previously described. The table below shows the results of this test for the product of the example at different concentrations in PTB in and against an unleaded base fuel.

The data show that as little as 5 PTB of the additive substantially eliminates rusting.

NACE Rust Rating of Oxazoline of Sarkosyl 0

Concentration	Rust Rating, %
10 PTB	Trace–1
10 PTB	Trace–1
5 PTB	Trace–1
5 PTB	Trace–1
Unleaded base fuel	50–100
Unleaded base fuel	50–100

These compounds also exhibited good carburetor detergency, being equivalent at 100 PTB in detergency to 1,035 PTB of a widely used commercial additive.

Hydrocarbon-Substituted Methylol Phenols

J.F. Pindar, J.M. Cohen and C.P. Bryant; U.S. Patent 4,273,891; June 16, 1981; assigned to The Lubrizol Corporation describe hydroxy aromatic compositions containing (a) a hydroxyl group bonded directly to a carbon of an aromatic nucleus, (b) a hydrocarbon-based substituent of at least about 50 aliphatic carbon atoms bonded directly to a carbon atom of an aromatic nucleus, (c) at least one methylol or lower hydrocarbyl-substituted methylol substituent bonded directly to a carbon atom of an aromatic nucleus, and not having any alkylene linkages between carbon atoms of two aromatic nuclei. These compositions are useful as additives for normally liquid fuels and lubricating oils and as intermediates for preparing other additives for fuels and lubricants.

Typical hydroxy aromatic compositions of the process are formed by reaction of formaldehyde with an alkenyl- or alkyl-substituted phenol wherein the alkyl or alkenyl substituent contains an average of about 50 carbon atoms.

The following are specific illustrative examples of the hydroxy aromatic compounds of the process. All parts and percentages in the examples are by weight and all temperatures are in degrees centigrade (°C) unless expressly stated to the contrary. Molecular weights are determined by vapor phase osmometry (VPO) or gel permeation chromatography (GPC). The textile spirits used in these examples is an aliphatic petroleum naphtha with a boiling range of about 63° to 79°C at 760 torrs.

Example 1A: Polyisobutenyl chloride (4,885 parts) having a viscosity at 99°C of 1306 SUS and containing 4.7% chlorine is added to a mixture of 1,700 parts phenol, 118 parts of a sulfuric acid-treated clay and 141 parts zinc chloride at 110° to 155°C during a 4-hour period. The mixture is then kept at 155° to 185°C for 3 hours before being filtered through diatomaceous earth. The filtrate is vacuum stripped to 165°C/0.5 torr. The residue is again filtered through diatomaceous earth. The filtrate is a substituted phenol having an OH content of 1.88%.

Example 1B: Sodium hydroxide (42 parts of a 20% aqueous sodium hydroxide solution) is added to a mixture of 453 parts of the substituted phenol described in Example 1A and 450 parts isopropanol at 30°C over 0.5 hour. Textile spirits (60 parts) and 112 parts of a 37.7% formalin solution are added at 20°C over a 0.8 hour period and the reaction mixture is held at 4° to 25°C for 92 hours. Additional textile spirits (50 parts), 50 parts isopropanol and acetic acid (58 parts of a 50% aqueous acetic acid solution) are added. The pH of the mixture is 5.5 (as determined by ASTM procedure D-974). The mixture is dried over 20 parts magnesium sulfate and then filtered through diatomaceous earth. The filtrate is vacuum stripped to 25°C/10 torrs. The residue is the desired methylol-substituted product having an OH content of 3.29%.

Example 2A: Aluminum chloride (76 parts) is slowly added to a mixture of 4,220 parts polyisobutenyl chloride having a number average molecular weight ($\overline{M}n$) of 1,000 (VPO) and containing 4.2% chlorine, 1,516 parts phenol, and 2,500 parts toluene at 60°C. The reaction mixture is kept at 95°C under a below-the-surface nitrogen gas purge for 1.5 hours. Hydrochloric acid (50 parts of a 37.5% aqueous hydrochloric acid solution) is added at room temperature and the mixture stored for 1.5 hours. The mixture is washed five times with a total of 2,500 parts water and then vacuum stripped to 215°C/1 torr. The residue is filtered at 150°C through diatomaceous earth to improve its clarity. The filtrate is a substituted phenol having an OH content of 1.39%, a Cl content of 0.46% and a $\overline{M}n$ of 898 (VPO).

Example 2B: Paraformaldehyde (38 parts) is added to a mixture of 1,399 parts of the substituted phenol described in Example 2A, 200 parts toluene, 50 parts water and 2 parts of a 37.5% aqueous hydrochloric acid solution at 50°C and held for 1 hour. The mixture is then vacuum stripped to 150°C/15 torrs and the residue is filtered through diatomaceous earth. The filtrate is the desired product having an OH content of 1.60%, $\overline{M}n$ of 1,688 (GPC) and a weight number average molecular weight ($\overline{M}w$) of 2,934 (GPC).

Example 3: Sodium hydroxide (8 parts of a 50% aqueous sodium hydroxide solution) and 145 parts paraformaldehyde are added to a mixture of 2,240 parts of a substituted phenol prepared as described in Example 2A except the polyisobutene used has a $\overline{M}n$ of 940 (VPO), and 1,271 parts diluent oil at 60°C and the mixture kept at 80°C for 22 hours. Acetic acid (6 parts of glacial acetic acid) is added and the mixture is held at 150°C for 2 hours and then filtered through diatomaceous earth. The filtrate is a 35% oil solution of the desired product having an OH content of 1.10%.

Example 4: Sodium hydroxide (32 parts of a 50% aqueous sodium hydroxide solution) and 290 parts paraformaldehyde are added to a mixture of 4,480 parts of the substituted phenol used in Example 3 and 3,099 parts diluent oil at 40° to 50°C; the mixture is kept at 80° to 85°C for a 14-hour period. Acetic acid (36 parts of glacial acetic acid) is added at 60°C. The mixture is held at 110° to 130°C for a period of 12 hours and then is filtered through diatomaceous earth. The filtrate is a 40% oil solution of the desired product having an OH content of 1.05%.

The compositions described here are useful in and of themselves as antirust and anticorrosion agents for fuels and lubricants. They are also useful as intermediates for the production of compositions that function in fuels and lubricants as detergents and dispersants for sludge formed in internal combustion engines.

These compositions can be employed in amounts preferably of from 0.1 to 10 pbw/100 parts of oil in a variety of lubricants based on diverse oils of lubricating viscosity, including natural and synthetic lubricating oils and mixtures thereof. These lubricants include crankcase lubricating oils for spark-ignited and compression-ignited internal combustion engines, such as automobile and truck engines, two-cycle engines, marine and railroad diesel engines, and the like. They can also be used in gas engines, stationary power engines, turbines and the like.

The following patents, abstracted elsewhere in this book, are for additives which show anticorrosion properties, as well as being useful detergents or dispersants:

4,134,846	4,147,641	4,231,758
4,144,034	4,203,730	4,231,759
4,144,035	4,204,841	4,259,086
4,144,036	4,210,425	

PIPELINE CORROSION INHIBITORS

Polymerized Fatty Acids plus Mannich Polymer

It is known that pipelines constructed of ferrous metals used to transport petroleum products, both refined and unrefined, corrode due to the fact that the petroleum products contain quantities of water and traces of corrosive impurities including oxygen. The problem is particularly severe when the pipelines are used to transport gasoline and other light petroleum distillate products. In order to protect the pipelines from this corrosive environment, it is common to treat the fluids transported by these pipelines with small quantities of corrosion inhibitors.

A preferred pipeline inhibitor is a blend of polymerized fatty acids formed by the polymerization of linoleic acid with oleic acid which produces a blend of dimer and trimer acids such that the trimer acid content is within the range of 9 to 36% and, preferably, 14%, with the balance being dimer acid. The preferred polymerized acids used as inhibitors of the type described should have an acid number of at least 110 and, preferably, greater than 150.

While these polymerized fatty acids give protection to the ferrous metals of pipelines transmitting petroleum products, the protection is not complete since they are incapable of preventing red rust. It would be of great benefit to the art if it were possible to find an additive which, when added to the polymerized fatty acid inhibitors, would improve their ability to inhibit corrosion in pipelines.

A.B. Gainer; U.S. Patent 4,197,091; April 8, 1980; assigned to Nalco Chemical Co. describes such an additive which, when used in an amount of from 0.5 to 5% in the described polymerized dimer/trimer fatty acid inhibitor and added to the hydrocarbon liquid to be transported by pipelines provides corrosion resistance of nearly 100%. The rust inhibitor consists of a major portion of a mixture of C_{36} dicarboxylic dimer acid and a C_{54} trimer acid, which mixture has an acid number of at least 110 and from 0.5 up to about 5% of a composition from the group consisting of (a) N,N'-disalicylidene-1,2-propylenediamine and (b) a Mannich polymer prepared by condensing 2 mols of p-dodecyl phenol with 1 mol each of formaldehyde and ethylenediamine.

The formula for the Mannich polymer (Formula A) is as follows in which the molecular weight is about 693.8 relative to a polystyrene standard.

Ingredients	Percent by Weight
Formaldehyde, 37°C inhibited	19.5
Ethylenediamine, 99%	7.3
p-Dodecyl phenol	31.5
Heavy aromatic naphtha, Exxon	41.7

Examples: To illustrate the effectiveness of the process, these compositions along with the mixed polymerized acids alone and a commercial inhibitor were tested using Military Test MIL-I-25017C, 8 Mar. 1971, superseding MIL-I-25017B, 22 Oct. 1962. Grade 1018 steel spindles are tested for rust formation in stirred isooctane-synthetic seawater mixtures.

The results of these tests using depolymerized isooctane as test fluid are as follows:

Composition Used	PTB Concentration	Rust (%)
Formula A	2	6, 12
45% dimer, 15% trimer, 2% formula A, 38% fuel oil	2	<0.1, 0
45% dimer, 15% trimer, 2% N,N'-disalicylidene-1,2-propylene diamine, 38% fuel oil	2	<0.1, 1
45% dimer, 15% trimer, 40% fuel oil	2	4, 4

Reaction Product of a Hydrocarbylsuccinic Anhydride and an Aminotriazole

Ashless detergents which are the reaction products of a hydrocarbylsuccinic anhydride and an aminotriazole such as 3-amino-1,2,4-triazole and 5-amino-1,2,4-triazole have been discussed in U.S. Patent 4,257,779.

The additives proposed by *R.L. Sung, J.J. Bialy, P. Dorn, W.P. Cullen and J.W. Nebzydoski; U.S. Patent 4,263,015; April 21, 1981; assigned to Texaco Inc.* are very similar, except that the hydrocarbyl radical has only from 6 to 30 carbon atoms when used for the purposes designated in this patent, rather than the 90 to 120 carbon atoms preferred for use in making the detergent additive.

This additive is effective as a rust inhibitor for motor fuels, fuel oils and lubricating oils.

Examples of suitable hydrocarbon-substituted succinic anhydrides include dipropenylsuccinic anhydride, tripropenylsuccinic anhydride, dodecenylsuccinic anhydride, tetrapropenylsuccinic anhydride, pentapropenylsuccinic anhydride and hexadecenylsuccinic anhydride. The hydrocarbon-substituted succinic anhydride and the aminotriazole are preferably reacted in approximately equimolar amounts.

Example 1: 100 g (1 mol) of tetrapropenylsuccinic anhydride and 100 g (1 mol) of 3-aminotriazole are dissolved in 1,200 ml of xylene. The reaction mixture was refluxed for about 6 hours followed by the removal of the solvent by distillation.

The reaction product recovered contained 15.7% nitrogen. IR spectroscopy indicated that the product was a mixture of the tetrapropenylsuccinamide of 3-aminotriazole and the alkenylsuccinimide of 3-aminotriazole.

Example 2: 142 g (1 mol) of pentapropenylsuccinic anhydride and 100 g (1 mol) of 3-aminotriazole are dissolved in 1,200 ml of xylene and refluxed for about 6 hours. The solvent is removed by distillation. The product is a mixture of the pentapropenylsuccinamide acid of 3-aminotriazole and the pentapropenylsuccinimide of 3-aminotriazole.

The reaction product is employed in a motor fuel composition in a concentration ranging from about 0.005 to 0.2% by weight.

The reaction product of Examples 1 and 2 were tested for corrosion-inhibiting properties in gasoline in the Colonial Pipeline Rust Test (described in U.S. Patent 4,185,965 above).

The base fuel employed in the tests was an unleaded grade gasoline having a research octane number of about 91. This gasoline consisted of about 24% aromatic hydrocarbons, 8% olefinic hydrocarbons and 68% paraffinic hydrocarbons and boiled in a range from about 90° to 375°F. A Base Blend was prepared from the foregoing base fuel and conventional antioxidant and metal deactivator in the amount of 4.5 PTB. The results are set forth in the following table:

Colonial Pipeline Rust Test

Run	Additive	Rust (%)
1	Base blend	50–100
2	Base blend + 10 PTB Ex. 1	Trace, trace
3	Base blend + 5 PTB Ex. 1	Rust-free, trace
4	Base blend + 1 PTB Ex. 1	Rust-free, trace

The foregoing tests show that there was a dramatic improvement in the rust-inhibiting properties of a motor fuel composition containing a range of concentrations of the rust-inhibiting additive.

Containing a Polymerized Unsaturated Aliphatic Monocarboxylic Acid

There is a need for a corrosion inhibitor for use in fuel storage tanks and pipelines where temperatures generally parallel outdoor ambient temperatures, maximum temperatures only occasionally exceeding about 100°F (38°C). The corrosion inhibitor should be effective at low concentrations and should not emulsify undesirable amounts of water. The two-component, acid/acid corrosion inhibitor described by *B.H. Garth and F.H. Schmidt; U.S. Patent 4,214,876; July 29, 1980; assigned to E.I. Du Pont de Nemours & Company* satisfies that need.

This corrosion inhibitor composition consists essentially of, by weight, (a) about 75 to 95%, of at least one polymerized unsaturated aliphatic monocarboxylic acid having about 16 to 18 carbon atoms per molecule, and (b) about 5 to 25% of a monoalkenylsuccinic acid in which the alkenyl group has 8 to 18 carbon atoms.

Preferred compositions contain most preferably from 80 to 85% of the polymerized monocarboxylic acid component, and most preferably 15 to 20% of the monoalkenylsuccinic acid component.

The polymerized unsaturated aliphatic monocarboxylic acids to be employed herein are those prepared from the corresponding monocarboxylic acids by methods which are well known in the art. Such polymerized acids generally contain 75% or more of dimer, trimer and higher polymerized acids and 25% or less of unpolymerized monocarboxylic acid.

For convenience, these acids may be referred to as Component A, or dimer acid, if a dimer acid is the major constituent of Component A.

Commercially available dimer acids include Empol Dimer Acids (Emery Industries) prepared by polymerizing linoleic acids and containing from 40 to 95% of dimer acids and from 4 to 25% of trimer acids. Commercial trimer acids include Empol Trimer Acids which contain from 40 to 95% of trimer acids and from 5 to 25% of dimer acids. Both types of compositions can contain up to 25% of monocarboxylic acids.

Because of their availability and low cost, mixtures of fatty acids called tall oil fatty acids are often used to produce dimer and trimer acid compositions. Polymerized tall oil fatty acids, such as Acintol FA-7002 (Arizona Chemical Company) can be used to prepare these compositions.

The contemplated monoalkenylsuccinic acids (Component B) are well-known in the art. The preferred monoalkenylsuccinic acid is dodecenylsuccinic acid, more preferably dodecenylsuccinic acid prepared from propylene tetramer.

These additives incorporated into hydrocarbon fuels in the range of about 0.0002 to 0.002% by weight (0.5 to 5 PTB) provide satisfactory corrosion-inhibiting properties. Concentrations higher than about 0.002% can be used but do not appear to provide further benefits. The most preferred concentration range is about 0.0004 to 0.0012% by weight (1 to 3 PTB).

The polymerized monocarboxylic acid Acintol FA-7002 was combined with dodecenylsuccinic acid in various weight ratios and dissolved in xylene to provide concentrates containing 79% by weight of the combination. The concentrates were added to depolarized isooctane in the concentrations indicated. The tests were run in duplicate.

For comparison purposes, similar concentrates were prepared using dodecenylsuccinic anhydride instead of the dodecenylsuccinic acid and similarly tested. The tests showed that the corrosion inhibitors of this process provide effective rust protection at very low concentrations (from 0.75 PTB up). The results also show that these corrosion inhibitors are markedly superior to similar combinations employing dodecenylsuccinic anhydride in place of dodecenylsuccinic acid. Thus, at an 81:19 ratio, twice as much of dodecenylsuccinic anhydride as dodecenylsuccinic acid is required to obtain no rusting. Similarly, at 86:14 and 91:9 ratios, the combinations containing the dodecenylsuccinic acid are greatly superior, per unit concentration, to the combinations containing the anhydride.

The results of the testing also show that the combination of Components A and B are synergistic.

COLD-END ADDITIVES–ETHYLENE POLYAMINES PLUS ALKANOLAMINES

As is well-known to boiler operators, sulfur-containing fuels present problems not only from a pollutional point of view, but also with respect to the life and operability of metallic equipment and parts which are in contact with the flue gases containing the sulfur by-products of combustion.

Upon combustion, the sulfur in the fuel is converted to sulfur dioxide and sulfur trioxide. In the flue gas, sulfur trioxide and water vapor are in equilibrium with sulfuric acid. Below about 450°F, essentially all of the SO_3 is converted to H_2SO_4 for typical flue gas compositions of oil-fired boilers. The resulting sulfuric acid condenses upon metal surfaces which are at temperatures below the acid dew point. Corrosion results from the attack of the condensed sulfuric acid on the metals.

The greater the sulfur content of the fuel, the more sulfuric acid will likely be produced. This is particularly the case in industrial and utility operations where low-grade oils are used for combustion purposes.

The basic area to which the following process is directed is referred to in the industry as the cold-end of a boiler, which is generally the path in the boiler system that the combustion gases follow after the gases have, in fact, performed their primary service of producing and/or superheating steam.

The use of alkanolamines alone as a cold-end additive is disclosed in U.S. Patent 4,134,728.

R.J. Sujdak; U.S. Patents 4,206,172; June 3, 1980; and 4,224,180; Sept. 23,

1980; both assigned to Betz Laboratories, Inc. has found that if a combination of ethylene polyamines and aliphatic, water-soluble alkanolamines is fed, preferably as an aqueous solution and in droplet form, to the moving combustion gases upstream of the cold-end surfaces to be treated and preferably at a point where the gases are undergoing turbulence, it will travel along with the gases as vapor and/or liquid droplets and deposit on the downstream cold-end surfaces.

In terms of a general formula, useful ethylene polyamines for this purpose could best be described by the following formula:

$$H_2N{-}CH_2({-}CH_2{-}NH{-}CH_2{-})_nCH_2{-}NH_2 \quad (1)$$

Since ethylenediamine is the lowest homolog in the series, the lower limit for n is 0. There seems to be no upper limit for n in Formula (1) other than that based on the commercial availability of the material. In any event, the highest homolog tested was poly(ethyleneimine) having the formula:

$$-\!\!\left(CH_2{-}CH_2{-}NH\right)\!\!-_n$$

which material had an average molecular weight of about 50,000 to 100,000, and, therefore, n was about 1,000 to 2,500.

The amount of additive used could vary over a wide range depending on the nature and severity of the problem to be solved and would be a function of the sulfur content of the oil.

The amount of additive will preferably comprise from about 0.0004 to 0.95 mol/bbl ethylene polyamine and from about 0.05 to 1 mol/bbl alkanolamine. The preferred relative proportions are from about 0.1 to 0.5 mol/bbl ethylene polyamine and from about 0.1 to 0.75 mol/bbl alkanolamine. Based on economic considerations, the total amount of active additive should not exceed about 1.1 mol/bbl. On a weight basis, the ratio of alkanolamine to ethylene polyamine is about 25:1 or less.

In order to assess the efficacy of the additives, various tests were conducted. Since the primary function of a cold-end additive is to eliminate or reduce corrosion caused by the condensation of sulfuric acid, techniques that measure corrosion were expected to yield the most direct information about product performance. Accordingly, the well-known method of quantifying the reduction in corrosion of a stainless steel air-cooled probe was used for determining efficacy as cold-end additives. The probe used was similar to a standard British Central Electricity Research Laboratories (CERL) acid deposition probe.

Flue gas constituents were allowed to condense on the probe for 45 minutes. The probe was then immediately washed with doubly distilled water and analyzed for iron and sulfate. Corrosion was measured by analyzing the probe washings for water-soluble iron.

The various additives tested were sprayed, using a standard atomizing spray nozzle arrangement, into the combustion gases at a point of turbulence located upstream of the air-cooled probe.

Immediately before base loading, the boiler was taken through a soot blowing cycle, and the burner tip was manually cleaned. The boiler was then baseloaded for 1 hour. Fuel oil of precisely the same composition must be fired over a given period of time to ensure reproducibility of baseline data throughout the period.

Example: The efficacy of various combined treatments was evaluated in tests using a boiler at a well-known oil refinery. The boiler was manufactured by Riley. During the tests, the boiler produced between 127,000 and 155,000 pounds of steam per hour, and it was baseloaded during each experiment. The generated steam was near 700°F at 600 psig pressure. Flue gas temperatures at the location of the corrosion probe ranged from 685° to 720°F. The materials tested were monoethanolamine (MEA) and triethylenetetramine (TETA), both obtained from Union Carbide. The results of these tests are reported in Figure 6.1 as a plot of concentration of iron in the probe washings, in ppm, against sampling temperature (°F). The sulfur content of the fuel oil was 0.69%.

Figure 6.1: Test for Corrosion Resistance

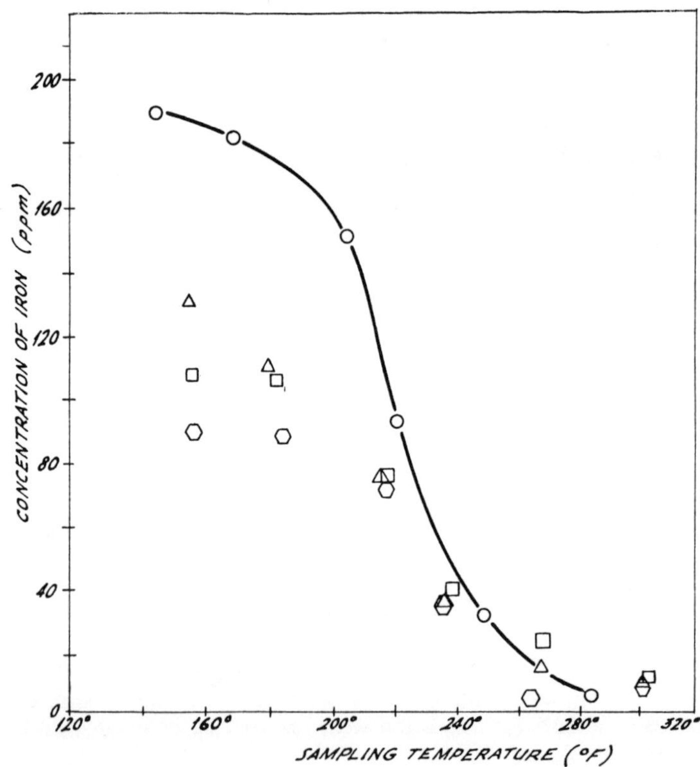

Source: U.S. Patent 4,206,172

These test results are to be interpreted according to the following legend:

Symbol	Mols of MEA	Mols of TETA	Total Mols of Amine
	..............(Per Barrel of Oil)..............		
Open circle	0	0	0
Open triangle	0	0.088	0.088
Open hexagon	0.141	0.0624	0.203

ANTIOXIDANTS

Alkenylsuccinic Acid or Anhydride/Aniline-Aldehyde Resin

S. Chibnik; U.S. Patent 4,153,564; May 8, 1979; assigned to Mobil Oil Corporation has provided compounds that act as emulsifiers, detergents and antioxidants when added to a lubricant or a fuel. They are prepared from (1) an alkenylsuccinic anhydride or acid and an aniline-aldehyde resin, or (2) the product of (1) and an aromatic triazole, e.g., benzotriazole, and aldehyde.

One class of additives may be made by a process summarized as follows:

$$\underset{(1)}{\text{R-CH-C(=O)-O-C(=O)-CH}_2\text{ (ring)}} + \underset{(2)}{\text{Ar(NH}_2\text{)-CH}_2\text{-[Ar(NH}_2\text{)-CH}_2\text{]}_x\text{-Ar(NH}_2\text{)}} \longrightarrow \underset{(3)}{\left[\text{succinimide(R)-N-Ar}\right]_n\left[\text{CH}_2\text{-Ar(NH}_2\text{)}\right]_m}$$

where Ar is an aromatic hydrocarbyl containing 6 to 18 carbon atoms, specifically derived, for example, from benzene, naphthalene and anthracene, R is an alkenyl group containing from 40 to 1,000 carbon atoms, preferably having 50 to 250 carbon atoms, n is from 1 to 3, x is from 0 to 4, preferably 0 to 2, and m is from 1 to 3.

The succinic anhydride used is prepared by conventional means. The alkenylsuccinic member is obtained by reaction of the anhydride with a polymer made from an olefin such as ethylene, propylene or butene or with a copolymer of these.

Another of the reactants, the aromatic amine-aldehyde resin, may be purchased. They can be prepared generally by reacting, at from about 60° to 120°C, an aromatic amine and an aldehyde in a ratio to yield the product (2) as set forth above.

Product (3), although itself effective as specified above, can be further reacted with an aromatic triazole and aldehyde.

The fuels in which these compounds may be used include the mixtures of hydrocarbons boiling in the gasoline range, usually from about 100° to 425°F. They can consist of straight or branched-chain paraffins, cycloparaffins, olefins and aromatic compounds or any mixture of such compounds, and may contain other additives such as antiknock compounds and the like.

Normally, the gasoline will contain from about 0.002 to 0.05% by weight of the reaction product of alkenylsuccinic anhydride with the amine-aldehyde resin, preferably about 0.01 to 0.04% by weight, or it will contain 0.01 to 0.04% by weight of the product of reaction between the abovementioned product and an aldehyde and an aromatic triazole.

Example 1: 41 parts of a mixture of products (Jefferson Chem. Co.) having the formula

$$C_6H_4(NH_2)-CH_2-\left[C_6H_3(NH_2)-CH_2\right]_x-C_6H_4(NH_2)$$

where x is either 0, 1 or 2, was reacted with 412 parts of polyisobutenylsuccinic anhydride (prepared using 1,300 molecular weight polyisobutene) at 150°C for 3 hours under a 5 mm vacuum. The product was diluted with a light mineral oil and was filtered.

Example 2: 12 parts of the reactant shown in Example 1 and 130 parts of the same polyisobutenylsuccinic anhydride were reacted for 3 hours at 150°C under a 2 mm vacuum. The molar ratio of anhydride to resin was 1:1. The product was diluted with 150 parts of light oil and was cooled to 95°C. 8 parts of tolyltriazole was added and a total of 3 parts of paraformaldehyde was added in equal portions over a 1-hour period. The reaction was completed by removal of volatiles at 150°C for 1 hour at 2 mm of pressure.

Example 3: Polyisobutenyl (500 molecular weight) succinic anhydride (100 parts) and 31 parts of the resin shown in Example 1 were reacted, and this product was reacted with 19 parts of benzotriazole and 6.3 parts of paraformaldehyde. Reaction conditions were similar to those of Examples 1 and 2.

Polyalkylene Amine plus Mannich Base

The deterioration due to oxidation and the like of distillate fuels, particularly in diesel fuel, manifests itself through the appearance of color and gums. The tacky oxidized fuel deposits adhere readily to injector parts and can cause injector sticking, nozzle-hole plugging and leakage past critical surfaces.

While many materials might effectively act as commercially successful dispersants for the gum, the field is severely limited to relatively few materials. Also, since the dispersant is an additive to the fuel, it must not significantly increase the deposits created in the combustion chamber, which interfere with the proper functioning of the piston.

O.L. Harle; U.S. Patent 4,166,726; September 4, 1979; assigned to Chevron Research Company describes a fuel composition for compression ignition engines which consists of a hydrocarbon boiling from 120° to 455°C and which contains 5 to 300 ppm of a polyalkylene amine and 5 to 300 ppm of a Mannich base.

The Mannich condensation reaction is well known in the art, and involves the condensation of the alkylphenol, an aldehyde and an amine. A particularly preferred alkylated phenol is dodecylphenol. The most preferred aldehyde reactant is formaldehyde, which may be used in its monomeric or its polymeric form, such as paraformaldehyde. A preferred amine is methylamine.

The condensation reaction will occur by simply warming the reactant mixture to a temperature sufficient to effect the reaction, e.g., from 75° to 175°C, for from 1 to 8 hours.

The polyalkylene amines which are suitable for use in the process are commercially available materials which are generally known for their detergent or dispersant properties. U.S. Patents 3,898,056, 3,438,757 and 4,022,589 show representative polyalkylene amines and methods of manufacture.

The alkylene amines include principally methylene amines, ethylene amines, propylene amines, butylene amines, pentylene amines, hexylene amines, heptylene amines, octylene amines, other polymethylene amines, and also the cyclic and the higher homologs of the amines such as piperazines and amino-alkyl-substituted piperazines.

The polyalkylene amine will generally have an average molecular weight in the range, preferably, of 1,000 to 1,500 and will have been reacted with sufficient amine to contain preferably from 0.8 to 1.2% by weight basic nitrogen.

The effectiveness of the additive combination of this formulation toward stabilizing diesel fuel from thermal degradation is shown by a test, in which the additive package and the diesel fuel are mixed until solution is complete. The resulting solution is filtered through a Whatman No. 1 filter paper. Then a 300-ml portion of the filtrate is transferred into 500-ml Pyrex bottles. Each bottle is covered with a piece of aluminum foil having a pinhole. The test samples are placed in an oven maintained at 105°C for 60 hours. At the end of this time, the bottles are allowed to cool to ambient temperature in the dark.

The sample bottle is shaken until all sediment is in suspension, and then it is filtered through a 5-micron-pore-size Millipore filter paper. The filter paper and precipitate collected thereon are dried in an oven at 90°C for 2 hours. The sample bottle is washed with a total of 50 ml of gum solvent (50% methanol/acetone). This solution is transferred to tared beaker and allowed to evaporate. The weight of the filter and gum residue is then determined.

In the above test, it is desired to limit or eliminate the residue due to thermal decomposition. Therefore, the smaller the residue value, the better the thermal stability of the test fuel.

A diesel fuel with no additive gave a residue of 41 ppm. The same fuel with 25 ppm of a polybutene amine and a Mannich base prepared from p-dodecylphenol, formaldehyde and methylamine in a 1:1:1 mol ratio gave a residue of only 14 ppm.

Metal Complexes of Thiobis(Alkylphenols)

According to *M. Braid; U.S. Patent 4,198,303; April 15, 1980; assigned to Mobil Oil Corporation*, nickel thiobis(alkylphenolates), e.g., nickel 2,2'-thiobis(4-t-octylphenolate), complexed with hydroxy materials such as alcohols and phenols impart antioxidant properties to various organic media such as lubricants, hydrocarbon fuels, etc., when incorporated therein. Synergistic or improved antioxidant compositions are provided when the hydroxy complexes are admixed with known antioxidants such as aryl amines and/or hindered phenols.

The complex has the following general structure:

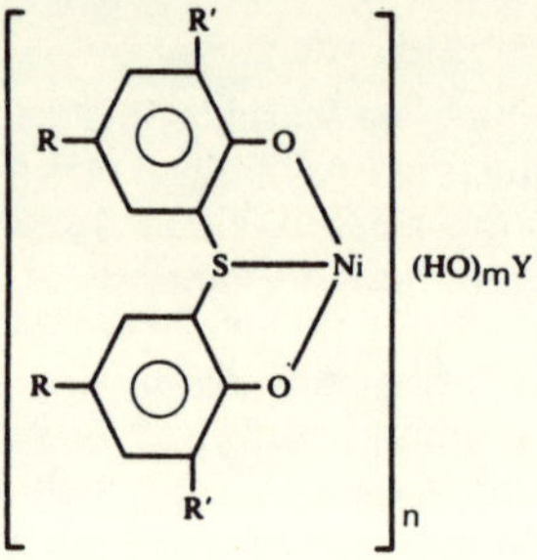

$$\left[\text{bis(R,R'-phenolato-O)-S-Ni complex} \right]_n (HO)_m Y$$

where R is either hydrogen or an alkyl group having from 1 to about 30 carbon atoms, R' is hydrogen or an alkyl group containing from 1 to 8 carbon atoms in any isomeric configuration except those in which a carbon atom bonded to a ring carbon atom is, in turn, bonded to more than two other carbon atoms, Y is a ligand such that YOH is preferably methanol, ethanol, n-propanol, isopropanol, n-butanol, isobutanol, or phenol.

Preferred aryl amine coantioxidants are phenyl naphthylamines. The preferred hindered phenolic compound is 4,4'-methylene-bis(2,6-di-tert-butylphenol). Generally the weight ratio of nickel complex to aryl amine and/or hindered phenol is from about 0.1-5.0 to 1.

The organosulfur-containing hydroxy-substituted nickel complexes can be effectively employed in an amount from about 0.01 to 5% by weight and preferably in an amount from about 0.1 to 2% by weight, of the total weight of the lubricant composition. As mentioned above, the nickel complexes may be incorporated in any organic media normally subject to oxidative degradation, especially lubricating viscosity hydrocarbon fuels and fuel oils which may be mineral oils or fractions thereof or synthetic oils.

Examples of the preparation of several of the nickel complexes which proved useful as antioxidants in mineral oil are given below.

Example 1: *Preparation of [2,2'-thiobis(4-tert-octylphenolato)]-2-propanol Nickel(II)* – 150 g of nickel 2,2'-thiobis(4-tert-octylphenol-phenolate) (purchased commercially and manufactured in accordance with U.S. Patent 2,971,940), melting range of 147° to 149°C, was added to about 650 ml of 2-propanol. The mixture was then heated while stirring. As the reaction temperature approached 85°C all of the solids dissolved. After refluxing for about ¼ to ½ hour solids began to precipitate. The reaction mixture became progressively more turbid and after 1.5 hours of refluxing the hot mixture was filtered and the solids collected and dried. 75.7 g of a light green colored solid with a melting point higher than 300°C were obtained. Elemental analysis of solids prepared in this way corresponded to the nickel 2,2'-thiobis(4-tert-octylphenolate) complex with 2-propanol, [2,2'-thiobis(4-tert-octylphenolato)]-2-propanol nickel(II).

Example 2: Nickel 2,2'-thiobis(4-tert-butyl-6-methylphenol-phenolate) prepared as described in U.S. Patent 2,971,940 (20 g) was dissolved in petroleum ether, boiling point 30° to 60°C. As the petroleum ether was allowed to boil off together with some of the alcohol, about 125 ml of 2-propanol was added to the boiling solution. At the end of the reaction period 50 ml of 2-propanol was added to bring the reaction mixture back to its original volume. The hot mixture was then filtered to collect the precipitated solids. The complex [2,2'-thiobis(4-tert-butyl-6-methylphenolato)] -2-propanol nickel(II) was then obtained as a light green solid, melting point >300°C.

Example 3: Nickel 2,2'-thiobis(4-tert-octylphenol-phenolate) prepared as described in U.S. Patent 2,971,940 (103 g) was dissolved in warm petroleum ether, boiling point 30° to 60°C (400 ml) and while stirring methanol (10 ml) was added at a rate sufficiently slow as to control the frothing of solvent generated by the resulting exothermic reaction. Additional methanol was then added while the mixture was heated to boil off the petroleum ether until solvent exchange was complete. The green solids which began to precipitate with the first methanol addition increased in amount throughout the reaction period and were collected by filtration of the hot methanolic reaction mixture. The complex [2,2'-thiobis(4-tert-octylphenolato)] methanol nickel(II) was thus obtained as a light green powdery solid melting higher than 300°C.

In related work by *M. Braid; U.S. Patent 4,211,663; July 8, 1980; assigned to Mobil Oil Corporation*, transition metal thiobis(alkylphenol-phenolates) are converted by treatment with aqueous alkali metal hydroxide to mixed alkali-transition metal complexes. For example, nickel(II) 2,2'-thiobis(4-tert-octylphenol-phenolate) upon treatment with aqueous potassium hydroxide is converted to potassium-containing derivatives with no observable loss or replacement of nickel.

It has also been found that di(transition metal)thiobis(alkylphenolates) may be converted to alkali-containing di(transition metal) complexes by similar treatment with aqueous alkali metal hydroxide. These derivatives have shown improved antioxidant properties compared with the original transition metal compound, and oxidation-inhibiting synergism has been observed in combinations with known antioxidants such as N-phenyl-1-naphthylamine.

These additive complexes are derived from the following general structure:

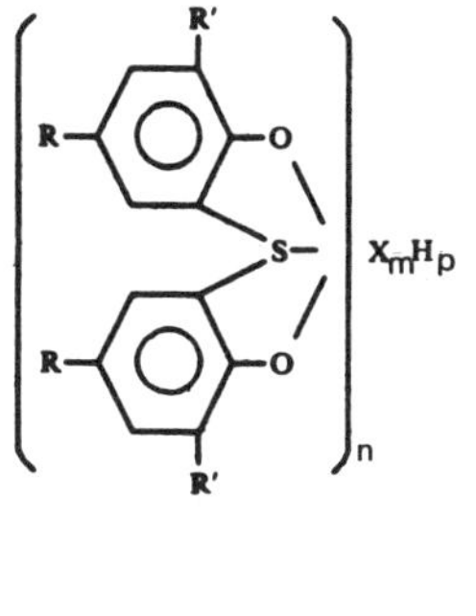

in which R is either hydrogen or an alkyl group having from 1 to about 30 carbon atoms or preferably 1 to 16 or 4 to 8 carbon atoms in any isomeric arrangement and R' may be the same as R except that isomers in which the carbon atom attached to the ring is connected to more than 2 carbon atoms are excluded. X is a transition metal selected from cobalt, iron, nickel and copper, and n = 1 to 3, m = 1 to 2, and p = 0 to 2.

These organic sulfur-containing alkali metal/transition metal complexes may be incorporated in any organic media normally subject to oxidative degradation, for example, oils of lubricating viscosity or greases prepared therefrom in which any of the abovementioned oils or fluids may be employed as vehicles.

Continuing his work on metal thiobis(alkylphenol-phenolato), *M. Braid; U.S. Patent 4,225,448; September 30, 1980; assigned to Mobil Oil Corporation* has found that the copper thiobis(alkylphenol-phenolates) such as 2,2'-thiobis(4-tert-octylphenol) copper complexes have the same proportion (2:1) of phenol to metal as the nickel complexes previously discussed but have different structures. Structures are as follows:

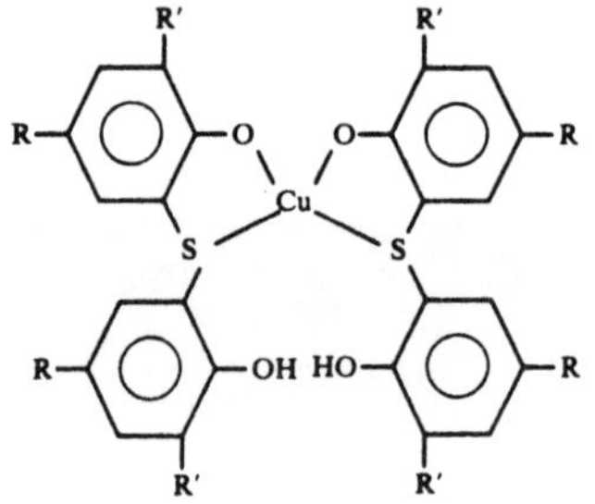

(1A) Cis

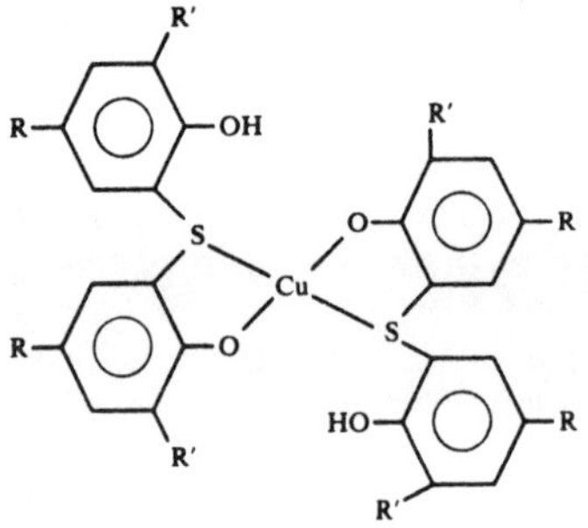

(1B) Trans

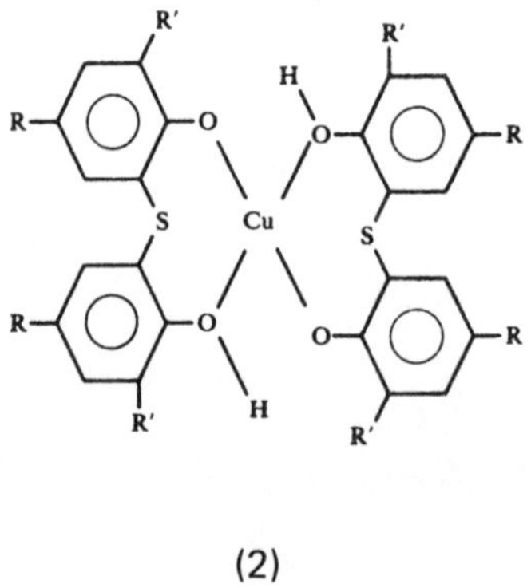

(2)

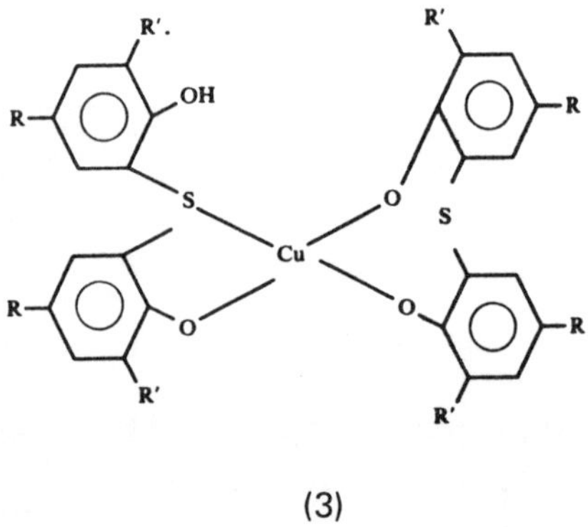

(3)

in which R is either hydrogen or an alkyl group having from 1 to about 30 carbons and, preferably 4 to 8 carbon atoms in any isomeric arrangement and R' is either hydrogen or an alkyl group having from 1 to 8 carbon atoms in any isomeric arrangement except that the carbon atom bonded (or attached) to the ring carbon atom is attached to no more than 2 other carbon atoms. Particularly preferred are alkyl groups wherein R is C_8H_{17} and 1,1,3,3-tetramethylbutyl.

Example 1: A mixture of 2,2'-thiobis(4-tert-octylphenol) (100 g) and copper(II) acetate monohydrate (22.8 g) in xylene (300 ml) was refluxed while stirring for about 1 hour while all of the water and some acetic acid was removed as an azeotropic distillate. Heating and stirring were continued for an additional 4.5 hours during which xylene and acetic acid were azeotropically distilled from the reaction mixture and fresh xylene was added concurrently to replace the distillate. At the end of this period acetic acid could no longer be detected in the distillate. Xylene solvent was removed from the reaction mixture by rotary evaporation. The dark brown semisolid residue was extracted with cyclohexane. The insoluble solids (41.9 g) were recrystallized from isooctane to afford the brown solid copper complex, melting point 150° to 155°C, for which the elemental analysis corresponded to a composition containing copper and 2,2'-thiobis(4-tert-octylphenol) in the ratio of 1:2. Isomer A, i.e., Structure (1A), (2) or (3).

Remaining in the crystallizing solvent are mixtures of the remaining isomers, e.g., if (1A) is the crystalline product, then (1B), (2) and/or (3) remain in the solvent and these are useful, as well.

Example 2: The mixture of Example 1 was further treated as follows. Evaporation of solvent from the cyclohexane extract left a brown solid residue which was treated with petroleum ether, boiling point 30° to 60°C, to separate a small amount of unreacted thiobis(alkylphenol). The mixture was filtered and the filtrate was stripped of the solvent to leave as a brown solid residue (54.6 g), an isomeric copper complex of 2,2'-thiobis(4-tert-octylphenol), melting point 108° to 112°C, with elemental analysis again corresponding to a ratio of 1:2. Isomer B, i.e., Structure (1B), (2) or (3).

Copper 2,2'-thiobis(4-tert-octylphenol) complex (Isomers A and B) were tested as antioxidants in lubricating oil along with the corresponding nickel complex. Isomer A showed improved oxidation inhibition over the nickel complex; Isomer B was not as effective as the nickel complex.

End-Capped Phenols with 1 to 10 Methylene-Bridged o-Hydrocarbyl Phenol Units

R.E. Malec; U.S. Patent 4,222,884; September 16, 1980; assigned to Ethyl Corporation provides an antioxidant composition which is normally a liquid making it easy to handle and to dissolve in organic substrates. It comprises a central segment of 1 to about 10 methylene-bridged o-hydrocarbyl phenol units end-capped with 3,5-dihydrocarbyl-4-hydroxybenzyl groups.

These antioxidants provide hydrocarbons such as gasoline, kerosene, diesel fuel, fuel oil and jet fuel with effective antioxidant protection. The most preferred of these antioxidants is a mixture of methylene-bridged compounds having the structure:

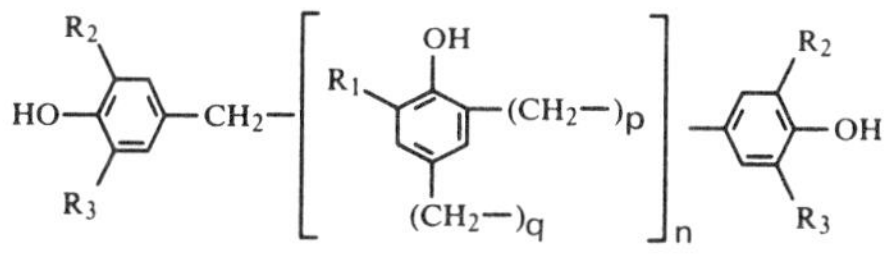

where R_1 and R_2 are tert-butyl groups, R_3 is either a methyl or a tert-butyl group, n is an integer from 2 to 5, p and q are 0 or 1, and p + q equals 1.

The antioxidants are used in the material to be protected in a range of from 0.05 to 5% by weight. Good results are usually achieved using from 0.1 to 3% by weight.

The additives are made by first condensing the o-hydrocarbyl phenol with formaldehyde using KOH (0.2 to 0.3 mol/mol of phenol) as catalyst. The reaction is preferably carried out in such a solvent as methanol, ethanol or isopropanol.

After the initial reaction has proceeded to produce the desired central segment, 2,6-dihydrocarbyl phenol is added to end-cap the condensation product.

The amount of 2,6-dihydrocarbyl phenol added should be an amount sufficient to end-cap the central segments. A preferred range is about 0.3 to 1.5 mols/mol of initial o-hydrocarbyl phenol.

In a highly preferred embodiment about 1.1 to 1.5 mols of formaldehyde/mol of o-hydrocarbyl phenol are used in the first stage and about 1.1 to 1.5 mols of formaldehyde/mol of 2,6-dihydrocarbyl phenol are used in the second stage.

Example: In a reaction vessel was placed 150 g (1.0 mol) of o-tert-butylphenol, 200 g methanol, 37.5 g (1.25 mols) of flaked paraformaldehyde and 14 g (0.25 mol) of potassium hydroxide. The mixture was stirred under nitrogen for 1.5 hours at 55° to 60°C forming the internal segments.

Following this, 227 g (1.1 mols) of 2,6-di-tert-butylphenol and 120 g (1.5 mols) of 38% aqueous formaldehyde solution were separately added over a 2-hour period while the reaction mixture was refluxed (70° to 77°C). The final mixture was heated an additional 2 hours at reflux. It was then diluted with heptane and acidified with 30 g of glacial acetic acid. The mixture was water-washed several times and then stripped of volatiles by heating under vacuum. The final product was a viscous yellow liquid weighing 429 g.

U.S. Patent 4,256,596, abstracted elsewhere in this book, discusses a composition which also exhibits good antioxidation properties.

-7-

COMBUSTION IMPROVERS AND FRICTION REDUCERS

FRICTION REDUCERS

The next six patents relate to the use of various molybdenum complexes as additives for hydrocarbon compositions such as gasoline, fuel oil and lubricating oils including greases, industrial oils, gear oils and lubricants for engines and other equipment having moving parts operating under boundary lubricating conditions.

There are many instances, as is well-known, particularly under boundary lubrication conditions where two rubbing surfaces must be lubricated, or otherwise protected, so as to prevent wear and to insure continued movement. Moreover, where, as in most cases, friction between the two surfaces will increase the power required to effect movement and where the movement is an integral part of an energy conversion system, it is most desirable to effect the lubrication in a manner which will minimize this friction. As is also well-known, both wear and friction can be reduced, with various degrees of success, through the addition of a suitable additive or combination thereof, to a natural or synthetic lubricant. Similarly, continued movement can be insured, again with varying degrees of success, through the addition of one or more appropriate additives.

While there are many known additives which may be classified as antiwear, antifriction and extreme pressure agents and some may in fact satisfy more than one of these functions as well as provide other useful functions, it is also known that many of these additives act in a different physical or chemical manner and often compete with one another, e.g., they may compete for the surface of the moving metal parts which are subjected to lubrication. Accordingly, extreme care must be exercised in the selection of these additives to insure compatibility and effectiveness.

The metal dihydrocarbyl dithiophosphates are one of the additives which are known to exhibit antioxidant and antiwear properties. The most commonly used additives of this class are the zinc dialkyl dithiophosphates which are conventionally used in lubricant compositions. While such zinc compounds afford excellent oxidation resistance and exhibit superior antiwear properties, it has heretofore been believed that these characteristics increased or significantly limited the ability to decrease friction between moving surfaces.

Known ways to solve the problem of energy losses due to high friction, e.g., in crankcase motor oils, include the use of synthetic ester base oils which are expensive and the use of insoluble molybdenum sulfides which have the disadvantage of giving the oil composition a black or hazy appearance.

Molybdenum Complexes of Hydroxy Amines

In light of the foregoing, the need for improved lubricating compositions that will permit operation of moving parts under boundary conditions with reduced friction is readily apparent. Similarly, the need for such a composition that can include conventional base oils and other conventional additives and can be used without the loss of other desirable lubricant properties, particularly those provided by zinc dialkyl dithiophosphates, is also readily apparent.

K. Coupland and C.R. Smith; U.S. Patent 4,164,473; August 14, 1979; assigned to Exxon Research & Engineering Co. have found disadvantages of the prior art lubricating additives can be overcome with a class of organomolybdenum complexes believed to be represented by the following formula:

```
            H2
            C
      H2C       X
      /         |
R—N - - - - - - Mo(X)2
      \         |
      H2C       X
            C
            H2
```

wherein R is a substantially hydrocarbyl group containing from 1 to 50, preferably 12 to 28, carbon atoms and X is selected from sulfur or oxygen. These complexes are particularly useful when used in combination with a sulfur donor, e.g., zinc dialkyl dithiophosphate. It is understood that the R substituent can contain substituted pendant hetero groups provided they do not detrimentally alter the hydrocarbon solubility of the molybdenum complex.

The hydrocarbon composition, therefore, comprises a major portion of a hydrocarbon, e.g., a lubricating oil and at least a friction reducing amount of a hydrocarbon soluble molybdenum complex and preferably a lubricity enhancing combination of: (a) the soluble molybdenum complex; and (b) an oil-soluble sulfur donor, preferably zinc dialkyl dithiophosphate, and, if desired, at least a sludge-dispersing amount of an oil-soluble dispersant, e.g., an ashless dispersant and at least a rust-inhibiting amount of a rust inhibitor. In practice, the lubricity enhancing combination is present in an amount sufficient to provide from about 0.005 to 0.2, preferably 0.03 to 0.15, optimally about 0.1, wt % molybdenum and at least about 0.25, e.g., 0.25 to 1, wt % sulfur donor, all weight percent being based on the total weight of the oil composition.

Example 1: Ethoduomeen T-13, a commercially available ethoxylated amine approximating in structure N,N'-tris(2-hydroxyethyl)-N-tallow-1,3-diaminopropane, (20.8 g), toluene (20.8 g) and molybdic oxide (5.23 g) were refluxed together with stirring. Water released in the reaction was collected by means of a

Dean and Stark receiver until a total of 0.65 cm^3 had been isolated. The organomolybdenum complex reaction product was diluted with light mineral oil (25 g) filtered and stripped of volatiles. The final product containing about 50 wt % of the complex product was a viscous amber liquid which analyzed for 3.77 wt % molybdenum.

Example 2: Ethoduomeen T-13 (20.41 kg), light mineral oil (24.49 kg) and ammonium heptamolybdate tetrahydrate (6.4 kg) were stirred together with heating at 150°C for 20 hours. Water and ammonia were allowed to be distilled from the reaction. During the last 4 hours a nitrogen sparge was applied to remove the last traces of volatile material. At the end of the reaction period there was isolated a dark red viscous oil containing about 50 wt % complex product and 7.0% molybdenum.

The product of Example 2 was evaluated by subjecting it to a study as described below. The molybdenum complex was added to a standard formulation for a lubricating oil in the amounts shown in the table.

This testing procedure measures the coefficient of friction between a journal and a 120° section of a journal bearing when lubricated by the test fluid lubricant. A journal (1.5 inch o.d.) is fitted on a shaft supported by two antifraction bearings. The bearing section is loaded against the journal by a load lever and a series of knife edges connected each to an arm. A bearing block, which holds the bearing section against the journal, is self-aligning thus eliminating any problem of edge loading. The frictional force between the bearing section and the journal is measured by a friction lever.

The shaft is coupled to a variable speed transmission which allows surface speeds of the journal to be varied from 0.3 to 16 cm/sec. Bearing loads can be increased from 0 to 3,000 psi by adding weights to the pan supported by the load lever. Infrared lamps keep the mass of the bearing at the desired test temperature (ambient to 130°C) as measured by a thermocouple in a thermowell.

The test lubricant is placed in a shallow pan and held such that the bottom face of the journal touches the oil. The tester and test lubricant are brought to 110°C. With only the weight hanger on the load lever the machine is started at 80 rpm. 36.6 kg of weights are added to the pan giving an effective contact pressure of 2,600 psi. Sufficient weight is placed on the load lever to overcome the friction force and keep it in the down position. The following speed cycle is carried out allowing five minutes of operation at each speed to allow lubrication conditions to stabilize. A torque measurement is made at each speed.

Speeds (rpm)	Torque (cm/sec)
80	16
60	12
30	6
20	4
10	2
7	1.4
3	0.6
1.5	0.3

The tester is then operated at constant speed, 30 rpm, for a 3-hour period. The speed cycle is repeated.

Between test lubricants, the tester is flushed with base case oil and run in the base case oil until its previous performance is repeated, taking the tester apart and cleaning it if necessary.

Test	Added Molybdenum Complex (wt %)	. . . . . Coefficient of Friction		. . . . Friction Reduction (%)	
		16 cm/sec	0.3 cm/sec	16 cm/sec	0.3 cm/sec
1	–	0.120	0.141	–	–
2	0.5	0.061	0.090	49.2	36.2
3	1.0	0.022	0.069	81.7	50.0

Molybdenum Complex of Thiobisphenols

In further work, *K. Coupland, C.R. Smith and J.M. Salva; U.S. Patent 4,248,720; February 3, 1981; assigned to Exxon Research & Engineering Co.* have developed a class of lubrication additives which consists of organomolybdenum complexes believed to be represented by the formula:

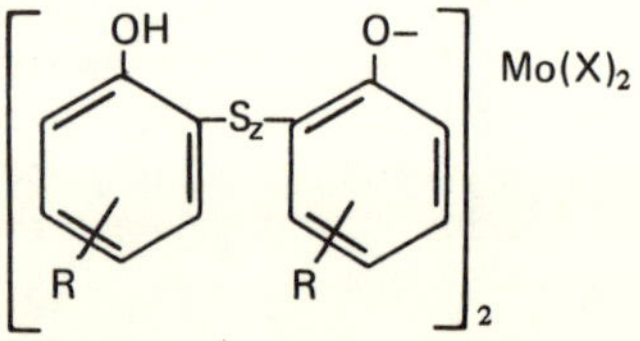

where z is 1 to 3, X is selected from sulfur or oxygen and R is a substantially hydrocarbyl group containing from 1 to 50, preferably 7 to 28, optimally 8 to 12 carbon atoms. These complexes are particularly useful when used in combination with a sulfur donor, e.g., zinc dialkyl dithiophosphate. It is understood that the R substituent can contain substituted pendant hetero groups provided they do not detrimentally alter the hydrocarbon solubility of the molybdenum complex.

A hydrocarbon composition is made comprising a major portion of a hydrocarbon such as a lubricating oil and at least a friction reducing amount of a hydrocarbon-soluble molybdenum complex and preferably a combination of (a) the molybdenum complex; and (b) an oil-soluble sulfur donor, preferably zinc dialkyl dithiophosphate, and, if desired, at least a sludge-dispersing amount of an oil-soluble dispersant, e.g., an ashless dispersant and at least a rust-inhibiting amount of a rust inhibitor. In practice, the lubricity enhancing combination is present in an amount sufficient to provide preferably from 0.03 to 0.15, optimally about 0.1, wt % molybdenum and at least about 0.25, e.g., 0.25 to 1, wt % sulfur donor, all weight percent being based on the total weight of the oil composition.

Whether the organomolybdenum complex is used alone or in combination with an active sulfur donor, its concentration may vary appreciably with the particular hydrocarbon. For example, when the molybdenum complex is used alone in a fuel such as gasoline, the concentration of the complex ranges from 10 to 1,000, preferably 20 to 50 weight parts per million based on the total weight of the fuel composition, whereas as a lubricant, it is used in combination with the active sulfur donor, which combination ranges from about 0.5 to 5, preferably 1 to 3 wt % based on the total weight of the lubricating oil.

Example 1: Nonyl phenol sulfide (550 g), diluent oil (550 g) and molybdic trioxide (84.2 g) were heated together with stirring at 100°C. Ethanolamine (72.5 g, 95% pure) was added dropwise and the temperature raised to 175°C for 4 hours. During the reaction, 22 cm^3 of volatiles were removed by distillation. The product (1,111 g) was isolated by filtration to yield a dark viscous oil containing 4.16 wt % Mo (representing a conversion of Mo of 93%).

Example 2: Nonyl phenol sulfide (550 g), diluent oil and molybdic trioxide (84.2 g) were heated together with stirring at 100°C. Ethylene diamine (70.3 g) was added to the mixture maintaining the temperature between 120° and 130°C. The temperature was raised to 175°C for 4 hours. During the reaction, volatile material was distilled from the reaction mixture. The product contained, after filtration, 4.4 wt % Mo (100% conversion).

Example 3: The procedure of Example 2 was followed substituting ammonium heptamolybdate tetrahydrate (103.2 g) as the source of molybdenum. The product contained 4.33 wt % Mo (96.8% conversion).

Example 4: The procedure of Example 1 was followed substituting ammonium heptamolybdate tetrahydrate (103.2 g) as the source of molybdenum. The product contained 4.36 wt % Mo (97.9% conversion).

Example 5: A lubricating oil composition was prepared by blending together the individual components, noted below, usually at a slightly elevated temperature, i.e., from about 45° to above 65°C to insure complex mixing. The final composition of Example 5 formulated into 10W/30 SE quality automotive engine oils was as follows:

Active Ingredient	Weight Percent
Mineral oil	94.9
Ashless dispersant	2.9
Magnesium sulfonate	0.2
ZDDP*	0.9
Rust-inhibitor	0.1
Viscosity index improvers	1.0
Silicone defoamer	0.01
Ashless antioxidant	–
Metal detergent-inhibitor	–

*Zinc dihydrocarbyl dithiophosphate such as zinc dinonyl phenol dithiophosphate.

This formulated blend was itself and with added Mo complexes subjected to testing under the testing procedure as described in the previous patent. Results are given in the table below.

Test	Lubricant Example	Added Mo Complex Example	Added Mo Complex Wt %	Coefficient of Friction 46 cm/sec	Coefficient of Friction 1 cm/sec	Friction Reduction (%) 46 cm/sec	Friction Reduction (%) 1 cm/sec
1	5	–	–	0.084	0.115	–	–
2	5	1	1.1	0.046	0.077	45.6	33.5
3	5	1	1.7	0.043	0.073	49.4	36.7

(continued)

Test	Lubricant Example	Added Mo Complex Example	Added Mo Complex Wt %	Coefficient of Friction 46 cm/sec	Coefficient of Friction 1 cm/sec	Friction Reduction (%) 46 cm/sec	Friction Reduction (%) 1 cm/sec
4	5	2	1.1	0.050	0.060	40.5	48.3
5	5	2	1.7	0.044	0.062	48.1	46.4
6	5	2	1.1	0.045	0.061	46.2	47.3
7	5	2	1.7	0.041	0.061	50.7	47.3
8	5	3	1.1	0.061	0.085	27.9	26.4
9	5	3	1.7	0.050	0.077	40.5	33.4

Alkanol or Phenol Solutions of Molybdenum-Thiobisphenol Complexes

P.W. Brewster; U.S. Patent 4,192,757; March 11, 1980; assigned to Exxon Research & Engineering Company has found that the molybdenum complexes of thiobisphenols described in the last patent, which are obtained as the solution reaction products of a hydrocarbyl substituted thiobisphenol, e.g., 2,2'-thiobis(4-isononyl phenol), with a molybdenum compound, e.g., molybdic oxide, in the presence of an amine, e.g., ethylene diamine are more readily produced in quantitative yields at lower temperatures and with a broader spectrum of amines when they are prepared in an alkanol solvent of a C_{8-50} alkyl substituted phenol, preferably nonyl phenol. These complexes are useful hydrocarbon additives, particularly when used in combination with an oil-soluble sulfur donor, e.g., a metal dialkyl dithiophosphate, which provides an additive combination for lubricants and fuels whereby the resulting lubricating composition exhibits an improved antifriction property.

When the complex is introduced into the lubricating oil in combination with the phenol, e.g., as the solution of the reaction, the modified lubricating oil exhibits a dynamic coefficient of friction comparable to if not somewhat better than that result obtained with the addition of only a common amount of organomolybdenum complex.

Example 1: Nonyl phenol sulfide (183 g) as ECA 9001, Solvent Neutral 150 mineral oil (183 g) and molybdic trioxide (28.1 g) were stirred together and then raised in temperature to 94°C at which time ethylene diamine (23.4 g) was thereafter slowly added over a 20-minute period. The temperature was raised with stirring to 121°C over 0.6 hour. While stirring at this temperature, the volatiles including water and ammonia were removed by gentle nitrogen sparging for 18 hours. After filtration, the resulting product solution, useful as a lubricating oil additive, had a viscosity of 177 SUS at 100°C and was black in color and contained about 4.3 wt % molybdenum and 1.9 wt % nitrogen.

Example 2: The procedure of Example 1 was used except for the following: the mineral oil replaced by 183 g of nonyl phenol (as ECA 9003, a $C_{9\,avg}$ nonyl phenol containing about 62 wt % monoalkyl and 33 wt % dialkyl phenol); use of densified MoO_3; going to 177°C over a 2-hour period while adding the ethylene diamine; and reacting at 177°C for 4 hours prior to filtration. The resulting filtered product solution (the filtration of which took less than 10 seconds to fully filter) analyzed for 4.5 wt % molybdenum.

A lubricating oil composition, as shown in the previous patent, was made up and the blend was tested alone and with the added molybdenum complex—prepared by this method and that of the previous patent—by the testing procedure described in U.S. Patent 4,164,473.

The tests show that this additive combination provides improved lubricity enhancement to lubricating oils when an active sulfur donor is present and that these combinations have utility as additives for lubricating oils.

While this additive combination provides frictional performance to a fully formulated lubricating oil comparable to that provided by the additive of the previous patent it is much easier to filter, thereby removing unwanted and deleterious reaction by-products. For example, where a quantity of the former (as shown by Example 2) filters through in about 10 seconds, a similar quantity of the latter would take from 0.5 to several hours, e.g., in Example 1 it took in excess of 1 hour. In this regard, a solvent mixture of up to an equal amount of mineral oil with the alkyl phenol solvent provides slower but still useful filtering periods of the solution reaction products.

Another advantage of the alkyl phenol as a solvent for the reaction of the hydrocarbyl phenol sulfide and molybdenum compound, preferably molybdic oxide (MoO_3), resides in the enhanced reactivity of the components, i.e., shorter reaction times and/or more heat stable complexes when such a solvent is used as compared with mineral oil solvent.

P.W. Brewster; U.S. Patent 4,201,683; May 6, 1980; assigned to Exxon Research & Engineering Co. has also found that these same molybdenum complexes may be prepared in a solvent of a C_{5-50} alkanol.

Among the alcohols useful in preparing these complexes are amyl alcohol, hexanol, heptanol, etc., through pentacontanol with the preferred alcohols being octanol through octadecanol. A highly suitable source of alcohols are the Oxo alcohols. Particularly suitable as a reaction solvent for this process is tridecyl Oxo alcohol.

The C_{5-50} alkanols are usefully present in the hydrocarbon composition in an amount of from about 0.25 to 5, preferably 1 part by weight per part by weight of the organomolybdenum complex.

Molybdenum Complexes of Lactone Oxazoline Dispersants

During the past decade, ashless sludge dispersants have become increasingly important, primarily in improving the performance of lubricants in keeping the engine clean of deposits and permitting extended crankcase oil drain periods while avoiding the undesirable environmental impact of the earlier used metal-containing additives.

Included within the published ashless sludge dispersants are the oil-soluble lactone oxazolines described in U.S. Patent 4,062,786.

J. Ryer, E.D. Winans, S.J. Brois and A. Gutierre; U.S. Patent 4,176,073; November 27, 1979; assigned to Exxon Research & Engineering Co. have found that lactone oxazoline dispersants can be reacted with a source of molybdenum to provide a molybdenum-containing ashless dispersant of improved thermal stability in hydrocarbons, preferably lubricating oils, and having the property of imparting enhanced lubricity to them. This has been accomplished by use of an aqueous-nonaqueous reaction medium. The operational embodiment of the process thus is a lubricating oil composition comprising a major proportion of mineral oil and a minor but friction-reducing amount of an oil-soluble molybdenum-containing lactone oxazoline lubrication oil dispersant having from 0.5 to 20 wt % molybdenum based on the weight of dispersant and further characterized by from one to two lactone rings, one to two oxazoline rings and at least one substantially saturated hydrocarbon group containing at least about 50 carbon atoms.

These additives are prepared from lactone oxazoline dispersants by reaction of the dispersant with an inorganic molybdenum compound in a binary solvent system comprising an aqueous component of the class consisting of water and ammonium hydroxide and a nonaqueous component such as mineral oil, tetrahydrofuran (THF) or a hydrocarbon boiling between 70° and 250°C. The volume ratio of aqueous to nonaqueous component ranges from 1:1,000 to 1:1, optimally 1:10.

It has further been found that a stable molybdenum complex can be obtained with little if any destruction of the ashless dispersant when complexing is effected at a temperature of 40° to 250°C, preferably from 50° to 200°C, in the binary solvent system.

It is preferred that the friction-reducing additive be present in the mineral oil in an amount to provide from about 0.01 to 2.0, optimally 0.05 to 0.5 wt % molybdenum in the oil (all weight percent being based on the total weight of the lubricating composition).

In preferred form, the molybdenum complex is that of a lactone-oxazoline dispersant derived from the reaction of 1 mol of C_{8-400} hydrocarbon substituted lactone acid material such as poly(isobutenyl) lactone acid wherein the hydrocarbyl substituent, e.g., the poly(isobutenyl) group has an $\overline{M}n$ ranging optimally from about 900 to 1,600 with one mol equivalent of a 2,2-disubstituted-2-amino-1-alkanol having 2 to 3 hydroxy groups and containing a total of 4 to 8 carbon atoms usefully by heating at a temperature of from 100° to 240°C until cessation of water evolution; the additive being complexed with from 1 to 2 molar equivalents of molybdic oxide, i.e., MoO_3 and containing from 0.5 to 20, optimally 5 wt % molybdenum.

Example: 224 g of a 50 wt % mineral oil solution of a poly(isobutyl) lactone oxazoline having an $\overline{M}n$ of 1,120 (0.05 mol), 8.1 g of $MoO_3{\cdot}H_2O$ (0.05 mol) and 10 cc of water were stirred and refluxed (ca 140°C) in 200 cc of xylene for 4 hours. During this time, water and xylene were removed by distillation. The reactants were freed from solid material by filtration and the filtrate stripped of volatile material by rotoevaporation. The product was a 50 wt % mineral oil solution of a dark brown oil containing 0.886 wt % molybdenum.

The practical exploitation of various types of molybdenum compounds and complexes as lubricant additives has been hindered not only by their insolubility and/or corrosiveness but also by low thermal stability.

K. Coupland, J.M. Salva and C.R. Smith; U.S. Patent 4,176,074; November 27, 1979; assigned to Exxon Research & Engineering Co. have developed a second method of reacting ashless oxazoline dispersants with a source of molybdenum by use of an aqueous-nonaqueous reaction medium. They have thus provided a lubricating oil composition comprising a major proportion of mineral oil and a minor but friction-reducing amount of an oil-soluble molybdenum-containing ashless oxazoline lubricating oil dispersant, the dispersant having from 0.5 to 20 wt % molybdenum based on the weight of dispersant and further characterized by from 1 to 2 oxazoline rings and a substantially saturated hydrocarbon group containing at least about 50 carbon atoms.

These materials are prepared from conventional ashless oxazoline dispersants by the reaction of the dispersant with a molybdenum source compound in a binary solvent system comprising an aqueous component of the class consisting of water and ammonium hydroxide and a nonaqueous component consisting of the class consisting of tetrahydrofuran (THF) and a hydrocarbon boiling between 70° and 250°C. The volume ratio of aqueous to nonaqueous component is preferably 1:100 to 1:4, optimally 1:10. The aqueous component can be considered a promoter for the molybdation of the oxazoline dispersant. Thus, for the purposes of this discussion, both the water and the ammonium hydroxide could be defined as an essential promoter of molybdation in a nonaqueous reaction medium. The temperature of the reaction, the type of oxazoline dispersant and the amounts of reactant are the same as those described in the previous patent.

Example: A mixture of 40 g of a 50 wt % mineral oil solution of a bisoxazoline dispersant obtained from the reaction of 1 mol of polyisobutenyl ($\overline{M}n$ of ~1,300) succinic anhydride with about 2 mols of tris(hydroxymethyl) aminomethane (THAM) to provide the bisoxazoline of polyisobutenyl succinic anhydride; 2.87 g of molybdic oxide ($MoO_3 \cdot H_2O$); 50 cc of toluene; and 5 cc of water were refluxed with stirring for 7.25 hours. During this time, water and xylene were removed by distillation. The reactants were freed from solid material by filtration and the filtrate stripped of volatile material by rotoevaporation. The product was a dark brown oil containing 3.56% molybdenum. This represents a conversion, based on molybdenum, of 79% on theory.

Alkyl Phosphate Esters

T.J. Griffin, Jr.; U.S. Patent 4,153,066; May 8, 1979; assigned to The Dow Chemical Company has found that certain metal salts of complex reaction products of a hydroxy ether and a phosphorus compound such as P_2O_5 are effective to reduce friction loss of refined oils and certain crude oils flowing through a confining conduit, such as in internal combustion engines.

The product employed is formed by reacting an essentially anhydrous hydroxy ether of the formula ROR_1OH wherein R is a C_{1-6} alkyl group, R_1 is a C_2 or C_3 alkylene group and the total carbon atoms of R_1 and R range from 3 to about 8 with a pentavalent phosphorus compound which is substantially free from acid groups such as Cl, F and the like. When the total carbon atoms in the hydroxy ether is 3 or 4, there is also reacted with the hydroxy ether and phosphorus compounds a long chain aliphatic monohydric alcohol containing at least 5 carbon atoms. A short chain aliphatic monohydric alcohol (C_{1-4}) can also be reacted therewith if desired. When the total carbon atoms in the hydroxy ether is 5 or more, there is reacted with the hydroxy ether and phosphorus compound either a long chain aliphatic alcohol (at least 5 carbons) or a short chain aliphatic monohydric alcohol (C_{1-4}) or a mixture thereof.

The abovedescribed compounds are reacted with a pentavalent phosphorus compound (preferably P_2O_5) for a period of time ranging from about 1.5 to 6 hours at a temperature ranging from about 70° to 90°C to form this complex reaction product. Reaction products having different selected characteristics can be prepared by reacting specific reactants and by varying the order in which they are reacted together.

As a friction reducing agent the reaction product is dispersed into the organic liquid along with an aluminum activator compound, preferably sodium aluminate, the reaction product and the activator being employed in amounts and a specific ratio to each other to impart to the organic liquid a desired reduction in friction loss.

Suitable hydroxy ethers which can be employed include, for example, ethylene glycol monomethyl ether, ethylene glycol monoethyl ether, ethylene glycol mono-n-butyl ether, ethylene glycol mono-n-hexyl ether, ethylene glycol monoisobutyl ether, propylene glycol monoethyl ether, propylene glycol monoisobutyl ether, propylene glycol monomethyl ether, mixtures thereof and other like compounds.

The mol ratio of the total of the short chain and/or long chain alcohol and the hydroxy ether to total phosphorus pentoxide ranges from about 2.8:1 to 7.0:1 with the most preferred ratio being about 3.64:1.

The reaction product of the hydroxy ether and the phosphorus pentoxide may be mixed with the aliphatic hydrocarbon–diesel oil, fuel oil, lubricating oils, and certain light crude oils–along with the sodium aluminate.

The reaction product and activator are employed in a total amount and weight ratio to each other to produce a product having the desired friction loss characteristics. These amounts and ratios will vary and are dependent on the reactants which are employed to make the reaction product, the exact activator, the organic liquid employed, and the desired degree of friction reduction. For example, less than about 8 gallons of the phosphate ester are employed per 1,000 gallons of organic liquid with from about 0.01 to 1.5 gallons of a 38% by wt of a sodium aluminate solution per 1,000 gallons of organic liquid. For different quantities of phosphate ester the amount of metal salt will vary proportionally.

Oil-Soluble Alkylene Glycol Ester Derivatives

J. Ryer, S. Brois and E.D. Winans; U.S. Patent 4,199,463; April 22, 1980; assigned to Exxon Research & Engineering Co. describe a class of oil-soluble alkylene glycol esters of partial esters derived from the reaction of organic acid materials such as dicarboxylic acids or anhydrides and an aldehyde/tris(hydroxymethyl)-aminomethane adduct or mixture which have utility as friction reducing additives for hydrocarbon fuels and lubricating oils.

These hydrocarbon-soluble alkylene glycol esters of partial, i.e., half, esters of 1-aza-3,7-dioxabicyclo[3.3.0] oct-5-yl methyl alcohols (hereafter designated DOBO) can be formed by the ethoxylation of the product of the reaction of organic dicarboxylic acids and anhydrides, with an aldehyde/tris(hydroxymethyl)aminomethane (THAM) adduct or aldehyde/THAM mixture. For liquid hydrocarbon compositions wherein the alkylene glycol esters have been found to be highly useful as friction reducing additives, the carbon chain length is from C_{6-50} carbon atoms for fuels such as gasoline and middle distillates, that is from 6 to 30 carbon atoms for gasoline and from 6 to 50 carbon atoms for middle distillate fuels, and from 6 to 150, preferably from 16 to 90, carbons for lubricating oils.

The composition thus comprises a major proportion of a liquid hydrocarbon of the class consisting of fuels and lubricating oils and a minor amount of a hydrocarbon-soluble alkylene glycol ester of a partial ester of 1-aza-3,7-dioxabicyclo-[3.3.0] oct-5-yl methyl alcohol, preferably in an amount from 0.001 to 20 wt % of the total composition.

The alkylene glycol esters are prepared by reacting 1 to 10 mols of an alkylene oxide such as ethylene oxide or propylene oxide with each mol of the half ester of the alkenylsuccinic anhydride/DOBO adduct, although it must be understood that the adduct could be catalytically reacted with a polyalkylene glycol to provide the derivative.

The dicarboxylic acid material/DOBO adduct, e.g., the alkenylsuccinic anhydride/DOBO adduct, is the essential precursor from which the alkylene glycol esters of half esters derivatives are obtained. The alkylene glycol ester derivatives can be characterized by the following formula:

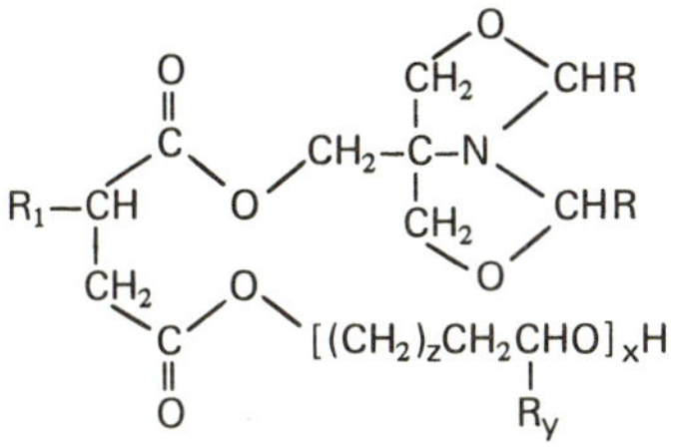

O
‖
C
O
CH_2
O
CHR
CH_2-C-N
R_1-CH
CH_2
CHR
O
CH_2
O
C
$[(CH_2)_zCH_2CHO]_xH$
‖
O
R_y

wherein X is 1 to 10, preferably 2 to 4; R_1 represents a hydrocarbyl group of 6 to 150 carbons; Z is 0 or 1; R is H, or C_{1-12}, preferably C_{3-8} hydrocarbyl group; and R_y is H or a CH_3 (methyl) group.

Example 1: *Preparation of aldehyde/THAM adducts (DOBO)* – 0.1 mol (12.1 g) of THAM was dissolved in an equal weight of water. To the resulting solution in a 250 ml Erlenmeyer flask equipped with magnetic stirrer was added 0.2 mol (6.0 g) of paraformaldehyde. The stirred mixture was heated to 70°C to effect dissolution of the paraformaldehyde and continued for 15 minutes at 70°C to produce 1-aza-3,7-dioxabicyclo[3.3.0] oct-5-yl methyl alcohol (DOBO) in quantitative yields. The product after evaporation of water and recrystallization from benzene melted at 60° to 61°C.

Example 2: *Half acid ester of 1-aza-3,7-dioxabicyclo[3.3.0] oct-5-yl methyl octadecenylsuccinate* – 0.5 mol of octadecenylsuccinic anhydride was added to a 1 ℓ round bottom flask and heated to 140°C for an hour to convert any partially hydrolyzed reactant to the anhydride form. After cooling the nitrogen-blanketed reactor to 100°C, 0.5 mol of DOBO was added in one portion. The alcohol reagent readily dissolved and the clear solution was heated to 174°C for about 2 hours. Infrared analysis showed that esterification was complete at this point.

Example 3: 1 mol of the product of Example 2 dissolved in an equal weight of heavy aromatic naphtha was placed in a stainless steel pressure vessel and 1 mol of ethylene oxide reacted with the product by heating at from 149° to 160°C for 2 hours at 30 to 40 psig and in the presence of 0.3% by wt of NaOH (present as a catalyst). The product recovered was a mixture of monoethylene glycol and diethylene glycol esters of the half acid ester of 1-dioxabicyclo[3.3.0] oct-5-yl methyl octadecenylsuccinate (approximately 1 weight proportion of the monoethylene glycol ester to 2 weight proportions of the diethylene glycol ester).

Example 4: The process of Example 3 was followed except that 2 mols of ethylene oxide was reacted with 1 mol of the product of Example 2. The resulting ethoxylated derivative was a mixture of monoethylene glycol, diethylene glycol and triethylene glycol esters of the half acid ester of Example 2 (the mixture containing about 90 wt % diethylene glycol ester, about 10 wt % ethylene glycol ester and a trace of triethylene glycol ester).

The product of Example 4 was evaluated in a lubricating oil composition by a ball-on-cylinder test. When 0.25 wt % of the ethoxylated product was added to the oil the coefficient of friction was reduced 63% and when 0.75 wt % was used, the friction reduction was 71%.

Sulfurized Fatty Acid Amide or Ester of Diethanolamine

B.T. Davis and J.B. Retzloff; U.S. Patent 4,236,898; December 2, 1980; assigned to Ethyl Corporation have produced hydrocarbon fuels of the gasoline boiling range which reduce friction between sliding metal surfaces in internal combustion engines. The reduced friction results from the addition to the hydrocarbon fuel of a small amount of a sulfurized fatty acid amide, ester or ester-amides of alkoxylated amine such as diethanolamine.

This additive is selected from sulfurized fatty acid esters, sulfurized fatty acid amides and sulfurized fatty acid ester-amides of an alkanol amine having the formula

$$H{-}N\begin{matrix} \diagup (R'{-}O)_n{-}H \\ \diagdown R'' \end{matrix}$$

wherein R' is a divalent aliphatic hydrocarbon radical containing 2 to 4 carbon atoms, n is an integer from 1 to 10, and R" is selected from hydrogen and the group $(R'{-}O)_n{-}H$.

The additives can be made by reacting a sulfurized fatty acid with an oxyalkylated amine (e.g., diethanolamine). Alternatively, sulfurized fatty acid amide can be made by reacting sulfurized fatty acid with ammonia or an alkanol amine (e.g., ethanolamine, diethanolamine) to form an intermediate which can be further oxyalkylated by reaction with an alkylene oxide (e.g., ethylene oxide, propylene oxide).

Another method is to first make the fatty acid ester, amide or ester-amide by reacting a fatty acid with an oxyalkylated amine (e.g., diethanolamine) and then reacting that intermediate with elemental sulfur at elevated temperature (e.g., 100° to 250°C).

Sulfurized fatty acids can be made by heating a mixture of fatty acid with elemental sulfur. Unsaturated fatty acids are preferred such as hypogeic acid, oleic acid, linoleic acid, elaidic acid, erucic acid, brassidic acid, tall oil fatty acids and the like. Sulfurized oleic acid, a commercial product, is most preferred.

The preferred amines used to make the additives are ethoxylated amines such as ethanolamine, diethanolamine, isopropanolamine and the like. These can be reacted to form both amides and esters.

Example: In a reaction vessel was placed 308 g (1 eq) of a commercial sulfurized oleic acid (Cincinnati Milacron), 105 g (1 mol) diethanolamine and a small amount of xylene. The mixture was heated under nitrogen to 185°C over 2 hours while removing water. The mixture was then stripped of solvent under vacuum leaving the product. It was analyzed for nitrogen. (Found: 3.48% total nitrogen, 0.93% basic nitrogen.) This shows a mixture of 73 wt % sulfurized oleamide of diethanolamine and 27 wt % sulfurized oleate ester of diethanolamine.

Other sulfurized fatty acids can be substituted for sulfurized oleic acid in the above example with good results.

The additives are used in an amount sufficient to reduce the sliding friction of metal surfaces in an internal combustion engine operating on a normally liquid hydrocarbon fuel containing the additive. A preferred range is about 0.05 to 0.5 wt %.

Tests have been carried out which demonstrate the ability of this fuel composition to significantly improve fuel economy. Tests were carried out in a 1978 U.S. production automobile having an inline-6-cylinder engine. The car was operated on a chassis dynamometer under controlled temperature and humidity conditions. Each test series consisted of three warmed-up shortened EPA city and full EPA highway cycles run in alternate fashion during which miles per gallon measurements were made. These tests were conducted before and after accumulating 500 miles on the car. A 45-minute warm-up at 55 mph preceded each test series. Miles per gallon measurements showed immediate fuel reduction as well as fuel reduction at the end of 500 miles. The shortened city cycle consisted of the first 3.6 miles of the 11.1-mile city cycle. The sequence of testing consisted of testing first clear Indolene fuel (baseline) followed by a test using the same fuel plus 0.1 wt % of a mixture prepared as described in the example. The results are reported in the following table in terms of miles per gallon percent improvement over baseline.

Test Cycle	Average Percent Improvement in mpg
Short city cycle	
Immediate*	1.19
After 500 miles**	0.26
Highway cycle	
Immediate*	1.25
After 500 miles**	0.63

*Average of 3 individual measurements.
**Average of 2 individual measurements.

N-Hydroxymethyl Aliphatic Hydrocarbylamides

R.L. Shubkin; U.S. Patent 4,243,538; January 6, 1981; assigned to Ethyl Corporation has provided fuel-efficient oil for internal combustion engines by including in the formulated oil a minor amount of an N-hydroxymethyl C_{12-36} aliphatic hydrocarbylamide. The additives can also be used in the engine fuel to reduce friction.

A preferred embodiment is a lubricating oil composition formulated for use in the crankcase of an internal combustion engine, the composition comprising a major amount of a lubricating oil and a minor friction-reducing amount of an oil-soluble aliphatic N-hydroxymethyl hydrocarbylamide having the structure

$$R-\overset{\overset{\displaystyle O}{\|}}{C}-\underset{\underset{\displaystyle Y}{|}}{N}-CH_2OH$$

wherein R is an aliphatic hydrocarbon group containing 11 to 35 carbon atoms and Y is selected from the group consisting of hydrogen, lower alkyls containing 1 to 4 carbon atoms and hydroxymethyl.

The most preferred additives are N-hydroxymethyl oleamide, N,N-bis(hydroxymethyl)oleamide and mixtures thereof.

Example: In a reaction vessel was placed 194 g oleamide, 210 g 37% aqueous formaldehyde, 10 g sodium carbonate and 400 ml n-butanol. The mixture was stirred under nitrogen at 60°C for four hours. Then 200 ml toluene was added and the aqueous phase removed. The organic layer was washed three times with water. The third wash emulsified and a small amount of sodium sulfate was added to break the emulsion. The organic layer was dried over calcium sulfate, filtered and volatiles removed under vacuum using a rotary evaporator. Product yield was 206.8 g which was mainly N-hydroxymethyl oleamide containing some N,N-bis(hydroxymethyl)oleamide.

Other similar additives can be made following the above general procedure by substituting other aliphatic hydrocarbyl amides having at least one hydrogen bonded to the amide nitrogen. Such amides include laurylamide, N-methyl tridecanamide, N-ethyl oleamide, eicosanamide, heptatriacontanamide, stearylamide, linoleamide, tall oil fatty acid amide and the like.

The additives are added to the lubricating oil in an amount which reduces the friction of the engine operating with the oil in the crankcase. A preferred concentration is about 0.1 to 1.0 wt %.

The more preferred lubricating oil compositions include zinc dihydrocarbyldithiophosphate (ZDDP) in combination with the additives.

Tests were carried out which demonstrated the friction-reducing properties of the additives. These tests have been found to correlate with fuel economy tests in automobiles. In these tests an engine with its cylinder head removed and with the test lubricating oil in its crankcase was brought to 1,800 rpm by external drive. Crankcase oil was maintained at 63°C. The external drive was disconnected and the time to coast to a stop was measured. This was repeated several times with the base oil and then several times with the same oil containing commercial oil formulated for use in a crankcase. The following table shows the percent increase in coast-down time caused by the additives.

Additive	Wt %	Increase (%)
N-hydroxymethyl oleamide	0.25	3.5

(continued)

Additive	Wt %	Increase (%)
N-hydroxymethyl oleamide	0.5	5.7
N-hydroxymethyl oleamide	1.0	6.5, 13*

*This result obtained using a different base oil.

Aliphatic Hydrocarbyl Sulfinyl or Sulfonyl Methane

A.G. Papay and J.P. O'Brien; U.S. Patent 4,246,125; January 20, 1981; assigned to Edwin Cooper, Inc. suggest a method of improving fuel mileage for automobiles by reducing engine friction, which they accomplish by the addition to the fuel or lubricating oil of a minor amount of an aliphatic hydrocarbyl sulfinyl or sulfonyl methane having the formula

$$R_1-\overset{\overset{O}{\uparrow}}{\underset{\underset{(O)_n}{\downarrow}}{S}}-CH_3$$

wherein R_1 is an aliphatic hydrocarbon group containing about 12 to 36 carbon atoms and n is 0 or 1.

When n is 0, the additives are aliphatic hydrocarbyl sulfinyl methanes. When n is 1, the additives are aliphatic hydrocarbyl sulfonyl methanes.

R_1 can be alkyl or alkenyl. Preferably, R_1 is an alkyl group. More preferably R_1 contains about 14 to 20 carbon atoms. Most preferably R_1 is an alkyl group containing about 16 to 18 carbon atoms. Examples of these are hexadecyl sulfonyl methane, octadecyl sulfonyl methane, hexadecyl sulfinyl methane, octadecyl sulfinyl methane and mixtures thereof.

The friction-reducing additives are readily made by oxidizing an aliphatic hydrocarbyl methyl sulfide. The aliphatic hydrocarbyl methyl sulfide can be made by reacting an olefinically unsaturated hydrocarbon with methyl mercaptan under UV radiation.

Example 1: In a reaction vessel was placed 166 g (0.7 mol) of a mixture of C_{16} and C_{18} monoolefin, predominantly straight chain α-olefin. Then 58 g of methyl mercaptan was added at 35°C over a 60-minute period while irradiating the glass vessel with UV from a mercury lamp. The product contained excess mercaptan.

In a reaction vessel was placed 198 g of the above mixture of octadecyl methyl sulfide and hexadecyl methyl sulfide. To this was added a solution of 94 g (0.83 mol) of 30% hydrogen peroxide dissolved in 79 ml methanol. The reaction was very exothermic and the temperature rose to 80°C after about 10 ml was added.

Addition was stopped and the reaction cooled with an ice bath to 60°C where it was stirred for a few minutes until reaction subsided. It was then cooled to 44°C and the rest of the hydrogen peroxide was added dropwise. Total addition time was 40 minutes. The mixture was stirred for 30 minutes at 70°C and then heated under vacuum to distill off methanol and water. Infrared analysis confirmed the remaining product to be mainly C_{16-18} alkyl sulfinyl methane.

Other aliphatic hydrocarbyl sulfinyl methanes can be made following the above general procedure using different olefins to make the aliphatic hydrocarbyl methyl sulfide.

Example 2: In a reaction vessel was placed 100 ml methanol and 126 g (0.42 mol) of the mixture of hexadecyl methyl sulfoxide and octadecyl methyl sulfoxide made in Example 1. To this was added dropwise a solution of 47 g of 30% hydrogen peroxide in 47 ml of methanol. Addition took 30 minutes at 40° to 50°C. The mixture was then heated to 70°C under vacuum to distill off water and methanol leaving a mixture of hexadecyl sulfonyl methane and octadecyl sulfinyl methane.

Other aliphatic hydrocarbyl sulfonyl methanes can be made following the above general procedure using different aliphatic hydrocarbyl methyl sulfides made from different olefinic hydrocarbons.

The additives are added to a lubricating oil in an amount which reduces the friction of an engine operating with the oil in the crankcase. A useful concentration is about 0.05 to 3 wt %. A more preferred range is about 0.1 to 1.5 wt %.

In a highly preferred embodiment such an improved motor oil also contains an ashless dispersant, a zinc dialkyldithiophosphonate and an alkaline earth metal salt of a petroleum sulfonic acid or an alkaryl sulfonic acid (e.g., alkylbenzene sulfonic acid).

These friction-reducing additives are also useful in liquid hydrocarbon fuel compositions. Fuel injected or inducted into a combustion chamber wets the walls of the cylinder. Fuels containing a small amount of the additive reduce the friction due to the piston rings sliding against the cylinder wall.

The additives can be used in both diesel fuel and gasoline used to operate internal combustion engines. Fuels containing about 0.001 to 0.25 wt % of the additives can be used.

Aliphatic Hydrocarbyl Sulfinyl or Sulfonyl Alkanol

In further work by *A.G. Papay and J.P. O'Brien; U.S. Patent 4,225,449; September 30, 1980; assigned to Edwin Cooper, Inc.* additives for reducing engine friction were developed which were selected from the group consisting of oil-soluble and fuel-soluble aliphatic hydrocarbylsulfinylalkanols and aliphatic hydrocarbylsulfonylalkanols having the structure

$$R_1-\underset{\underset{(O)_p}{\downarrow}}{\overset{\overset{O}{\uparrow}}{S}}-R_2-OH$$

wherein R_1 is an aliphatic hydrocarbon group containing about 12 to 36 carbon atoms, R_2 is a divalent aliphatic hydrocarbon group containing 1 to 4 carbon atoms and p is 0 or 1.

The most preferred additives are: 2-(hexadecylsulfinyl)ethanol; 2-(octadecylsulfinyl)ethanol; 2-(hexadecylsulfonyl)ethanol; and 2-(octadecylsulfonyl)ethanol.

The additives are readily made by conventional methods by first making an aliphatic hydrocarbylthioalkanol. These can be made by reaction of an olefinically unsaturated hydrocarbon with a mercaptoalkanol under UV radiation. This is then oxidized with hydrogen peroxide to form the sulfinyl or sulfonyl derivative depending upon the degree of oxidation. Generally, an equal mol amount of hydrogen peroxide will form the sulfinyl derivative and two mols of hydrogen peroxide per mol of hydrocarbylthioalkanol will form the sulfonyl derivative. Intermediate amounts will form mixtures which are also useful.

Example: *Preparation of hydrocarbylsulfinylalkanols* – In a UV cell was placed 30 g of a mixture of 50 wt % α-hexadecene and 50 wt % α-octadecene and 10 g of mercaptoethanol. The mixture was stirred 30 minutes under UV radiation leaving 40 g of a mixture of 2-(octadecylthio)ethanol and 2-(hexadecylthio)ethanol.

In a second reaction vessel was placed 75 g of the above mixture, 100 ml methanol, 3 g sodium bromide and 5 ml acetic acid. Then 32 g of 30% hydrogen peroxide was added dropwise at 45° to 50°C over a 30-minute period. The methanol was then distilled off and about 200 ml of heptane was added. The heptane was distilled off to remove water. The sodium bromide precipitate was then filtered at about 75°C. The filtrate was cooled forming a precipitate. This was filtered off to yield 45 g of 2-(C_{16-18} alkylsulfinyl)ethanol.

Other similar sulfinyl derivatives can be made following the above general procedure by using different aliphatic hydrocarbylthioalkanols.

The additives are added to lubricating oil in an amount which reduces the friction of an engine operating with the oil in the crankcase. A preferred concentration is about 0.1 to 1.5 wt %.

In a highly preferred embodiment the motor oil also contains an ashless dispersant, a zinc dialkyldithiophosphonate and an alkaline earth metal salt of a petroleum sulfonic acid or an alkaryl sulfonic acid (e.g., alkylbenzene sulfonic acid).

Tests were conducted which demonstrated the friction-reducing properties of the additives.

LFW-1 Test: In this test a metal cylinder is rotated around its axis 45° in one direction and then 45° in the opposite direction at a rate of 120 cycles per minute. A metal block curved to conform to the circular contour of the cylinder presses at a fixed load against the periphery of the cylinder. Test lubricant is applied to the rubbing surface between the cylinder and the block. Torque transmitted to the block from the oscillating cylinder is measured. The greater the torque the greater the friction. Results are given in terms of "percent improvement" which is the percent reduction in torque compared to that obtained with the test oil without the test additive. The test oil is a fully formulated oil of SAE SE quality. Test results are given in the following table:

Additive	Percent Improvement
C_{16-18} alkylsulfinylethanol (0.15%)	14
C_{16-18} alkylsulfinylethanol (0.15%) plus dimethyl octadecylphosphonate (0.2%)	20
C_{16-18} alkylsulfonylethanol (0.3%)	11
C_{16-18} alkylsulfonylethanol (0.3%) plus dimethyl octadecylphosphonate (0.2%)	14
Dimethyl octadecylphosphonate (0.3%)	11
Dimethyl octadecylphosphonate (0.5%)	11.5

The above results show that both the sulfinyl and sulfonyl additives are very effective in reducing friction and that their effectiveness is improved by use in combination with a phosphonate.

N-Aliphatic Hydrocarbyl Hydroxyalkyl Sulfinyl or Sulfonyl Succinimide

In still further work, *A.G. Papay and J.P. O'Brien; U.S. Patent 4,237,020; December 2, 1980; assigned to Edwin Cooper, Inc.* developed another group of compounds for decreasing friction in automotive engines. According to their results, engine friction is lowered by adding to the engine crankcase oil or fuel a small friction-reducing amount of an N-C_{10-36} aliphatic hydrocarbyl hydroxyalkylsulfinylsuccinimide or N-C_{10-36} aliphatic hydrocarbyl hydroxyalkylsulfonylsuccinimide.

A preferred embodiment is a lubricating oil composition containing a friction-reducing amount of oil-soluble additive having the formula

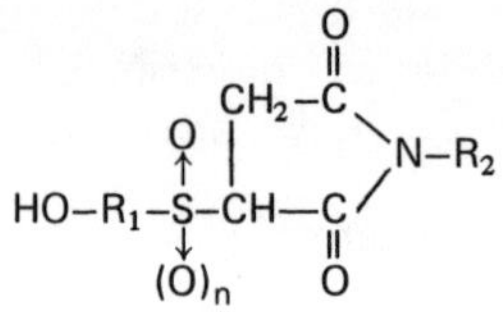

```
                        O
                        ‖
                   CH2—C
            O      |       \
            ↑      |        N—R2
   HO—R1—S—CH——C  /
            ↓           ‖
           (O)n         O
```

wherein R_1 is a divalent aliphatic hydrocarbon group containing 1 to about 6 carbon atoms, R_2 is an aliphatic hydrocarbon group containing about 10 to 36 carbon atoms, and n is 0 or 1.

Highly preferred additives are: N-oleyl[(2-hydroxyethyl)sulfinyl] succinimide and N-oleyl[(2-hydroxyethyl)sulfonyl] succinimide.

Example: In a reaction vessel was placed 196 g of maleic anhydride and 900 ml acetone. The mixture was stirred under nitrogen to dissolve the maleic anhydride. Then 534 g of oleylamine were added dropwise over one hour at 17° to 21°C. Then 220 g triethylamine were added dropwise over 30 minutes at 21° to 22°C. Then 156 g of 2-mercaptoethanol were added over a 30 minute period at 22° to 26°C. The reaction mixture was heated to 100°C to distill off acetone and then to 160°C under 30 inches Hg vacuum to remove triethylamine and water yielding 853 g of N-oleyl[(2-hydroxyethyl)thio] succinimide.

In a reaction vessel was placed 413 g of the above intermediate and 400 g of methanol. The mixture was stirred at 40°C under nitrogen to form a solution and then 91 g of 30% hydrogen peroxide was added over a 30 to 45 minute period at about 55°C. Solvent was distilled off up to 100°C and then 30 inches Hg vacuum applied to complete removal of volatiles leaving 422 g of N-oleyl-3-[(2-hydroxyethyl)sulfinyl] succinimide.

The corresponding sulfonyl additive can be made following the above procedure using twice the amount of hydrogen peroxide. Likewise, other mercaptoalkanols may be used to form the hydroxyalkylsulfinyl or sulfonyl substituents. Alternatively, different hydrocarbyl amines can be substituted forming the corresponding N-hydrocarbyl group.

The additives are added to lubricating oil in an amount which reduces the friction of an engine operating with the oil in the crankcase. A preferred concentration is about 0.1 to 1.5 wt %.

Tests were conducted which demonstrated the friction-reducing properties of the compounds.

SAE-2 Friction Machine Test: In this test a heavy flywheel is rotated at 1,440 rpm. A series of 9 clutch plates are then brought to bear axially at a defined load against the flywheel. The flywheel is connected to the rotating plates. The stationary plates are connected to a device which measures rotational torque. The time from initially applying pressure through the clutch plate until the rotating plate ceases to rotate is measured. Also, the rotational torque measured at the stationary plates is plotted against time. Torque rises initially to a value referred to as "dynamic torque" and then rises finally to a value called "static torque." The clutch plates are immersed in test lubricant. A reduction in friction is indicated by (1) an increase in time required to bring rotation to a halt, and (2) a decrease in dynamic and static torque. Results are reported in percent time increase (percent improvement) and percent reduction in dynamic and static torque compared to that obtained using the same oil without the test additive.

The test oil is a fully formulated oil of SAE SE quality. Test results are given in the following table. In the table the test additive was used at 0.3 wt % and the phosphonate coadditive at 0.2 wt %.

	SAE 2 (%)		
Additive	Time	Dynamic	Static
N-oleyl-(2-hydroxyethyl)sulfinylsuccinimide	11	10	20
N-oleyl-(2-hydroxyethyl)sulfinylsuccinimide + dimethyl octadecylphosphonate	14	15	31

The results show a significant reduction in friction which is further improved by use of the phosphonate coadditive.

Reaction Product of Molybdenum Compound, Phenol plus an Amine

J.L. Karn; U.S. Patent 4,266,945; May 12, 1981; assigned to The Lubrizol Corporation has prepared molybdenum-containing compositions by the reaction of an acid of molybdenum or salt thereof, phenol or aldehyde condensation product therewith, and a primary or secondary amine. The preferred amines are diamines such as tallow-substituted trimethylene diamine and their formaldehyde condensation products. An optional but preferred ingredient in the reaction mixture is at least one oil-soluble dispersant. These molybdenum-containing compositions are useful as additives in lubricants and fuels, and are especially useful in lubricants when combined with compounds containing active sulfur.

As will be apparent from the summary above, these compositions are prepared from a mixture comprising three essential reactants. Reagent A is at least one acid of molybdenum or salt thereof, most often one in which the molybdenum is hexavalent. Illustrative acids are active forms of molybdic acid (H_2MoO_4) and the isopolymolybdic acids.

The salts of the molybdenum acids are preferred for use as reagent A, especially the ammonium and alkali metal salts.

Reagent B may be at least one phenol; that is, at least one compound containing a hydroxy radical bound directly to an aromatic ring. Preferred are phenols containing at least one alkyl substituent containing about 3 to 100, and especially about 6 to 20 carbon atoms, such as heptylphenol, octylphenol, dodecylphenol, tetrapropenealkylated phenol and octadecylphenol.

Reagent B may also be a condensation product of one of the abovedescribed phenols with at least one lower aldehyde containing not more than 7 carbon atoms. Suitable aldehydes include formaldehyde, acetaldehyde, propionaldehyde, the butyraldehydes, the valeraldehydes and benzaldehyde. Also suitable are aldehyde-yielding reagents such as paraformaldehyde, trioxane, methylol, methyl Formcel and paraldehyde. Formaldehyde and the formaldehyde-yielding reagents are especially preferred.

Reagent C may be an aliphatic amine containing at least one primary or secondary amino group. The organic radicals attached to the amino group are aliphatic hydrocarbon-based radicals.

The amines may be mono- or polyamines; polyamines are preferred. The especially preferred polyamines are diamines having the formula $R^3NH{-}R^4{-}NH_2$. In these diamines, R^3 is an aliphatic hydrocarbon radical having at least about 6 and usually about 6 to 20 carbon atoms, and R^4 is a divalent aliphatic hydrocarbon radical having about 2 to 8 and preferably about 2 to 6 carbon atoms and is usually a C_{2-4} alkylene radical. Examples of especially preferred diamines of this type are those sold under the trademark Duomeen. In these diamines, R^3 contains 16 to 18 carbon atoms and may be, for example, oleyl (Duomeen O) or radicals derived from tallow fatty acids (Duomeen T), coconut fatty acids (Duomeen C), etc.

Component C may also be a condensation product of at least one amine as described hereinabove with at least one of the lower aldehydes previously described with reference to reagent B. The use of amine-aldehyde (especially formaldehyde or a formaldehyde-yielding reagent) condensation products as reagent C is preferred, but is usually unnecessary when reagent B is a phenol-aldehyde condensation product.

The preparation of condensation products, also useful as reagent C, is illustrated by the following examples.

Example 1: A mixture of 100 parts of Duomeen T, 3.2 parts of lime, 14.3 parts of water, 30.3 parts of paraformaldehyde and 49 parts of mineral oil is heated at 99° to 105°C until substantially all water is driven off. It is then cooled and filtered; the filtrate is a 70% solution in oil of the desired condensation product.

Example 2: A mixture of 100 parts of Duomeen T, 14.1 parts of water and 30.2 parts of paraformaldehyde is heated under reflux as water is removed by distillation. When no more water is evolved, the temperature is increased to 148°C and the mixture is blown with nitrogen until all water has been removed. The condensation product is then cooled and filtered.

The molybdenum-containing compositions may be prepared by merely blending the abovedescribed reagents, a small amount of water, and, if desired, a substantially inert, normally liquid organic diluent such as benzene, toluene, xylene, petroleum naphtha, mineral oil, ethylene glycol monomethyl ether or the like and heating to a temperature within the range of 80° to 200°C, preferably about 90° to 150°C, until reaction is substantially complete. After completion of the reaction, volatile materials are ordinarily removed by vacuum stripping, blowing with an inert gas or the like and the composition is filtered.

In a preferred embodiment, there is also present in the reaction mixture leading to the molybdenum-containing composition an oil-soluble basic nitrogen-containing ashless dispersant. The amount of dispersant used is generally about 1 to 10% and preferably about 3 to 6% by wt, based on the total of components A, B and C; these percentages are exclusive of inert diluents.

There may also be used in the lubricants other additives in combination with the molybdenum-containing compositions (and, optionally, active sulfur compounds). Such additives include, for example, detergents and dispersants of the ash-producing or ashless type, corrosion- and oxidation-inhibiting agents, pour point depressing agents, auxiliary extreme pressure agents, color stabilizers and antifoam agents.

The molybdenum-containing compositions described herein which are free of sulfur compounds may also be used as fuel additives to decrease friction and fuel consumption.

Generally, these fuel compositions contain about 1 to 50,000, preferably about 4 to 10,000 pbw of the molybdenum-containing composition per million parts of fuel.

The fuel compositions can contain, in addition to the molybdenum-containing composition, other additives which are well known to those of skill in the art. These can include antiknock agents such as tetraalkyl lead compounds, lead scavengers such as haloalkanes (e.g., ethylene dichloride and ethylene dibromide), deposit preventors or modifiers such as triaryl phosphates, dyes, cetane improvers, antioxidants such as 2,6-ditertiarybutyl-4-methylphenol, rust inhibitors such as alkylated succinic acids and anhydrides, bacteriostatic agents, gum inhibitors, metal deactivators, demulsifiers, upper cylinder lubricants, antiicing agents and the like.

The additives can be added directly to the fuel or lubricant. Preferably, however, they are diluted with a substantially inert, normally liquid organic diluent such as mineral oil, naphtha, benzene, toluene or xylene, to form an additive concentrate. These concentrates usually contain about 20 to 90% by weight of the molybdenum-containing composition and may contain, in addition, one or more other additives known in the art.

The following patents, covered in other sections of this book are also noted here, since the additives they describe are useful friction reducers as well as corrosion inhibitors, detergents, etc.: 4,209,411; 4,257,780; 4,266,944.

COMBUSTION IMPROVERS

Picric Acid with Ferrous Sulfate

H.M. Webb; U.S. Patent 4,145,190; March 20, 1979; assigned to Natural Re-

sources Guardianship International, Inc. describes an energy-saving fuel additive for jet, gasoline, and diesel engines, including the use as an additive for domestic heating and light industrial oils (No. 2 and 3) and bunker or residual fuels (No. 4, 5 and 6) which comprises as active ingredients a catalytic mixture of a major proportion of picric acid and a minor proportion of ferrous sulfate.

A preferred solvent suitable for use is a combination of an alkyl benzene, such as toluene, and a lower alcohol, such as isopropanol. The combination specially may include a minor amount of nitrobenzene as well as a particulate reducer such as a long chain tertiary amine (Primene 81R).

The additive mix or concentrate, denoted MSX Mix, useful for bulk addition is as follows for one gallon:

	....For 1 U.S. Gallon....	
	Preferred	**Range**
Ferrous sulfate, g	1.4	0.08-1.4
Picric acid (trinitrophenol), g	45.0	2.8-45.0
Toluene, kg	2.4	2.4-1.0
Isopropyl alcohol, kg	1.0	1.0-2.4
Nitrobenzene, g	2.7	0.08-2.7
Long chain amine (tertiary dodecylamine), g	1.7	0.2-1.7
Water	.... Balance	

The MSX Mix is utilized for dosage to fuels in the ratio of 1:1,000 to 1:2,000 with a preferred dosage of 1:1,600 parts by volume.

Example: *MSX Mix formulation* – Toluene and isopropyl alcohol were mixed together. The trinitrophenol (picric acid) was introduced to this mixture and stirred gently. It dissolved completely when left overnight. The nitrobenzene was added with a slight stir. The ferrous sulfate was dissolved in a small amount of hot water (a maximum of ½ gallon for 100 gallon mix) and added to the mixture.

The product was allowed to stand overnight. It was inspected for any sediment settling, after quality control tests were made and the product passed. It was released for ultimate packaging.

The water usually present with the trinitrophenol (picric acid) was taken into consideration in the formulation of this product.

As a result of utilization of this composition, it has been found by tests that improvements in fuel economy between 12.5 and 15.5% urban and up to 27% highway conditions have been experienced. The variable range is due to make, condition, size of the vehicle, coupled with the variations in road conditions that drivers have at city versus highway driving, etc. It can be further stated that a mean average mileage improvement for all tests is about 20%.

Reaction Product of a Polyamine and an Acrylate Ester

T.C. Shields; U.S. Patent 4,240,804; December 23, 1980; assigned to Sherex Chemical Company, Inc. has developed a group of additives for gasoline which are designed to improve fuel combustion and to reduce the deposition of carbon in combustion areas of an engine.

These additives comprise the reaction product of a polyamine (including diamine), or oxygen-containing polyamine, preferably an ether polyamine, and an alkyl ester of acrylic or alkyl acrylic acid. When the polyamine is a diamine, then the alkyl portion of the alkyl acrylate must be eight or more carbon atoms. A preferred formulation is one wherein the reaction product of the polyamine and the acrylate ester is combined with an amine which can be: (1) the polyamine parent of the ester; (2) an oxygen-containing polyamine parent of the ester; (3) other oxygen-containing monoamines; (4) other oxygen-containing polyamines; (5) other monoamines or polyamines; (6) mixtures of the above. Some of these combinations exhibit synergistic properties.

Polyamines which can be utilized include, but are not limited to:

4-aza-8-oxaeicosyl amine (tridecyl ether diamine)
4-aza-8-oxa-16-methylheptadecyl amine (isodecyl ether diamine)
4,8-diazahexacos-17-enyl amine (tallow triamine)
4-azahexadecyl amine (coco diamine)
4-azadocos-13-enyl amine (oleyl diamine)

Suitable esters include, but are not limited to, ethyl acrylate, and 2-ethylhexyl acrylate.

The amine and the acrylate ester are permitted to react, preferably in a ratio of one mol to one mol at room temperature. Ordinarily, little if any heating or cooling of the reacting mixture will be required and the reaction will proceed readily when the reactants are thoroughly mixed. No byproducts are formed and the end product is a clear one-phase liquid. The product so obtained is added to the liquid hydrocarbon fuel, for example gasoline, preferably in a ratio of between 2 and 60 pounds per 1,000 barrels (PTB).

A preferred composition for use as a hydrocarbon fuel additive is a mixture of polyamines and the reaction product of polyamines and acrylate esters. For example, a mixture of the reaction product of tridecyl ether diamine and 2-ethylhexyl acrylate, when admixed with tridecyl ether diamine or another polyamine or oxygen-containing amine or polyamine (as previously defined–see notations) results in an additive having enhanced properties.

Example: This example illustrates the synergistic effect obtained by the use of a combination of an ether polyamine and the reaction product of an ether polyamine and an acrylic or methacrylic ester.

A commercially available gasoline additive (hereinafter designated "G") is available as a "package" containing 18% tridecyl ether diamine and 12% dipropoxylated tallow amine. The remainder of this package is alcohols, lubricating oils, and small amounts of other agents. This package constitutes a balanced gasoline additive package possessing not only gasoline detergency performance properties but also antiicing, water shedding, anticorrosion and other properties needed for modern fuel additives. The 2-ethylhexyl acrylate adduct of tridecyl ether diamine was mixed with additive G in a ratio of 1 to 4 pbw, respectively, to make additive G-1. The 2-ethylhexyl acrylate adduct of isodecyl ether diamine was mixed with additive G in a ratio of 1 to 4 pbw, respectively, to make additive G-2. The 2-ethylhexyl acrylate adduct of tallow triamine was mixed with additive G in a ratio of 1 to 4 pbw, respectively, to make additive G-3. These mixtures were then tested along with two commercial additives, X and Y in leaded and unleaded gasoline following a newly revised CRC engine test procedure. Additive X is the traditionally used superior performance reference material for testing engine detergency performance. Additive Y is a commercially available engine detergent additive. The additive mixture, G-1, was tested at varying concentrations as shown in the table below.

	Amount (PTB)	Gasoline Type	Deposit Reduction (%)
Additive X	50	Leaded	90
Additive Y	20	Leaded	48
Additive Y	50	Leaded	78
Additive G	24	Leaded	38
Additive G	30	Leaded	46
Additive G-1	20	Leaded	21
Additive G-1	30	Leaded	89
Additive G-2	30	Leaded	84
Additive G-3	30	Leaded	74
Additive Y	11.1	Unleaded	63
Additive G	11.1	Unleaded	28
Additive G-1	11.1	Unleaded	94
Additive G-1	3.7	Unleaded	28

From the data in the table it is readily apparent that a mixture of polyamines or ether polyamines and their reaction products with 2-ethylhexyl acrylate perform in leaded gasoline as well as additive X, and in a lower concentration, and better than additive G at equal concentrations. In unleaded gasoline, the additive of this process performs significantly better than the Y or G additives at equal concentrations.

Tertiary Diamines plus Ethanol

A.F. Kaspaul; U.S. Patent 4,244,703; January 13, 1981; assigned to California-Texas Oil Company describes fuel additives which improve the combustion process in internal combustion engines thus improving fuel economy. The fuel additives improve air/fuel distribution prior to and during combustion, the fuel being particularly better vaporized prior to combustion due to the action of the additives.

The octane ratings of fuels are also increased by the use of these additives, thereby allowing the utilization of efficient high compression engines which need not be burdened with a plurality of energy-wasteful pollution control devices in order to reduce polluting emissions.

The additives are tertiary diamines which can be represented by the general formula

$$\begin{matrix} R \\ R_1 \end{matrix} \!\!>\! N{-}(CH_2)_n{-}N \!<\!\! \begin{matrix} R \\ R_1 \end{matrix}$$

in which R is an alkyl group and particularly a methyl group; wherein R_1 is hydrogen or an alkyl group and particularly methyl; and wherein n is an integer between 1 and 6.

The preferred tertiary diamine is known both as tetramethyldiaminepropane and N,N,N',N'-tetramethyl-1,3-propanediamine. While the tertiary diamines can be used per se as fuel additives, it is preferred that the diamines be mixed with an anhydrous alcohol, particularly ethanol, prior to admixture with the fuel. A one to one ratio by weight is preferred. The diamine/anhydrous alcohol mixture can further be admixed with a substantially one to one mixture of isopropyl alcohol and diacetone alcohol, the diamine being preferably present in the resulting admixture in a concentration which is approximately 10% of the concentration of either the isopropyl alcohol or the diacetone alcohol. The admixture of the diamine, anhydrous alcohol, isopropyl alcohol, and diacetone alcohol is the preferred additive and is admixed with a fuel such as gasoline in a preferred concentration range of between 0.5 and 4.0 ml of diamine to 20 gallons of fuel.

During make-up of the preferred additive, the diamine is fixed mixed with an equal part of anhydrous ethanol, the diamine/ethanol mixture then being added to an equal parts mixture of isopropyl alcohol and diacetone alcohol. The ratio of the diamine/ethanol mixture to the isopropyl alcohol/diacetone alcohol mixture is preferably between 0.5 to 0.025.

The additives may be mixed with fuel in bulk either at the refinery, at a distribution center, or at a point of sale. The additives can also be mixed with fuel in a gas tank of a vehicle by the operator of the vehicle. It is further contemplated that the additives could be metered into the carburetor fuel or induction air from a rechargeable reservoir. The additives could also be metered into induction air via an active air filter.

A number of road tests utilizing a variety of vehicles and test tracks were conducted and are summarized as follows, an additive comprised of 1:1 tetramethyldiaminemethane and anhydrous alcohol being used at a concentration of 4 ml per 20 gallons of gasoline:

Vehicle	Track	Improvement (%)
Lincoln Continental Town Coupe, 1977	Miscellaneous and Solvang Runs	4
Lincoln Continental	Miscellaneous and Solvang Runs	7
Oldsmobile Cutlass, 1968	Westlake Village Santa Monica Westlake Village	20
Buick Skylark, 1971	Malibu West Solvang Malibu West	12

The Oldsmobile was used for commuting and was operated mostly on fast-moving surface streets and freeways. Fuel economy improved each time the additive was added to a full tank of gasoline, the first time at 105,000 miles, 6%; second time at 113,400 miles, 8.2%; third time at 118,300 miles, 24%; then a steady 20%. Without the additive, fuel economy was around 15 mpg from 105,000 miles to about 120,000 miles.

It is seen from the foregoing that a family of fuel additives is provided which increases the thermal efficiency of gasoline-operated engines, fuel economy and emission control being particularly increased.

Iron(II) Chelate and Picric Acid/Organic Base Complex

M.T. Scholtz; U.S. Patent 4,265,639; May 5, 1981 describes fuel additives to promote combustion which are particularly useful in gasoline and diesel fuels. The additives include active catalysts and stabilizing ingredients. The active catalytic ingredients are iron(II) and picric acid or picrates–both substances already known for their use in fuels. However, these substances are stabilized in this formulation, the iron being used as an iron(II)/ethylenediaminetetraacetic acid (EDTA) chelate and the picric acid being complexed with an amine.

Particular amines are primary, secondary or tertiary monoamines of the formula

$$\begin{matrix} R_1 \diagdown & \\ & N{-}R_3 \\ R_2 \diagup & \end{matrix}$$

where R_1, R_2 and R_3 are hydrocarbyl groups having 1 to 12 carbon atoms, and more particularly 2 to 8 carbon atoms, or hydrogen, provided that the amine contains at least 3 carbon atoms. Amines containing 12 or less carbon atoms are preferred. These hydrocarbyl groups may be fully aliphatic or aromatic in character or they may have both aliphatic and aromatic constituents. Aliphatic groups or constituents may be saturated or unsaturated and may be characterized as having a straight-chain, branched-chain or cyclic structure. Particular amines include cyclohexylamine, 2-ethylhexylamine, n-propylamine, n-butylamine, triethylamine, diethylamine, dicyclohexylamine and aniline, cyclohexylamine being preferred.

According to one method, additives may be prepared by the steps of (1) combining EDTA and amine in a suitable solvent such as methanol; (2) adding picric acid to a suitable solvent such as methanol to make a solution of picric acid; (3) adding a source of +2 iron to the picric acid solution of step (2) with stirring to obtain a green iron picrate solution; and (4) adding portions of the solution of step (1) to the green iron picrate solution of step (3) with agitation until the green color disappears and a clear yellow solution is obtained. The color change of step (4) is rapid, taking place immediately upon sufficient addition of the solution of step (1).

According to yet another method, additives may be prepared by the steps of (1) combining EDTA and amine in a suitable solvent such as methanol; (2) dissolving picric acid in the solution of step (1); and (3) adding a source of +2 iron to the solution of step (2) with stirring to give a red orange solution which on standing (e.g., without stirring for about 12 to 100 hours) goes to a clear yellow solution.

According to a further method, additives may be prepared by (1) combining EDTA and amine in a suitable solvent such as methanol; (2) adding a +2 iron source to the solution of step (1) with stirring to get a chelated iron solution; and (3) adding a picric acid solution to the solution of step (2) to get a clear yellow solution.

According to a fourth method, additives may be prepared by the steps of (1) combining the EDTA and the amine in a methanol solution; (2) adding picric acid to the solution of step (1) and dissolving; (3) adding a source of +2 iron to the solution of step (2) and stirring to suspend and dissolve to give a red orange solution; and (4) adding isopropyl alcohol to the solution of step (3) to obtain a clear organic solution which goes yellow upon standing (e.g., without stirring for about 12 to 100 hours).

Example: *Additive stability in gasoline* – Gasoline containing 650 ppm of a solution having a 1:140:6.4:80 ratio of iron:picric acid:EDTA:cyclohexylamine, the concentration of the solution being 60.3 mg/ℓ with respect to iron, is prepared. The gasoline takes on a pale yellow color imparted by the additive which is yellow. There is no significant color change and no visible evidence of precipitation, gum formation or other breakdown of either the gasoline or combustion catalyst after several weeks of observation. When this gasoline is refluxed for one hour, there is again little evidence of significant breakdown. When this gasoline is exposed to trace amounts of water there is relatively little visible evidence of precipitation or other deterioration in either the gasoline or the combustion catalyst, even after several weeks.

Comparative Example: Gasoline is prepared in an identical manner as in the above example except that both the EDTA and the cyclohexylamine of the additive are omitted. The green iron(II) picrate additive immediately changes color upon addition to the gasoline and the gasoline takes on a brown coloration upon such addition. Upon storage of this treated gasoline in a closed bottle for several weeks in a darkened cupboard a brown/green deposit is adhered to the glass at the bottom of the container. In the presence of light the formation of gummy deposits and precipitated material occurs more rapidly.

Performance tests demonstrated an increase in miles per gallon in the range of 4% attributable to the use of these additives. The additives were found to be effective at both speeds of 35 and 55 mph.

β-Carboxyethyl Polysiloxane

T. Minezaki; U.S. Patent 4,272,254; June 9, 1981 has found that the combustion efficiency of fuels such as gasoline can be improved by adding thereto an organic silicon compound having at one end a fatty acid polysiloxane.

The silicon compound can be synthesized by the following steps: First, trichlorosilane ethyl cyanide is produced by reacting acrylonitrile with trichlorosilane. Then, a hydrolysis is effected to produce trichlorosilane ethyl carboxylic acid. Then, by the action of thionyl chloride, trichlorosilane propionyl chloride is produced which is then hydrolyzed into β-carboxyethyl polysiloxane, i.e., an organic silicon compound. The process is expressed by the following formulas:

(1) $SiHCl_3 + CH_2CHCN \rightarrow Cl_3SiCH_2CH_2CN$

(2) $Cl_3SiCH_2CH_2CN \xrightarrow[\text{hydrolysis}]{} Cl_3SiCH_2CH_2COOH$

(3) $Cl_3SiCH_2CH_2COOH + SOCl_2 \rightarrow Cl_3SiCH_2CH_2COCl_3$

(4) $2Cl_3SiCH_2CH_2COCl_3 \xrightarrow[\text{hydrolysis}]{} \begin{array}{l} \;\;\;| \\ \;\;\;O \\ \;\;\;| \\ O{=}Si{-}CH_2CH_2COOH \\ \;\;\;| \\ \;\;\;O \\ \;\;\;| \\ O{=}Si{-}CH_2CH_2COOH \\ \;\;\;| \\ \;\;\;O \\ \;\;\;| \end{array}$

This silicon compound added to a hydrocarbon fuel can assist the oxidation of each component in the combustion system. More specifically carbon monoxide is changed to carbon dioxide, while the hydrocarbon is decomposed into water and carbon dioxide. Simultaneously, the nitrogen oxides are decomposed and oxidized. In addition, since the organic silicon is rich in oxygen, it improves the combustion efficiency by decreasing the carbon content in the exhaust gas, thereby suppressing the production of smoke. Usually, 0.2 to 2.5 mg of β-carboxyethyl polysiloxane is added to one liter of hydrocarbon fuel.

The concentrations of carbon monoxide and hydrocarbon in the exhaust emissions of an automobile engine using a fuel containing β-carboxyethyl polysiloxane added to gasoline were measured and the results are as shown:

	CO Concentration (%)	HC Concentration (ppm)
Before addition	3.5	300
After addition	0.5	180

The automobile was a Nissan Gloria, 6 cylinders, with automatic transmission and an L20 engine (1973).

The idling speed of the engine was increased as a result of addition of β-carboxyethyl polysiloxane. It was therefore, necessary to effect such a slow adjust to adjust the speed of the engine by changing the air-fuel ratio of the mixture. The fuel consumption rate was decreased from 5.2 km/ℓ to 7.5 km/ℓ as a result of the slow adjust and the idling adjust.

-8-

ANTISTATS AND OTHER ADDITIVES

ANTISTATIC AGENTS

Reaction Product of α-Olefin/Maleic Anhydride Copolymer with Amines

The low electrical conductivity of many volatile organic liquid compositions has presented the problem of controlling static buildup, particularly during handling and transportation, for the purpose of insuring safe and effective distribution without the concomitant danger of ignition or explosion. For example, volatile organic liquids such as hydrocarbon fuels, (e.g., gasoline, jet fuels, turbine fuels and the like), or light hydrocarbon oils employed for such purposes as solvents or cleaning fluids for textiles, possess a very low degree of electrical conductivity.

In the use of such fluids, electrostatic charges, which may be generated by handling, operation or other means, tend to form on the surface, and may result in sparks, thus resulting in ignition or explosion. These hazards may be encountered merely in the handling or transportation of such organic liquids and even in operations, such as centrifuging, in which a solid is separated from a volatile liquid, during which electrostatic charges can accumulate.

Various materials have heretofore been proposed for incorporation into such organic liquid compositions for increasing their electrical conductivity and thus reduce the aforementioned dangers of ignition and explosion. Such materials, however, have not been proved to be sufficiently effective in increasing the desired electrical conductivity of these fluids and, in many instances, have been found to be too costly for the relatively small degree of increased protection which they are capable of providing.

In accordance with the process of *W.J. Cheng and D.B. Guthrie; U.S. Patent 4,163,645; August 7, 1979; assigned to Petrolite Corporation*, improved liquid hydrocarbon compositions are provided containing, in an amount sufficient to impart antistatic properties, the Lewis acid reaction product of an amine derivative of an olefin/maleic anhydride copolymer. Incorporation of this antistatic agent in the composition imparts increased electrical conductivity to an extent greater than that heretofore realized with many other antistatic materials.

In general, the additive is used in amounts of 0.1 to about 200, and preferably from about 1 to 10 pounds, per thousand barrels of the total volume of the liquid composition. This is equivalent to about 0.33 to 660 ppm and preferably from about 1.0 to 33 ppm.

When a copolymer derived from an α-olefin and maleic anhydride is reacted with an amine, one may obtain the amine salt, the amic acid, the amine salt of the amic acid, etc., or mixtures thereof.

It has now been found that this reaction, when carried out in the presence of a Lewis acid, proceeds at a faster rate and produces a unique product which is more fluid in solution and more effective as an antistatic agent.

In general, the copolymer is prepared by reacting an α-olefin with maleic anhydride in essentially equimolar ratios. A wide variety of α-olefins can be employed in preparing the α-olefin-maleic anhydride copolymer. Representing the α-olefin is the following idealized formula, $RCH_2{=}CH_2$, where R is a hydrocarbon group such as an alkyl group, $H(CH_2)\!-_n$, where n is a number such as from about 0 to 50, but with an optimum of about 10 to 18.

The amount of Lewis acid employed may depend on various factors such as the particular Lewis acid, the particular copolymer, the particular amine, etc., employed. Based on the weight of the copolymer, with Lewis acids such as BF_3, one may employ from about 0.01 to 300 mol percent, but with an optimum of about 40 to 60 mol percent. The preferred Lewis acid is BF_3 in the form of a complex such as an etherate.

The amines utilizable in forming the amic acids and salts thereof are the primary and secondary aliphatic amines having between about 4 and about 30 carbon atoms per molecule or mixtures thereof.

A very useful and readily available class of primary amines are the tertiary-alkyl and secondary-alkyl primary monoamines in which a primary amino ($-NH_2$) group is attached to a tertiary carbon atom or secondary carbon atom, respectively; and to mixtures thereof.

Example 1: *Without Lewis acid* – To a polymerization reactor are charged tetradecene-1 196 g, maleic anhydride 98 g, and aromatic solvent 50 g. The contents are heated to 130° to 150°C and 0.8 g di-tert-butyl peroxide is added. The reaction temperature is controlled at 130° to 160°C for 12 hr. The polymer is diluted with aromatic solvent to 50%. The amine (Primene 81-R, 404 g) is added at 75° to 80°C and is stirred for 6 hr at 100° to 110°C. The product is diluted to the desired concentration with aromatic solvent. The measure of antistatic activity in conductivity units at 100 ppm in fuel oil is 1,910 picomhos/meter.

Example 2: *With Lewis acid* – The polymer solution of Example 1, 26.4 g and 51 g of aromatic solvent is charged to a reactor and heated at 100°C until homogeneous. The Lewis acid (BF_3 etherate at 47% BF_3, 1.3 g) is charged. The amine (Primene 81-R, 18.2 g) is added at 87° to 106°C. The resulting solution is stirred at 100° to 110°C for 6 hr. The product is diluted to the desired concentration with aromatic solvent. The measure of antistatic activity in conductivity units at 100 ppm in fuel oil is 2,315 picomhos/meter.

Porphyrin Compound plus a Metal Alkyl Sulfosuccinate

K.W. Willcox; U.S. Patent 4,171,958; October 23, 1979; assigned to Phillips Petroleum Company has found that the conductivity of liquid hydrocarbons or halohydrocarbons having a boiling point in the range of –12° to 400°C can be considerably raised by the addition of from 0.1 to 500 ppm of a porphyrin. The conductivity is further raised by the addition to the above mixture of 0.5 to 100 ppm of a calcium or sodium alkyl sulfosuccinate containing 8 to 16 carbons. The porphyrin may have the formula 1 or 2

(1) (2)

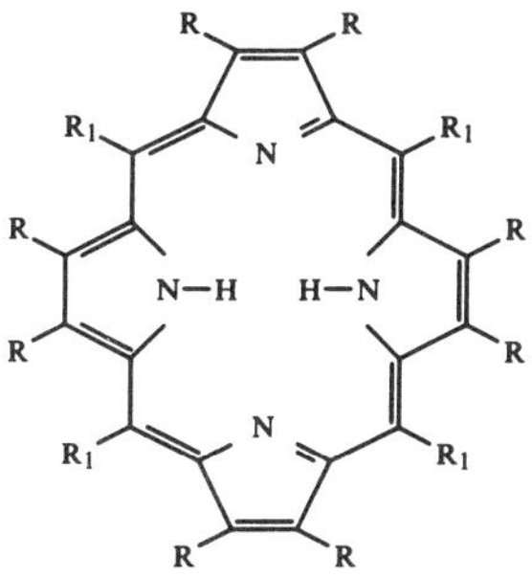

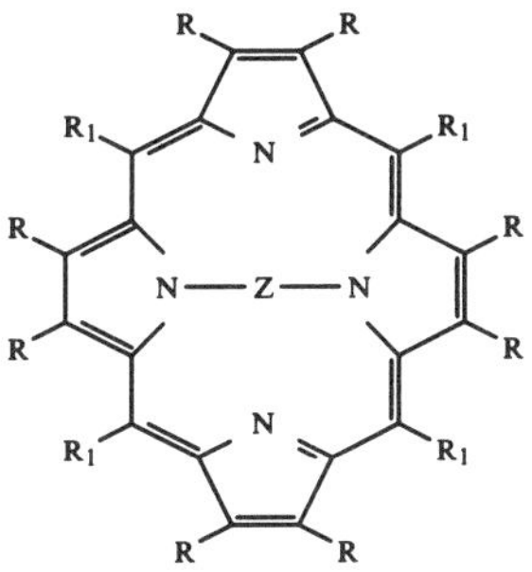

or be a compound of either of the foregoing structures having the hydrogen-free nitrogen atoms protonated, where each R is hydrogen, an alkyl group of 1 to 8 carbon atoms or a carboxy group, each R_1 is hydrogen, an alkyl group of 1 to 4 carbon atoms, an aromatic hydrocarbyl group of 6 to 10 carbon atoms or such an aromatic group substituted with an alkyl group of 1 to 12 carbon atoms, and Z is M or M=O, where M is Cu, Zn, Cr, Fe, Ni, Co or V.

Examples of effective porphyrins are diprotonated α-β-γ-δ-tetraphenylporphyrin; octaethylporphyrin or its diprotonated compound; copper, nickel or vanadyl octaethylporphyrin or α-β-γ-δ-tetraphenylporphyrin; diprotonated tetraisopropyl tetraphenylporphyrin, etc. The first of these is preferred. The preferred commercially available alkyl sulfosuccinate is sodium dioctylsulfosuccinate.

Mixtures Containing an Aminomethylene Sulfonic Acid

There is a definite need for antistatic additives which provide cost-effective activity in a broad spectrum of fuels. *J.R. Spence; U.S. Patent 4,252,542; Feb. 24, 1981; assigned to Standard Oil Company (Indiana)* reports that this objective may be attained by use of an additive which is: (1) an aminomethylene sulfonic acid; (2) a crosslinked aminomethylene sulfonic acid; or (3) a mixture of an aminomethylene sulfonic acid and an amine compound adduct of a maleic anhydride-olefin copolymer.

The aminomethylene sulfonic acids are produced by reacting an amine compound, e.g., a substituted succinimide containing free primary or secondary nitrogen atoms or a compound of the formula RNH_2 or $R_1NHCH_2CH_2CH_2NH_2$ wherein R and R_1 have 1 to 25 carbon atoms, with a carbonyl compound and sulfur dioxide.

Amine compounds which can be used to prepare the aminomethylene sulfonic acids are primary monoamines, polyamines, preferably alkylene polyamines and polyalkylene polyamines.

Amines having 12 to 25 carbon atoms and polyamines having 4 to 6 amine nitrogens are preferred for maximum antistatic activity in the noncrosslinked antistatic agents.

Amines having the structure $R_1NH\text{-}CH_2CH_2CH_2\text{-}NH_2$ include Duomeen CD, T, O, and L-15.

The preferred adducts of acyclic hydrocarbyl succinic anhydride and polyamines are based on diethylenetriamine, triethylenetetramine, and tetraethylenepentamine. These polyamines are especially preferred due to the low cost, the high activity of the resulting antistatic additives and the reactivity of the amino nitrogens. Acyclic hydrocarbyl succinic anhydrides having an aliphatic carbon chain from about 12 to 25 carbon atoms are preferred.

The crosslinked aminomethylene sulfonic acid antistatic agents are produced by reacting an amine compound with a polyfunctional crosslinking agent and then reacting the crosslinked amine with a carbonyl compound and sulfur dioxide, or by reacting the noncrosslinked aminomethylene sulfonic acid with a polyfunctional crosslinking agent.

An effective antistatic composition can be prepared by forming a composition containing about 1 to 100 parts by weight of an aminomethylene sulfonic acid and about 1 part of an amine compound adduct of a maleic anhydride-olefin copolymer. Suitable maleic anhydride-olefin copolymers are made by reacting approximately equimolar amounts of a 1-olefin and maleic anhydride under conventional free radical conditions. These polymers and their synthesis are well known. Useful olefins are α-olefins, such as ethene, propene, 1-butene, 1-pentene, 1-hexene, 1-heptene, 1-octene, 1-octadecene, 1-eicosene, etc.

Further detail of the production of the maleic anhydride-olefin copolymer and the subsequent amine compound adduct are found in U.S. Patent 3,003,858. The preferred maleic anhydride-olefin copolymer has a molecular weight from about 2,000 to 8,000 preferably 4,500 to 6,500 and is made from 1-octadecene and maleic anhydride. This copolymer is preferred for the low cost, high activity, and ease of preparation of the components.

The effective mixture of the crosslinked aminomethylene sulfonic acid and the amine compound adduct of the maleic anhydride-1-olefin copolymer contains about 1 to 100 parts of the polymeric-type sulfonic acid for each part of the amine compound maleic anhydride-olefin copolymer adduct. These mixtures are advantageously made in a hydrocarbon carrier fluid such as a light-fraction kerosene, lubricating oil, toluene, xylene, C_{9+} aromatics, etc.

The aminomethylene sulfonic acid or mixture with the polymer adduct in a solvent can be blended into the fuel of interest in concentration of from about 0.1 to 10.0 ppm of fuel based on the neat polymer.

A field of specific applicability of these additives is in the improvement of organic liquid compositions in the form of petroleum distillate fuel oils having an initial boiling point from about 75° to 135°F and an end boiling point from about 250° to 1000°F.

Particularly contemplated among the fuel oils are Nos. 1, 2 and 3 fuel oils, used in heating and diesel fuel oils, gasoline, turbine fuels and the jet combustion fuels, as previously indicated.

Example 1: 30.0 g of polybutenyl succinic anhydride, molecular weight about 660, was added to 5.65 g of triethylenetetramine in 50 ml of a C_{9+} aromatic solvent and heated to 180°C for 30 min. The solution was cooled to 35°C and 100 ml of solvent was added. 6.394 g of toluene diisocyanate was added slowly and the solution was heated to 160°C for 2 hr. The solution was cooled to 85°C and 1.1845 g of paraformaldehyde was added. The solution was held at 85°C and stirred until clear, approximately 30 min. 2.6 g SO_2 was sparged through the solution for 2.5 hr at 85°C.

Example 2: Example 1 was repeated except with 25.48 g polybutenyl succinic anhydride, (MW 578, 0.044 mol); 5.48 g triethylenetetramine (0.037 mol), 6.53 g toluene diisocyanate (0.37 mol), 1.1253 g formaldehyde (0.037 mol) and 2.4 g sulfur dioxide (0.037 mol).

Example 3: 10 g of a maleic anhydride 1-octadiene copolymer, MW 5,600, was reacted with 4.17 g n-butylamine in 200 ml of toluene at 100°C for 2 hr in a 500-ml round-bottom flask with a reflux condenser, then at reflux for 2 hr until no water or amine distilled. The resulting solution was 6.88% by weight product in toluene.

Testing showed the beneficial antistatic activity of the aminomethylene sulfonic acid; in some cases only 13 ppm provided adequate conductivity. Crosslinking increased the activity. Further testing showed the great increase produced by the addition of the amine/maleic anhydride-octadecene copolymer adduct to the aminomethylene sulfonic acid.

The conductivity of No. 2 heating fuel with 10 ppm of the compound of Example 2 gave a conductivity of 262 picosiemens/meter at 740°F; that of 1.0 ppm of the compound of Example 3 was 175 picosiemens/meter; but the combination of 1 ppm of the compound of Example 3 plus 10 ppm of the compound of Example 2 gave a conductivity of 1,002, more than double the expected conductivity of 437.

Polymeric Amines plus α-Olefin/Acrylonitrile Copolymers

M.I. Naiman, J.I. Knepper and N.E.S. Thompson; U.S. Patent 4,259,087; March 31, 1981; assigned to Petrolite Corporation have found that when certain α-olefin/acrylonitrile copolymers are used in conjunction with certain polymeric polyamines, synergistically effective antistatic agents for hydrocarbon fluids are obtained.

The preferred olefins for preparing the copolymer are 1-alkenes of 10 to 20 carbons, the most preferred being the 1-decene copolymer. The acrylonitrile is one with a formula ($CH_2{=}C(R)CN$, where R is hydrogen or an alkyl such as methyl. The copolymers are prepared by complexing the acrylonitrile with a Lewis acid such as $AlCl_3$, $ZnCl_2$, AlR_nCl_{3-n} and then polymerizing the complexed material with a terminal olefin using a free radical initiator.

A wide variety of polymeric polyamines can be employed in conjunction with the α-olefin/acrylonitrile copolymers to yield these compositions. One polyamine component of the antistatic composition is a polymeric reaction product of epichlorohydrin with an aliphatic primary monoamine or N-aliphatic hydrocarbyl alkylene diamine.

Mixtures of aliphatic primary amines can also be used, and are preferred since mixtures of primary amines derived from tall oil, tallow, soybean oil, coconut oil, cottenseed oil and other oils of vegetable and animal origin are commercially available and at lower cost than individual amines. The above mixtures of amines generally contain alkyl and alkenyl amines of from about 12 to 18 carbon atoms, although sometimes an individual amine mixture, depending upon the source, contains small amounts of primary amines having fewer or more carbon atoms. A preferred example of a commercially available mixture of primary monoamines is hydrogenated tallow amine which contains predominantly hexadecyl- and octadecylamines with smaller amounts of tetradecylamine.

Polymeric polyamines as described above are commercially available. One such suitable product which is believed to be a polymeric reaction product of N-tallow-1,3-propylenediamine with epichlorohydrin is Polyflo 130.

The most preferred olefin-N-alkylmaleimide copolymer is octadecane-N-(N-cyclohexyl-2,4-diamino-2-methylpentane)maleimide copolymer.

The preferred diamine is N-cyclohexyl-2,4-diamino-2-methylpentane, having the formula:

$$NH_2{-}C(CH_3)_2{-}CH_2{-}CH(CH_3){-}NH{-}C_6H_{11}$$

Example: *Preparation of the copolymer of octadecene-N-(N=cyclohexyl-2,4-diamino-2-methylpentane)maleimide* — 35 g of octadecene-maleic anhydride copolymer, 10 cc amyl alcohol and 100 cc xylene were added to a 500-ml flask equipped with stirrer, thermometer and Dean Stark condenser for water removal. The mixture was stirred at 120° to 140°C for 0.75 hr. To this solution was added 19.8 g (0.1 mol) N-cyclohexyl-2,4-diamine-2-methylpentane. The reaction mixture thickened quickly and then thinned as the temperature was raised to reflux. Water and amyl alcohol were azeotropically distilled off. At the end of 5 hr 90 to 95% of the theoretical amount of water was collected. The mixture was cooled to 50°C and the clear homogenous solution was diluted with xylene to make a 20% active solution. The in situ esterification step with amyl alcohol as solvent and reactant was used for the imide synthesis to eliminate and reduce possible crosslinking with the difunctional amine and improve homogeneity in the presence of the aromatic solvent.

The described compositions, when added in concentrations of 1 to 100 ppm to fuel oils, increase the conductivity of the fuel substantially. Depending on the nature of the fuel and the structure of the specific nitrile copolymer the increases in conductivity will vary. However in all cases there is a substantial increase.

EXHAUST EMISSION REDUCERS

Compounds of Si and/or Ca, Mg, Mn, etc. plus Ash-Containing Resin

Combustion of a petroleum fuel containing sulfur, sodium and vanadium compounds results in corrosion, slagging and clogging, with ash deposits on equipment utilizing the fuel.

An effective means adapted for preventing the detrimental effects resides in the use of mineral additives introduced into the flue gases resulting from combustion of petroleum fuel.

More technologically simple, however, is the method of introducing such additives along with the fuel in the form of a composition prepared in advance or during the fuel supply for combustion or during the process of fuel combustion.

This is accomplished by *V.I. Chikul, B.I. Tyagunov. Y.S. Ulanen, O.S. Chikul, K.E. Zeger, L.Z. Tamarkin, R.A. Petrosian, V.E. Chmovzh, A.S. Smirnov, V.I. Polyakovskaya and P.A. Trufanov; U.S. Patent 4,180,385; December 25, 1979* with a fuel composition incorporating a high-boiling fuel and an additive containing compounds of the elements such as Si and/or Ca, and/or Mg, and/or Al and/or Mn, in amounts of ~ 7 kg/ton of fuel, the additive, including an ash-containing resin resulting from thermal processing of a solid fuel.

In this process for producing a fuel composition the additive is prepared by thermal processing a solid fuel and formation of a dust-containing vapor-gas mixture, followed by condensation of a resin from the resulting vapor-gas mixture. This resin is discharged together with the dust deposited therewith. The vapor-gas mixture delivered to condensation is incorporated with compounds of elements such as Ca and/or Mg, and/or Al, and/or Fe and/or Mn in the form of salts, oxides, oxychlorides or hydroxides.

In this process preparation of the additive is combined with the solid fuel processing, whereby preparation of the additive is substantially simplified and made less expensive.

The process makes it possible to reduce the range and consumption rates of mineral raw materials in preparing the additive preparation; furthermore, the necessity of fine grinding the raw materials is avoided; the necessity of using an expensive multicomponent intermediate liquid is also avoided; energy consumption is substantially reduced.

An apparatus for implementation of the process is fully described and shown in a line drawing in the patent.

The apparatus operates in the following manner: A finely ground dried fuel such as shale is continuously fed into a reactor. Also continuously supplied into the reactor is a solid heat carrier agent, such as ash of the previously processed shale. The ash is heated to a temperature of from 700° to 1100°C. The resulting mixture of shale and heat carrier agent is kept inside the reactor for a period required to heat the shale to a temperature of from 450° to 650°C and to perform complete thermal decomposition thereof. The mixture of the heat carrier agent and solid residue from the thermal decomposition of the shale is continuously discharged from the reactor.

Volatile products of the shale thermal decomposition, i.e., the mixture of resin vapors, water vapors and gas and the dust entrained therewith are delivered from the reactor to a scrubber wherein condensation of heaviest fractions of the resin and deposition of a substantial portion of the dust occurs. To ensure the condensation of resins and facilitate the removal of dust-laden resin, the cooled resin of the previously processed shale is also continuously supplied into the scrubber.

The vapor-gas mixture is delivered from the scrubber to a condenser. The resin circulating and condensed in the scrubber along with the deposited dust is also fed into the condenser. Additional condensation of the resin in the amount required to produce the additive is effected in the condenser and the dust remaining in the vapor-gas mixture is entrapped. The residual vapor-gas mixture, practically free from the dust, is delivered to a further application via an outlet pipe.

The resin mixture fed from the scrubber and condensed in the condenser is delivered to a vessel from which it is fed into the scrubber again.

The mixer is continuously fed with a solution of magnesium chloride. In the mixer the magnesium chloride solution is mixed with the circulating resin and heated to a temperature of from 110° to 250°C.

Temperature of the resin fed to the mixer is adjusted by varying operating conditions of the cooler, thus ensuring the required heating of the solution of magnesium chloride.

In the scrubber, water with magnesium chloride dissolved therein is evaporated and magnesium chloride precipitated from the solution as a solid product is suspended within the resin in the form of fine dust.

The dust-containing resin is the final additive ready for use and is fed by means of a pump as it is accumulated to the required amount or continuously, for use in the fuel composition.

Organic Magnesium and Manganese Compounds to Reduce Sulfur Emissions

R.C. Diehl and J.L. Walker; U.S. Patent 4,202,671; May 13, 1980; assigned to Calgon Corporation have developed a means for conditioning fuel oils with the use of organic magnesium and organic manganese compounds which reduce fuel oil requirements and sulfur trioxide emissions.

Suitable organic magnesium compounds include oil-soluble magnesium sulfonates and magnesium carbonates. A particularly preferred organic magnesium compound is (Hybase n-400), a magnesium sulfonate and magnesium carbonate in oil of the formula:

$$(C_{22}H_{45}C_6H_4SO_3)_2Mg + MgCO_3$$

Suitable organic manganese compounds include manganese salts of mixed aliphatic acids, as for example, compounds of the formula:

$$(C_nH_{2n+1}COO)_xMn$$

wherein n is 4 to 20 and x is 2 or 4. The organic magnesium and organic manganese compounds may be present in a weight ratio of from 0.1 to 10 parts organic magnesium compound per part organic manganese compound. In use, the organic magnesium and organic manganese compounds are blended together in fuel oil to provide an oil-soluble composition having a concentration of from 1 to 10 weight percent magnesium and 1 to 15 weight percent manganese. This composition may be fed to the combustion chamber in a concentration of 10 to 10,000 mg/ℓ, preferably 75 to 1,500 mg/ℓ of fuel oil.

The following example will illustrate the fuel conditioning composition.

Example: A composition containing 40% by weight No. 2 fuel oil, 10% by weight of a 12% solution of manganese salts of mixed aliphatic acids (Nuxtra Manganese 12) and 50% by weight of an organic magnesium compound (Hybase n-400) was fed to a boiler for 6 weeks at a concentration of 1 gal per 1,500 gal of a high sulfur oil. Combustion efficiency was increased and fuel consumption decreased by 2.3%; sulfur trioxide emissions were reduced by 75% and the rate of corrosion was reduced.

Irradiation with Low Radioactivity

Y. Ito; U.S. Patent 4,205,959; June 3, 1980; assigned to Michie Ito; Japan has devised a method for treating the liquid fuel for use in an automobile engine with low radioactivity in order to decrease the amount of noxious ingredients in the exhaust gas and to reduce fuel consumption.

As the means for irradiation of the liquid fuel, an ore produced in the region of Ina, Nagano-ken, Japan, called monazite is used. This ore is broken into pieces having a diameter of from 3 to 5 mm, cleaned, and soaked in the fuel for the necessary amount of time. Any other convenient method of applying a low amount of radiation could also be used.

Monazite contains the following components: Ce_2O_3, 39 to 74%; ThO_2, 1 to 18%; ZnO_2, 1 to 7% and Y_2O_3, 1 to 5% in the form of $CeYPO_4$ and $ThSiO_4$.

Example 1: Washed monazite pieces are soaked in gasoline at 50 g/ℓ of gasoline for 5 hr. Results of the tests on the exhaust gas of Gasoline A (nonirradiated) and Gasoline B (irradiated as described) were as follows:

Components	Gasoline A Weight of Exhaust Components per 1 km, g	Gasoline B Weight of Exhaust Components per 1 km, g	Reduction (%)
CO	7.23	6.86	5
HC	2.52	2.51	2
NO_x	1.52	1.29	15
CO_2	184.5	187.7	–

Addition of Wax Oxidates to Diesel Fuel

Diesel fuels used in diesel engines give off in the exhaust of the engine particulates which recent tests show to be harmful pollutants. These particulates include not only those that exist as visible smoke when the diesel engine is overloaded or when the engine is worn or dirty, but also those that are invisible and emerge from partly loaded clean diesel engines. The Federal Environmental Protection Agency recently determined that diesel-powered automobiles emit unacceptably high levels of air pollution which must be reduced to ward off a possible health hazard.

The proposed standards would allow six-tenths of a gram per mile for 1981 model cars, to be reduced to two-tenths of a gram per mile by the 1983 model year.

The main object of the process developed by *W.M. Sweeney and H.G. Sprague;*

U.S. Patent 4,222,746; September 16, 1980; assigned to Texaco Inc. is to provide diesel fuel compositions which emit, during use, reduced amounts of particulate combustion products.

The fuel used in this process comprises a hydrocarbon diesel composition containing from 0.1 to 1.5 weight percent of a wax oxidate characterized by a ratio of neutralization number to saponification number below about 0.40 and Saybolt Universal viscosity at 210°F higher than 1,600 sec. Optionally a synergistic amount (from 0.1 to 1.5 weight percent) of a fuel-soluble organometallic compound such as alkylcyclopentadienyl manganese tricarbonyl, ferrocene and iron pentacarbonyl can be added to the fuel.

The wax oxidates used preferably are produced by reacting mineral oils or petrolatum with air in the presence of a catalyst at a temperature between 270° and 400°F at a pressure below 20 psig and at an air rate between 15 and 35 cu ft of air per pound of reactant per hour. A more complete explanation of the method of preparation and examples thereof are set forth in U.S. Patent 2,705,241.

The process is further illustrated in detail in the following examples.

Example 1: ½% by weight of a wax oxidate prepared from a light solvent neutral oil is added to a diesel fuel and reduces the particulates from a diesel engine by about 25% as compared with the engines burning the same fuel without the oxidate. The addition of 0.1 weight percent of the wax oxidate produced 3% reduction.

Example 2: Proceeding as in Example 1 but using 1.5 weight percent of a wax oxidate washed with alkali to remove acidic materials gave the same results.

Example 3: This example shows that wax oxidates act synergistically with iron pentacarbonyl in reducing particulate emission.

A diesel fuel containing ½ weight percent iron pentacarbonyl and ½ weight percent of a wax oxidate produced from a light solvent neutral oil reduces particulate emission by 40% as compared with the pure fuel.

The same fuel with ½ weight percent iron pentacarbonyl alone reduces such emissions by 10% while the addition of wax oxidate alone reduces the emissions by 25% as compared with the base fuel.

TRACER COMPOUNDS

Chlorohydrocarbons and Chlorocarbons

There are occasions when it is desirable to give gasoline or other fuels from a given source a distinctive characteristic, a label, such that it can be identified and distinguished from any similar fuel from other sources. Some such occasions arise in cases of suspected theft from a storage location. Other occasions involve investigations of intended or unintended commingling of similar fuels from different sources, or investigations of the path or time delay in distribution of fuels, as from refinery to customer. Still other occasions arise when spills or leaks of fuel of uncertain origin result in contamination of earth or water.

It would be desirable, particularly when fuel spills or leakage may occur, to have available an inexpensive tracer material which could be added to all the fuel from a given source, (i.e., marketing station, refinery, or company) on a continuing basis, without giving unwanted side effects. Then if a spill or leak releases fuel of unknown origin, as when gasoline found percolating through the earth may have come from any of several underground tanks or pipelines in the vicinity, a test for tracer in the recovered fuel would determine its source.

In selecting a suitable tracer, several factors must be taken into consideration. Among the major ones are: cost, ease of detection, stability, solubility and compatibility with the fuel, inertness to air, water and normal soil components, corrosivity, volatility and toxicity. Balancing all of these factors against each other, *J.L. Keller; U.S. Patent 4,141,692; February 27, 1979; assigned to Union Oil Company of California* has found that the chlorohydrocarbons and chlorocarbons having at least 3 chlorine atoms and at least 2 carbon atoms per molecule, and having an atomic ratio of Cl/C of at least 1 to 3 appear to present an optimum combination of required properties.

Firstly, they are easily detectable in minute quantities by conventional gas chromatographic methods, using an electron capture detector. They can be selected according to boiling point so as to give a chromatogram peak readily distinguishable from the peaks resulting from other fuel components. Their boiling point can also be selected so as to fall within or above the upper three-quarters of the boiling range of the fuel, thereby minimizing evaporative loss of tracer from spills or leaks. They are suitably inert to air, water and soil components, as well as conventional fuel components, and they are noncorrosive. They are relatively nontoxic, as compared to compounds such as CCl_4 and $CHCl_3$. Finally, a considerable variety of suitable members are commercially available at low cost. No other class of compounds is known which meets all these qualifications.

Preferred tracer compounds for use herein contain from 3 to about 8 chlorine atoms and 2 to 10 carbon atoms per molecule, and have a Cl/C atomic ratio between 0.5 and 3. From the standpoint of gas chromatographic detectability, it is further preferred that at least 2 chlorine atoms per molecule be bonded to the same or adjacent carbon atoms, and/or that at least one chlorine atom, preferably at least 2, be bonded to an olefinic carbon atom. Exemplary preferred tracer compounds for gasolines are as follows:

	Boiling Point (°C)	Melting Point (°C)
1. Trichloroethylene	87.2	-73
2. 1,1,2-Trichloroethane	113.5	-37
3. Tetrachloroethylene	121	-22
4. 1,1,2,2-Tetrachloroethane	146	-36
5. Pentachloroethane	162	-29
6. Hexachloroethane	186*	187
7. 1,2,4-Trichlorobenzene	213	17
8. 1,2,4,5-Tetrachlorobenzene	240-46	138-140
9. Pentachlorobenzene	275-77	85-86

*777 mm.

In general, a wide range of concentrations can be utilized, from about 0.01 to 10 mg/ℓ, but to insure against loss of detectability through aging or other factors while at the same time maintaining reasonable economy, the concentration range of about 0.2 to 4 mg/ℓ of the tracer compound is preferred.

Example: Three gasoline samples containing 0.5 mg/ℓ of compound 8 above were treated as follows: Sample A was extracted 5 times in succession with ½ its volume of water. Sample B was similarly extracted with 1% H_2SO_4. Sample C was similarly extracted with 1% aqueous NaOH solution. Three other samples, E, F and G, each containing 1 mg/ℓ of compound 3 were treated as were samples A, B and C respectively. Chromatographic analysis gave the following results. The data demonstrate that the tracers are stable and not measurably affected by acid or alkalis, and are detectable in extremely small quantities.

Sample	Tracer	Treatment	Chromatogram Peak Height (mV)
A	Tetrachlorobenzene	Water extracted	0.6
B	Tetrachlorobenzene	1% H_2SO_4 extracted	0.6
C	Tetrachlorobenzene	1% NaOH extracted	0.6
D	Tetrachlorobenzene	None	0.6
E	Tetrachloroethylene	Water extracted	13
F	Tetrachloroethylene	1% H_2SO_4 extracted	13
G	Tetrachloroethylene	1% NaOH extracted	13
H	Tetrachloroethylene	None	13

Tests on gasoline containing 0.5 mg/ℓ of tracers 8 and 3 which had been percolated through a 32 ft column of dry earth showed that the adsorptive capacity of earth for the tracer is very small.

Morpholino-Substituted Naphthylaminopropanes

Another need for markers or tracers for organic liquids such as petroleum fuels is to make it possible to identify different kinds of fuels which are priced and taxed in different ways.

For example, gasoline used for off-highway, nonvehicular purposes such as mining, lumbering or fishing, is commonly taxed at lower rates than that for highway vehicular use. Further, certain grades of oil are used interchangeably for heating oil or for diesel motor fuel. These situations can lead to abuse of the tax laws and cheating by unscrupulous persons.

R.B. Orelup; U.S. Patent 4,209,302; June 24, 1980; assigned to Morton-Norwich Products, Inc. has provided a group of compounds as markers for fuels. The two preferred compounds are 1-(4-morpholino)-3-(α-naphthylamino)propane and 1-(4-morpholino)-3-(β-naphthylamino)propane.

The marker is added to a petroleum fuel in a concentration as low as about 0.5 to 1 ppm of fuel, in which it dissolves without imparting any color to the fuel or otherwise indicating its presence to the naked eye. Thereafter, the marked fuel is extracted with a relatively small portion of an aqueous acidic solution which removes and concentrates the marker in the aqueous phase. The aqueous phase containing the marker is separated from the fuel and treated with a small quantity of a stabilized solution of diazotized 2-chlor-4-nitroaniline whereupon a characteristic pink coloration develops instantly. This procedure of marker extraction and treatment with stabilized diazotized 2-chlor-4-nitroaniline is hereinafter referred to as the detection procedure.

A control, unmarked fuel carried through the same detection procedure, either does not develop any color, or assumes a pale yellow to brownish hue, depending upon the nature and components of the original fuel.

The marker is usually added to the fuel at a concentration of about 12 ppm, at which concentration it imparts no coloration to the fuel. At 12 ppm, the marker gives an intense bluish pink coloration when carried through the described detection procedure. At this level, if the marked fuel is admixed with an unmarked fuel in a concentration as little as 4 parts of marked fuel per 96 parts of unmarked fuel, the resulting concentration of marker in the mixture will be 0.5 ppm, a concentration which is easily detected by the described detection procedure.

The presence of the marker by the above-described detection procedure may be confirmed by extracting the aqueous acidic extract with an immiscible polar solvent, such as amyl alcohol or hexyl Cellosolve, which extracts the marker and changes its shade to purple. This polar solvent solution of the marker also may be spotted on silica gel-coated plates for thin layer chromatography. These techniques serve to distinguish the marker from any natural fuel component colors which may mask the marker color at very low concentrations. The aqueous acidic solution which is used to extract the marker from the marked fuel may be a solution of methanesulfonic acid or hydrochloric acid. A convenient methanesulfonic acid solution is one containing 2% by weight of methanesulfonic acid. A 2 to 3% solution of hydrochloric acid in water may also be used. The specific acid used in the aqueous acidic solution may be varied as desired or needed in accordance with the nature of the fuel under consideration.

An important element in the detection of the marker is the diazotized 2-chlor-4-nitroaniline solution. Normally, diazotized aromatic amines are relatively unsatble, and when used for preparation of azo dyes, are reacted in the azo coupling as soon as possible after diazotization is complete.

It has been found when 2-chlor-4-nitroaniline is diazotized in a nonaqueous medium, particularly glacial acetic acid, the resulting pale yellow diazo solution is very stable. For example, a standardized solution has been observed to retain more than 75% of its initial activity when stored 6 months at 18° to 35°C in a brown bottle, and to be still highly active after one year. Since testing of marked fuel samples desirably are to be performed by enforcement officers in the field, rather than under laboratory conditions, an effective diazo reagent must be stable for lengthy periods of time under ambient temperature conditions.

The markers described above have moderate fuel solubilities per se. For convenience in handling, storage and metering into fuels, these markers may be converted to a permanent liquid state, soluble in all proportions in petroleum fuels and colorless at use concentrations, by admixture with fatty acids and solvents. For example, a typical formulation would be the following: Marker I 34%, oleic acid 39%, and xylene 27%.

MISCELLANEOUS ADDITIVES

Boron Compound plus Amine as Biocide

It is well known that hydrocarbon fuels when kept in storage vessels such as at oil refineries, distribution terminals, and in smaller fuel tanks almost invariably contain small amounts of water. Some species of bacteria and fungi normally found in soil and groundwater have been found also to exist in this water, deriving their nutrition from the hydrocarbon phase as well as from trace elements found in the water. This metabolism is responsible for the consumption of some amount of petroleum product, tank corrosion, and general product deterioration by forming rust, hydrogen sulfide, gums and filamentous material at the interface between the water and the hydrocarbon.

It is the primary objective of *J.W. Amick; U.S. Patent 4,166,725; September 4, 1979; assigned to Standard Oil Company (Indiana)* to provide a biocidal agent which is fuel-soluble so it is carried along as fuel is transferred from one storage vessel to another. In this manner effective control of microbiological growth may be accomplished by adding the agent to fuel at a single point in the distribution system.

This objective is accomplished in a hydrocarbon fuel composition which comprises a major portion of a mixture of hydrocarbons boiling in the range from about 70°F to 750°F containing a minor amount of one or more oil-soluble boron compounds having biocidal properties and one or more oil-soluble amine compounds having biocidal properties.

Some boron compounds which are useful are condensation products of an alkali metal borate and ethylene glycol, diethylene glycol, triethylene glycol, 1,4-butanediol, 2,3-butanediol, and 1,5-propanediol.

The amine compound component may be any oil-soluble amine compound having biocidal properties. Included within this group of compounds are various alkyl and aryl substituted monoamines and various alkyl and aryl substituted diamines. Some amine compounds which are encompassed by this process have the following general formula:

$$(H)_2N{-}R_2{-}N(R_3)(R_4)$$

where R_2 is a C_{1-4} alkyl group, and R_3 and R_4 are selected from the group consisting of hydrogen and C_{1-30} alkyl groups. R_3 may be hydrogen and R_4 may be C_{6-24} alkyl groups.

In general, effective results will be obtained by using each biocidal agent of the kind disclosed herein in an amount between about 0.0005 and about 0.05%. Normally concentrations of about 0.001 to 0.025% will be preferred so as to obtain a good balance between economy and effectiveness of results.

Starter Fluid for Very Cold Weather

F.E. Kaiser, Jr.; U.S. Patent 4,177,040; December 4, 1979; assigned to U.S. Aviex Corp. has devised an engine-starting compound used for starting internal combustion engines, such as diesel engines, in extremely cold weather.

The compound, consisting of a combination of a lower alkyl ether, a CO_2 or nitrous oxide pressurizer, a hydrocarbon propellant of the class consisting of butane, isobutane and propane, and a petroleum distillate with a 5 to 8 carbon chain is utilized for the starter fluid. In this combination, the petroleum distillate with its flash point between 10° to 30°F will prolong the initial fire of the starting fluid. This will provide additional Btu to increase the internal temperature of the engine cylinder due to the longer period of burn with the petroleum distillate continuing to burn after the ether and propane have been exhausted. By so prolonging the burning time of the starter fluid the fuel of the internal combustion engine is more easily ignited thus reducing the amount of time required to crank the engine and also the amount of starter fluid needed to obtain engine fuel ignition is reduced.

The starting fluid is utilized in internal combustion engines of both the diesel and gasoline type. The fluid in its mixed form is available for use in a sealed pressure-charged, valved aerosol-type container with the fluid generally being under a pressure between 60 psi and 130 psi. The starting fluid is introduced into the cylinders of the internal combustion engine which upon cranking of the engine causes the initial ignition or firing of the fluid followed by the ignition of the fuel within the cylinder of the engine.

The starting fluid is a mixture of a lower alkyl ether, such as diethyl ether or isopropyl ether, a hydrocarbon propellant of the type consisting of butane, isobutane and propane, a pressurizer which is preferably CO_2, nitrous oxide, or a combination of CO_2 and nitrous oxide, and a petroleum distillate of the type having 5 to 8 carbon chains or an isomer thereof. Such petroleum distillate will be either pentane, hexane, heptane, or octane or mixture thereof.

The proportion of the components of the fluid mixture is with the ether being 35 to 65% by volume, the hydrocarbon propellant from 4 to 8% by volume, the pressurizer being 3 to 5% by volume and the petroleum distillate being 36 to 52% by volume.

A preferred starting fluid is of the following components and proportions: diethyl ether, 40%; propane, 6.7%; CO_2, 3.3%; and petroleum distillate, 50%.

The ether and hydrocarbon propellant are introduced into the aerosol container in liquid form and remain so until expelled from the container.

Multicomponent Demulsifier for Hydrocarbons

As is well known, the performance of various petroleum products and other hydrocarbons can be adversely affected by the presence of water therein. As is also well known, water can be dispersed in such products during processing. For example, the presence of water is often found in gasoline storage. Morevoer, seawater is often present with gasoline or petroleum products when the same are transferred from tankers and/or barges to stationary, land storage facilities. Such contamination is, of course, not, per se, bad and, indeed, long-range deleterious effects can often be avoided by allowing sufficient settling time and thereafter decanting or otherwise separating the hydrocarbon and water phases.

Due to the large volumes stored and transported, however, the provision of adequate settling time is not, generally, practical. Moreover, when a hydrocarbon-water mixture, such as gasoline and water is subjected to high shear agitation, such as can occur in refinery blending and/or high shear pumping, such as can be encountered in transferring from tanker or barge to land storage or from tank to tank while in storage, relatively stable oil-in-water and water-in-oil emulsions can be formed. Then, upon settling the two emulsions will separate and decanting will permit separation of the two emulsions, but the hydrocarbon phase will contain water and the water phase will contain oil.

The problems associated with the formation of stable emulsions have become more common in recent years and particularly since the advent of the use of additives possessing surfactant properties, which have been used with increasing frequency in mineral oils, solvents and fuels, as dispersants, oxidation and rust inhibitors, antiicing agents, pour point depressants, detergents, etc.

Heretofore, several methods have been proposed for breaking oil-in-water and water-in-oil emulsions. For the most part, these prior art processes involve the use of demulsifiers and the use of various alkoxylated alkylphenol-formaldehyde resins and various quaternary ammonium halides have been contemplated for this purpose. The prior art processes heretofore developed do not, however, function quickly and water is often contained in various hydrocarbons when the same reached the ultimate consumer. Moreover, significant quantities of hydrocarbon are lost when the water phase is discarded. Also, when the amount of water present during blending or other processing is relatively large, dirt will be entrained in the hydrocarbon reaching the consumer.

A.A. Zimmerman, and J. Ryer; U.S. Patent 4,209,422; June 24, 1980; assigned to Exxon Research & Engineering Co. have found that the foregoing and other disadvantages of the prior art demulsifying systems can be avoided or at least significantly reduced with a multicomponent demulsifier comprising:

(a) an alkoxylated alkylphenol-formaldehyde resin wherein the alkyl may be one or more alkyl groups having from about 3 to 12 carbon atoms and the alkoxylation of the resin is effected using about 1 to 100 mols of one or more alkylene oxides per mol of OH radicals in the resin, the alkylene oxide having from about 2 to 6 carbon atoms, and

(b) at least two alkoxylated polyamines wherein the number of nitrogen atoms in the polyamines differ by at least two and wherein the alkoxylation of the polyamines is effected using from about 1 to 100 mols of one or more alkylene oxides per atom of nitrogen in the polyamine and wherein the alkylene oxides have from about 2 to 5 carbon atoms.

The amount of alkoxylated alkylphenol-formaldehyde used varies from about 1.5:1 to about 3:1 pbw of the formaldehyde resin per part by weight of either the higher or lower nitrogen content alkoxylated polyamine. The amount of polyamines vary from about 0.5:1 to 2:1 pbw of higher nitrogen content alkoxylated polyamine per part by weight of lower nitrogen content alkoxylated polyamine.

A quaternary ammonium compound can be substituted for one of the alkoxylated polyamines, and particularly the polyamine having the larger number of nitrogen atoms.

Particularly preferred are the dialkyldimethyl quaternary ammonium salts wherein the alkyl groups each contain from about 12 to about 14 carbon atoms.

Example: In this example, 2,500 ml of a premium quality gasoline were combined with 2.5 ml water, 0.8 PTB of an alkoxylated p-tert-amylphenol-formaldehyde resin, 0.3 PTB of an alkoxylated diamine and 0.3 PTB of an alkoxylated higher nitrogen content polyamine. The p-tert-amylphenol-formaldehyde resin used in this experiment was alkoxylated with a mixture of ethylene and propylene oxides containing about 4 mols of ethylene oxide per mol of OH groups in the base polymer and about 8 mols of propylene oxide per mol of hydroxy groups in the base polymer. The alkoxylated diamine and the alkoxylated higher nitrogen content polyamine was identical to those used in the previous examples. The water phase added to the gasoline has a pH of 4.

The gasoline, demulsifiers and water were combined in a pump-around system and continuously circulated using a centrifugal pump, (for example, a single-stage centrifugal pump 316 S.S. MRT seal explosion-proof 115 volt 60 cycle one-phase motor). Samples were periodically withdrawn and the time to filter through in a 0.5 μ filter determined. In this example, the time required to filter was less than 100 seconds. Without the disclosed demulsifier package the time for filtration was in excess of 200 seconds.

Dyes Prepared by Diazotization and Coupling

G. Hansen, F. Merger, G. Nestler and G. Zeidler; U.S. Patent 4,210,414; July 1, 1980; assigned to BASF AG, Germany describe certain dye mixtures which are particularly useful for coloring motor fuels, heating oils, etc. because of their ready solubility in aromatics. The dyes are mixtures having the formula:

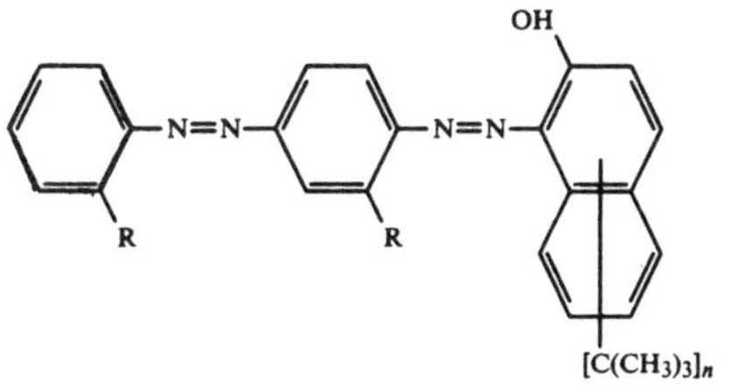

OH

N=N N=N

R R

$[C(CH_3)_3]_n$

where R is hydrogen or methyl and n is a number from 0 to 6. Preferably, n is 0, 1 or 2. Particularly preferred mixtures comprise the following dyes

(a) R = H, n = 0
(b) R = CH_3, n = 0
(c) R = H, n= 1 and
(d) R = CH_3, n = 1,

which still contain certain amounts of dyes with n = 2 and very small amounts with n > 2.

The dye mixtures in particular contain at least 10% and at most 90% of one of the dyes (a) to (d); mixtures which contain the dyes in about equal amounts are preferred. In mixtures with unequal amounts, the dyes where R = CH_3 are preferred.

The dye mixtures may be prepared by synthesizing the individual dyes and mixing them or, more advantageously, by carrying out the synthesis with a mixture of diazo components and coupling components. The method of preparation of the dyes, by diazotization and coupling, exhibits no unusual features.

Surprisingly, the dye mixtures of the process exhibit excellent solubility in solvents which are conventionally used for the preparation of stock solutions, for example, for coloring fuel oil. Specific examples of such solvents are toluene, xylene and high-boiling, (e.g., 180° to 250°C) aromatic mixtures (for example, Shellsol and solvent naphtha).

The dye mixtures are therefore exceptionally useful, for example, for preparing stock solutions containing a large amount of dye, e.g., 3 to 10%.

In the example which follows, parts and percentages are by weight, unless stated otherwise.

Example: 19.7 parts of 4-aminoazobenzene and 22.5 parts of 4-amino-3,2'-dimethylazobenzene are stirred with a mixture of 350 parts of water and 60 parts of 10 N hydrochloric acid for about 3 hr at room temperature and after adding 200 parts of ice, a concentrated aqueous solution of 13.8 parts of sodium nitrite is introduced. The diazotization is complete in from 3 to 4 hr at from 0° to 5°C; the excess nitrite is then removed in the conventional manner with amidosulfonic acid. Thereafter, the suspension of the diazonium salt mixture is added dropwise to an ice-cooled emulsion of 15.1 parts of β-naphthol and 21.0 parts of tert-butylnaphthol in 200 parts of water, 12 parts of KOH and 100 parts of ice.

Sodium acetate solution is then added until the pH reaches a value of from 3 to 4. After stirring overnight, the coupling is complete. The product is filtered off, washed neutral with water and dried under reduced pressure at 60°C.

Metal Deactivator for Aviation Fuel

The additives developed by *E.A. Allen, P.C. Phillips and T. Smithson; U.S. Patent 4,233,035; November 11, 1980; assigned to National Research Development Corporation, England* are organic compounds which are capable of chelating metals and may be used to improve the high-temperature thermal stability of hydrocarbon fuels, especially aviation fuels.

Aviation fuels commonly contain metals such as iron, zinc and especially copper, picked up from reagents used in their production or from containers, pipelines etc. Concentrations of such metals as low as 0.1 ppm adversely affect the high-temperature stability of fuels, apparently by acting as oxidation and polymerization catalysts. The high temperature stability of such metal-containing fuels may be improved by the addition of so-called metal deactivators which are generally metal-chelating organic compounds.

The metal deactivator used in this process is: (A) an oxime having the formula

$$R^1-\underset{\underset{NOR^3}{\parallel}}{C}-R^2$$

where R^1 is an alkyl or a benzenoid aromatic group, R^2 is an ortho-hydroxy benzenoid aromatic group or an α-hydroxy group with the formula $-CH(OH)-R^4$ where R^4 is an alkyl or benzenoid aromatic group; and R^3 is hydrogen or alkyl group; or (B) a ketimine of a β-diketone and a diamine and having the formula:

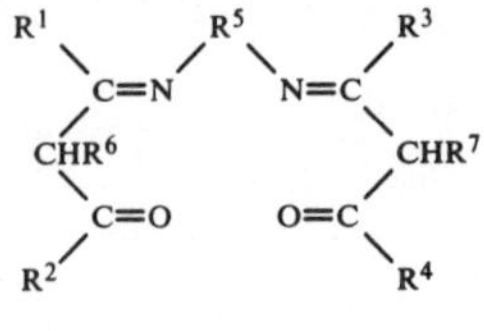

```
R1          R5          R3
  \        /  \        /
   C=N          N=C
  /                \
CHR6                CHR7
  \                /
   C=O          O=C
  /                \
R2                  R4
```

where R^1, R^2, R^3 and R^4 represent alkyl or benzenoid aromatic groups, R^6 and R^7 represent alkyl groups or hydrogen atoms and R^5 represents an alkylene group, a cycloalkylene group, a benzenoid aromatic group or an alkyl-substituted derivative thereof: or (C) an imine of an ortho-acylphenol and a diamine having the formula

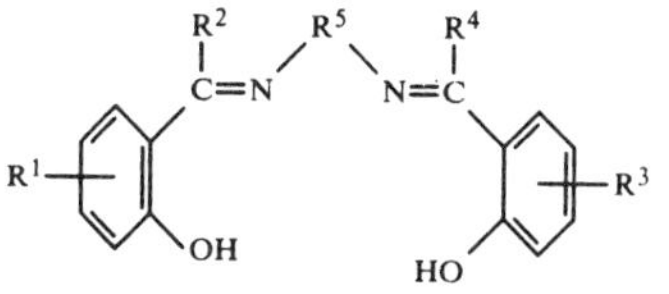

wherein R^1, R^3 represent alkyl groups, R^2, R^4 represent alkyl or benzenoid aromatic groups or hydrogen atoms and R^5 represents an alkylene group, a cycloalkylene group, a benzenoid aromatic group or an alkyl-substituted derivative thereof. The term benzenoid aromatic group as used herein includes phenyl groups, substituted, especially alkyl-substituted, phenyl groups, hydrocarbon polycyclic aromatic groups based on benzene and derivatives, especially alkyl-substituted derivatives, of the polycyclic aromatic groups.

The alkyl groups present in these metal deactivators, either as groups R^1 to R^5 in the above formula or as substituents in the benzenoid aromatic group, may be straight or branched chain alkyl groups and may vary in length over a wide range from C_1 to at least C_{20}. While short chains of C_{1-10} will generally be easier to introduce, longer chains may be advantageous in increasing the solubility of the deactivator in the hydrocarbon fuel.

The metal deactivators will normally be added to the fuel in low concentrations typically 5 to 10 ppm (by weight) although higher or lower concentrations may sometimes be desirable. They form complexes with the trace metals in the fuel and the solubility of these complexes in the particular fuel used governs the suitability of each deactivator for use in that fuel. The metal deactivators are especially advantageous in aviation fuels, such as AVTUR (aviation turbine fuel similar to ASTMS specification JET A/1 or UK Defence Standard D. Eng. D 2494) where copper is the predominant trace element. Hence for use in such fuels the deactivator should preferably form a copper complex having a solubility in the fuel, at room temperature, of at least 10 ppm and preferably at least 40 ppm by weight. This solubility may be influenced by other additives in the fuel and should therefore be verified on the actual fuel mix to be used.

Especially suitable metal deactivators for AVTUR-type aviation fuels include 2-hydroxy-5-tert-butyl benzophenone oxime and 2-hydroxy-5-nonyl acetophenone oxime.

Among the ketimines, especially suitable metal deactivators for AVTUR-type aviation fuels include bisbenzoylacetone propylenediamine.

Especially suitable imine metal deactivators for AVTUR-type aviation fuels include N,N'-di(5-tert-octylsalicylidene) ethylenediamine and N,N'-bis(2-hydroxy-5-methylbutyrophenone) ethylenediamine.

Example: 2-hydroxy-5-tert-butyl benzophenone oxime was prepared as follows: 9.2 g 5-tert-butylphenol and 10 g benzoic acid were codissolved in 50 cc tetrachloroethane. 10 g boron trifluoride was passed into the solution after which the mixture was heated for 5 hr at 100°C, and then poured into 500 cc 10% aqueous sodium hydroxide. The organic layer was separated and the solvent removed by steam distillation leaving crude 2-hydroxy-5-tert-butyl benzophenone which was purified by distillation. 25.4 g 2-hydroxy-5-tert-butyl benzophenone (prepared as above), 13.9 g hydroxylamine hydrochloride, 75 cc ethanol, and 75 cc pyridine were mixed together and refluxed for 3 hr. After cooling the mixture was poured into water and extracted with ether. The ether extract was washed with dilute hydrochloric acid and water, dried and evaporated. The residual oil was distilled under vacuum to give 14.2 g 2-hydroxy-5-tert-butyl benzophenone oxime.

Fuel Containing Insecticide Dispersed in Engine Exhaust Fumes

W.H. Nichols, Jr.; U.S. Patents 4,173,094, November 6, 1979; and 4,240,802; December 23, 1980 has devised a process and composition for dispensing toxic agents such as insecticides by means of the exhaust of an internal combustion engine. This is accomplished by using a fuel composition containing a small but effective amount of an insecticide, fungicide or herbicide.

The insecticide may be malathion, present in the fuel in the amount of from 2.5 to 10% by weight of the mixture. The fuel may be a gasoline-kerosene mixture for running a lawnmower.

Among the insecticides believed to be suitable for incorporation into a gasoline or gasoline-kerosene fuel mixture are the following halogenated hydrocarbon compounds: aldrin, chlordane, dieldrin, DDT, (dichlorodiphenyltrichloroethane), heptachlor, and mirex.

Also suitable are organophosphorus compounds such as malathion. Other insecticides likewise believed to be effective include halogenated hydrocarbons such as BHC ((benzene hexachloride), TDE (1,1-dichloro-2,2-bis(p-chlorophenyl)-ethane, methoxychlor, toxaphene, CPCBS, (4-chlorophenyl-4-chlorobenzene sulfonate), CPBS, (4-chlorophenylbenzene sulfonate), BPIPS [2-(p-tert-butylphenoxy)isopropyl-2-chloroethyl sulfite], carbon tetrachloride, methyl bromide, ethylene dibromide, ethylene dichloride, tetrachloroethane and chloropicrin; and organophosphorus compounds such as TEPP (tetraethyl pyrophosphate), parathion, paraoxon, TPAM (diethylthiophosphoric acid ester of 7-hydroxy-4-methylcoumarin), schradan, dimefox, mipafox, Systox and EPN (O-ethyl O-p-nitrophenyl phenylphosphonothioate).

In a series of tests, mixtures containing from about 5 to 20% by volume of a 50% by weight solution of malathion in a gasoline and gasoline-kerosene fuel were employed in two different lawn mowers. The lawn mowers were operated for periods of about 1½ hr (the time necessary to empty a fuel tank) five times over a period of three months without apparent adverse effect upon the lawn mower engines.

The lawn mowers were employed at normal grass-cutting intervals during the period July through September 1977 in Sarasota, Florida. In each case, visual observations of the effect of the exhaust from the engines on insects, including mosquitoes and other flying and crawling insects, were made. Close examination showed that insects were killed by the exhaust fumes, mosquitoes and several varieties of bugs being found dead in the treated area. Live insects were trapped and exposed to the exhaust fumes gathered in a plastic bag and were killed upon such exposure.

Operator observations comparing mowing the lawn with the same lawn mowers powered by gasoline fuels not containing a toxic agent showed that mosquitoes and bugs were not killed. This observation was reinforced by the fact that, when mowing with a conventional gasoline fuel, the operator was bitten by mosquitoes to a considerable extent but when employing a fuel mixture in accordance with the process during similar hours and under similar conditions there was noticeably less mosquito activity, in fact, virtually none at all on the operator.

-9-

FUEL COMPOSITIONS AND SUPPLEMENTS

GASOLINE-ALCOHOL COMBINATIONS

Gasoline plus Methyl Alcohol plus Ammonia

According to *M. Kanao; U.S. Patent 4,166,724; September 4, 1979* a fuel can be made for use in an internal combustion engine which eliminates or reduces the presence of nitrogen oxides, hydrocarbons and carbon monoxide in its exhaust gas. The fuel consists of gasoline and light oil having therein an additive mixed in the proportion of 10 liters of the fuel to about 3 to 5 ml of the additive, the additive comprising a mixture of about 15 to 16% of liquefied ammonia and about 85 to 84% of methyl alcohol.

Methyl alcohol can absorb ammonia very well. The alcohol can contain 40% of ammonia by weight at 10°C and at a pressure of 1 atmosphere. NH_3 may be added, from the start, by bleeding it directly into petroleum fuels at room temperature under normal pressure or under high pressure. Alternatively, it may be first dissolved in a suitable solvent, and then added directly in a different form without being mixed in the fuels.

Mixture Solubilized with Ethyl or Methyl tert-Butyl Ether

The following two patents relate to fuel mixtures for use in internal combustion engines. More particularly, they relate to solubilizing ethanol containing water in gasoline by means of an additive which provides additional octane rating to the resulting blend and has no adverse effect on its storage stability, water-shedding or corrosion properties.

In accordance with the process described by *F.S. Bove and S. Herbstman; U.S. Patent 4,207,076; June 10, 1980; assigned to Texaco Inc.* from 9 to 12% of crude alkyl (preferably ethyl) tert-butyl ether is blended in with a fuel consisting of 70 to 84% gasoline and 5 to 20% of 95% (or wet) ethanol (ETOH).

For obvious economic reasons the ethyl tert-butyl ether (ETBE) used here preferably is of the type referred to as "crude ETBE bottoms" which is the higher boiling constituent left in the reactor when excess grain alcohol is reacted with isobutylene over a sulfonated resin. This material is a mixture of tert-butyl alcohol, ethanol and hydrocarbons as well as ethyl tert-butyl ether. This material has been found to solubilize grain alcohol in gasoline in all proportions thereby allowing a wide latitude in the precise amount of ethanol which can be blended with the gasoline. In addition, the presence of this material in the blend considerably increases its octane rating.

Various fuel blends were compared for solubility, octanes, and oxidation stability using 95% ETOH, a Brazilian-type base fuel simulated from a U.S. unleaded gasoline and prepared ETBE.

The addition of ETBE served to improve the solubility properties of the 95% ETOH by acting as a cosolvent. In fact, if the gasoline contained 12% by vol ETBE (estimated ETBE blending concentration basis conversion of all available isobutylene) the 95% ETOH would be soluble at all practical concentrations. It appears that both fuel composition and temperature greatly affect 95% ETOH solubility, whereas when 100% ethanol was used, no solubility problems were encountered. For example, although at room temperature the 95% ETOH was not soluble in the Brazilian-type gasoline (contains 12.5% aromatics and 11.0% olefins) at any concentration studied (up to 32% by vol), it was soluble in a U.S.-type lead-free gasoline (contains 32% aromatics and 9.5% olefins) at 18% by vol and greater. However, even in the higher aromatic fuel, phase separation occurred at 48°F. Neither the U.S. nor the Brazilian-type gasoline posed any solubility problems when 100% ethanol was used.

The octane upgrading potential using ETBE mixture and 95% ETOH was very favorable. The results indicate that using either pure or 95% ETOH provides blending octanes of 113-131.

Peroxide buildup in the neat ETBE after 4 weeks storage was evaluated. The ETBE samples were stored in metal (steel) cans and aerated three times weekly for four weeks at ambient temperatures.

The active oxygen of freshly prepared ETBE was 0.0055 wt % compared to 0.025 wt % for the aerated samples. Under these relatively severe aerating conditions there was a 4 to 5-fold increase in peroxide formation. The 0.025 wt % active oxygen content is equivalent to about 0.25 wt % ether hydroperoxide content in the ETBE.

Addition of a phenylenediamine-type antioxidant inhibits peroxide formation at 30 pounds per 1,000 barrels (ptb) in the neat ETBE. This would be equivalent to 3 ptb in gasoline containing 10% by vol ETBE. It is possible that lower dosages of antioxidant are sufficient to inhibit peroxide formation.

The anticorrosion property of a fuel containing 12% by vol ETBE and 10% by vol of 95% ETOH was evaluated in the Nace Test, in which a steel spindle is suspended in a stirred mixture of 90% fuel and 10% distilled water for 3.5 hr at 100°F. The spindle is then visually rated for rust. The test fuel provided excellent corrosion protection and was equivalent to a fuel containing 18% by vol of 95% ETOH. Since the Brazilian gasoline containing 95% ETOH could not be evaluated for corrosion protection due to solubility problems previously identified, a U.S.-type base fuel was used to obtain a corrosion protection comparison between the gasoline/ETBE/95% ETOH and gasoline/95% ETOH blends.

In similar work by *F.S. Bove, W.M. Sweeney, and S. Herbstman; U.S. Patent 4,207,077; June 10, 1980; assigned to Texaco Inc.* 12 vol % of methyl-tert-butyl ether (MTBE) is blended in with a fuel consisting of 70 to 90 vol % gasoline and 5 to 20 vol % of 95% (or wet) ethanol. Pure (at least 99% purity) MTBE has been found to solubilize grain alcohol in gasoline in all proportions thereby allowing a wide latitude in the precise amount of ethanol which can be blended with the gasoline. In addition, the presence of this material in the blend considerably increases its octane rating.

This method of stabilization is generally applicable to hydrocarbon mixtures in the gasoline boiling range of about 90° to 420°F. These mixtures essentially have no lubricity value and are obtained by separating an appropriate boiling fraction from a hydrocarbon distillate obtained in the refining of crude oil.

The addition of methyl-tert-butyl ether (MTBE) was shown to solubilize the water present in grain alcohol when that material is used in gasoline mixtures.

Fuel Supplement of Methyl and Ethyl Alcohols, Xylene plus Alkali Metal Hydroxide

S.B. King; U.S. Patent 4,231,756; November 4, 1980 describes a gasoline and petroleum fuel supplement for use in internal combustion engines which results in or causes more complete combustion of the fuel in the engine and a reduction in the overall amount of pollution emitted from the engine exhaust.

The fuel supplement is a formulation of chemicals which may be combined with gasoline and/or water to provide more complete combustion when used with gasoline in the present day internal combustion engine. The mixture and ratio between the ingredients and the amount of gasoline is determined by construction of the motor, weight of the vehicle and conditions of operation.

The supplement provides increased gasoline mileage of up to 50% or more. It produces gaseous vapor which causes the blow-by vapors in the engine to burn when they become united in the motor. Consequently, the normally harmful dangerous and wasted hydrocarbons and other gases as well as the inert nitrogen gases burn more cleanly during combustion.

The fuel supplement is formed of a combination of essential ingredients in the following relative proportions: 250 to 1,500 ml of methyl alcohol, 100 to 800 ml of xylene, 250 to 1,500 ml of ethyl alcohol, 200 to 800 mg of potassium hydroxide, and 200 to 800 mg of sodium hydroxide. The sodium hydroxide and potassium hydroxide may be added to the other ingredients in solid form in the above-stated proportions or may be added in the form of an aqueous solution. When an aqueous solution of the hydroxides is used, the solutions may comprise, for example, from about 150 to 600 g/ℓ of the respective hydroxides. Obviously, the size of the batch of fuel supplement produced is a matter of choice so long as the relative proportions of ingredients are maintained as stated above.

In addition, from about 10 to 60 cc of cobalt chloride, sodium peroxide, sodium bromide and/or sodium oxide, may be added to the above supplement to reduce the pollution in the engine.

When the above ingredients forming the supplement are mixed together, the total mixture is then mixed either with gasoline or with water. When the supplement is mixed with water, the final product comprises one-third to two-thirds by volume supplement and the remainder water. When the supplement is mixed with gasoline, the product comprises from about 70 to 90% by vol of supplement and from about 10 to 30% by vol of gasoline.

Either of these mixtures may be injected or otherwise added to the carburation system in an internal combustion engine. Alternately, the supplement may be added directly to the gasoline in the fuel tank. It has been found that approximately 1 oz of supplement per gallon of fuel achieves the desired results.

Example 1: A fuel supplement was used in a 1977 Plymouth with a 318 V8 engine having a 4,220 lb registered weight. The supplement was formed by mixing 900 ml of methyl alcohol, 200 ml of xylene, 900 ml of ethyl alcohol, 500 mg potassium hydroxide dissolved in 15 ml of water, and 500 mg of sodium dissolved in 15 ml of water. The supplement was vaporized and the gaseous vapors were added through the carburetor to the combustion chamber. The mileage increased from 15 mi/gal, without using the supplement, to 27 mi/gal, on the average, using the supplement.

Example 2: Example 1 was repeated, except that 33 cc of sodium bromide, industrial grade (Fisher Scientific Laboratories, St. Louis, Missouri), was added to the supplement. A similar increase in mileage was evidenced, along with a decrease in the pollutants leaving through the exhaust.

Example 3: Example 1 was repeated, except that 50 cc of cobalt chloride, industrial grade (Fisher Scientific Laboratories, St. Louis Missouri), was added to the supplement. A similar increase in mileage was evidenced, along with a decrease in the pollutants leaving through the exhaust.

Fuel Supplement Containing a Lower Alkanol plus Alkali Metal Hydroxide

In other work reported by *S.B. King; U.S. Patent 4,255,158; March 10, 1981* a similar fuel supplement is suggested, with a different combination of ingredients.

This fuel supplement is formed of a combination of essential ingredients in the following relative proportions: 250 to 3,000 ml of a lower alkanol, such as methyl alcohol, ethyl alcohol, n-propyl alcohol, isopropyl alcohol or mixtures thereof, and 0.75 to 120 g of an alkali metal hydroxide, such as sodium hydroxide, potassium hydroxide, lithium hydroxide or mixtures thereof. The sodium hydroxide and/or potassium hydroxide and/or lithium hydroxide may be added to the lower alkanol ingredients in solid form in the above-stated proportions or may, in the alternative, be added in the form of an aqueous solution. When an aqueous solution of the hydroxides is used, the solution may comprise, for example, from about 150 to 4,000 g/ℓ of the respective hydroxides. Obviously, the size of the batch of fuel supplement produced is a matter of choice so long as the relative proportions of ingredients are maintained as stated above.

When the above ingredients forming the supplement are mixed together, the total mixture is then mixed either with gasoline or with water. In a preferred embodiment, the above ingredients are mixed with distilled or deionized water. When the supplement is mixed with the distilled or deionized water, the final product comprises from about ¼ to ¾ by volume supplement and the remainder water. When the supplement is mixed with gasoline, the product comprises from about 70 to 95% by volume of supplement and from about 5 to 30% by volume of gasoline.

Either of these mixtures may be injected or otherwise added to the carburation system of an internal combustion engine, for example, at the PCV valve, carburetor intake manifold or to each cylinder. A carburetor intake manifold converter may also be used to inject and vaporize the supplement. Alternately, the supplement may be added directly to the gasoline in the fuel tank. It has been found that adding to the fuel tank approximately 1 oz of supplement per gallon of fuel achieves the desired results.

Example: A fuel supplement was added in a 1968 Pontiac Le Mans Sedan having a 350 V8 engine weighing 3,620 lb registered weight. The supplement was formed by mixing approximately 33% by volume of supplement with 66% distilled water. The supplement was prepared by mixing 1,000 ml of methyl alcohol, 1,000 ml of ethyl alcohol, 7.5 g of sodium hydroxide and 7.5 g of potassium hydroxide. The mixture was added slowly to the distilled water in the above proportions. The supplement was added to the intake manifold through the PCV line and the supplement was vaporized and the gaseous vapors were added through the carburetor to the combustion chamber using the intake manifold converter. The mileage increased from 15 mi/gal, without using the supplement, to 25 to 30 mi/gal with the supplement.

Methanol, Alkyl-Substituted Benzene, Heptane and Chlorinated Benzene

Conventional gasoline fuels vary in their natural (without additives) octane (or antiknock) values depending on the nature of the petroleum crudes from which they are prepared and the processing to which the crudes are subjected. Gasoline fuels of low natural octane values are less costly than fuels of higher natural octane values and can be made in higher yields from the same amount of petroleum crude.

To obtain adequate engine performance from gasoline fuels of low natural octane values, it is customary to blend them with additives which increase their octane values. Tetraethyllead is a particularly effective additive for increasing the octane value of gasoline fuels, providing increased octane values at much lower additive levels than other materials.

Tetraethyllead, however, produces lead-containing emissions which are harmful to the environment and its use as a gasoline additive is being phased out. It is essential that other antiknock fuels and other antiknock additives be found that can be used in motor vehicles without requiring carburation adjustment.

It is known that methanol, added to gasoline, increases its octane value. Its low cost and potential abundance are also attractive. However, methanol has only limited solubility in gasoline fuel fractions and only small amounts of methanol can be added to gasoline when methanol is the sole antiknock additive.

Aromatic compounds, and particularly lower alkyl-substituted benzenes, such as toluene and the xylenes, are known to increase the solubility of methanol in gasoline and have the added benefit of being, themselves, useful to increase the octane number of gasoline.

D.G. Parker and J.D. Milligan; U.S. Patent 4,242,100; December 30, 1980; assigned to Tri-Pak, Inc. have, therefore, provided a motor fuel composition which comprises about 50 to about 70 parts by weight of methanol, about 10 to about 30 parts by weight of at least 1 alkyl-substituted benzene having 7 to 10 carbon atoms, about 0 to about 30 parts by weight of normal heptane, and about 2 to 10 parts by weight of a chlorinated benzene.

The composition may be used as a motor fuel by itself, or it may be blended with conventional gasoline fractions in proportions usually in the range of about 25:75 to 75:25 by weight as an extender for the gasoline fraction and/or to enhance the octane number of the gasoline fraction. In either case, an internal combustion engine can operate with such a fuel without changing its normal carburation settings.

These compositions are fuels of exceptionally high octane number, well over 100. They are also capable of raising the octane numbers of relatively low octane gasolines to values of about 100 when blended therewith.

The alkyl-substituted benzene may, for example, be toluene, any of the xylenes (or a mixture of xylenes), ethyl benzene, any of the trimethyl benzenes, cumene, p-cymene, or mixtures of these compounds. The preferred alkyl-substituted benzene component is a mixture of toluene and xylene (mixed xylenes). In the total composition, the preferred range for toluene is from about 5 to 15 parts by weight and the preferred range for xylene is the same. Most preferably the toluene and xylenes are present in about equal weights.

The preferred amount of methanol in the composition is from about 55 to 65 parts by weight.

The chlorinated benzene can be mono- or polychlorinated, e.g., chlorobenzene, the o-, m-, and p-dichlorobenzenes, and the like. Preferred is o-chlorobenzene, and a preferable weight range for this constituent is about 3 to 7 parts by weight. It is believed that the chlorinated benzene acts as a solubilizer to extend the solubility range of methanol in the alkyl benzene or heptane-alkyl benzene system so that the fuel composition remains as a single phase with methanol contents higher than those which could be tolerated without the chlorinated benzene component.

If desired, and particularly for winter use in cold climates, a small amount (from a trace to about 5 wt %) of a butane fraction may be added to the composition. The butane may be added by bubbling butane vapor through a mixture of the other components of the composition.

Example: A fuel composition was prepared by blending:

	Parts by Weight
Methanol	55
Xylene, mixed	10
Toluene	10
n-Heptane	20
o-Dichlorobenzene	5

Butane was bubbled through the mixture to dissolve therein a small amount (less than about 1 part by weight) of butane. The mixture had a vapor pressure of 4.9 lb/in^2 at 100°F and an API gravity of 44.4 at 60°F. It contained no free sulfur and no mercaptan sulfur and had a corrosion rating 1A, equivalent to no corrosion.

The octane rating of the above composition was 110.0. When blended at a 50:50 weight ratio with a commercial gasoline having an octane number of 91.9, the blend had an octane rating of 101.7. All antiknock tests were by The Research Method (ASTM Method D-908).

Dry Single-Phase Gasohol

As is well known, liquid hydrocarbons may be combined with certain water-miscible alcohols. Typical of such products is gasohol, a mixture of gasoline and absolute ethanol. It is found that if such mixtures are formulated from alcohols which contain water in amounts as small as 0.1 to 5 vol %, the resulting composition separates into two phases.

W.M. Sweeney and S. Herbstman; U.S. Patent 4,261,702; April 14, 1981; assigned to Texaco Inc. describe a process for forming dry compositions from mixtures of hydrocarbons with wet alcohols, particularly one for the preparation of dried, stable gasohol from gasoline and 95 wt % ethanol.

The process comprises:

(1) mixing a liquid hydrocarbon fuel and aqueous alcohol thereby forming a two-phase wet hydrocarbon fuel including an upper hydrocarbon-rich phase and a lower phase containing alcohol and water;

(2) separating the lower phase containing alcohol and water from the upper hydrocarbon-rich phase;

(3) reacting the lower phase containing alcohol and water at pH below 7 with a ketal, an acetal, or an ortho-ester which reacts with the water thereby forming a substantially water-free composition;

(4) blending this water-free composition and the upper hydrocarbon-rich phase thereby forming a single phase dry hydrocarbon product; and

(5) recovering the single phase dry hydrocarbon product.

In the process, the charge hydrocarbon fuel, preferably dry gasoline, is mixed with the aqueous water-miscible alcohol which is miscible with the hydrocarbon fuel.

The liquid hydrocarbon charge may be mixed with typically 5 to 10 vol % of a water-miscible alcohol which is miscible with the hydrocarbon fuel. In one embodiment 90 volumes of gasoline may be mixed with 10 volumes of 95 wt % ethanol. In another embodiment 80 volumes of diesel fuel may be mixed with 20 volumes of 95 wt % ethanol.

Mixing is preferably effected by passing the two charge components through a packed bed of inert materials, at 25° to 125°F, i.e., ambient temperature of 75°F.

The uniformly mixed two-phase composition is passed to a settling (or separation) operation wherein the two phases separate. The upper hydrocarbon layer is recovered from the settling (or separation) operation as a haze-free substantially dry composition which may be withdrawn.

The lower aqueous layer is withdrawn and passed to a reaction operation to which there is added as a reactant a ketal, an acetal, or an ortho-ester. Mixtures of these components may be employed.

The ketal may be characterized by the formula $R_2C(OR')_2$; the acetal may be characterized by the formula $RCH(OR')_2$; and the ortho-ester may be characterized by the formula $RC(OR')_3$.

Preferably R and R' are lower alkyl, C_{1-10} and more preferably C_{1-4} alkyl. Illustrative ketals may include 2,2-dimethoxy propane, butane or pentane, 2,2-di(cyclohexoxy)propane, etc. Illustrative acetals may include dimethoxy ethane, propane and n-butane, 1,1-diethoxy-n-butane, etc. Illustrative ortho-esters may include methyl ethyl orthoformate, methyl orthobutyrate, and n-propyl orthoacetate. The preferred ketal reactant is 2,2-dimethoxy propane.

The reaction may be carried out in a mixing vessel, but is preferably in a packed bed through which the lower aqueous phase and the added reactant pass as reaction is effected in the presence of acid catalyst.

The acid catalyst which may be employed in small-to-trace amounts may be an inorganic acid such as sulfuric acid, hydrochloric acid, etc. or an organic acid such as p-toluenesulfonic acid, etc.

It is a particular feature of the process that it may be possible to use solid acid composition bearing protons to catalyze the reaction of water with ketal or acetal or ortho-ester. Typical of such solid acids are resins such as reticular sulfonated styrene-divinyl benzene copolymer cation exchange resins, e.g., Amberlyst 15 (Rohm and Haas) having a hydrogen ion concentration of 4.9 meq/g of dry resin and a surface area of 42.5 m^2/g.

In the preferred embodiment, the pelletted solid resin acid catalyst may be used in the form of a gravity packed bed through which the reactant and the aqueous layer pass.

The amount of reactant present will preferably be 1 to 10 wt % greater than the equivalent amount of water in the aqueous layer of water. In the case of the acetal and ketal, the equivalent amount of reactant is 1.0 mol of reactant per mol of water; and in the case of the ortho-ester, the equivalent amount thereof is also 1.0 mol per mol of water.

Reaction is typically effected at 25° to 200°F, say 100°F, and the reaction effluent typically contains (1) alcohol, (2) ketone or aldehyde, (3) water in amount less than about 2 wt %, and (4) reactant. The reaction effluent is a single phase substantially water-free composition.

Reaction effluent, typically 1 to 10 parts, 5 parts being preferably blended with 90 to 99 parts, e.g., 95 parts of the upper hydrocarbon-rich phase recovered from the separation operation to form a single phase substantially water-free product.

In the case of gasohol, the net product may be a single phase substantially water-free gasohol product.

ALCOHOL FUELS FOR DIESEL ENGINES

It has been shown in Brazilian Patent Application P17700392 that alcohols, such as methanol and ethanol, can be substituted for conventional petroleum derived diesel fuels for burning in diesel engines, when used in combination with an ignition accelerator, such as ethyl nitrate or nitrite. Reportedly, the addition of alkyl nitrate or nitrite accelerators to the alcohol achieves a level of auto-ignition sufficient to operate in diesel engines. Unfortunately, these fuel compositions, devoid of any petroleum derived products, are notably deficient in lubricity or lubricating properties with the result that engine wear from the use of these fuels in internal combustion reciprocating diesel engines is a serious problem. Of particular concern are wear problems associated with the fuel injector mechanisms used in such engines. Wear problems have also been encountered in diesel engines operating on light diesel fuel oils.

A series of patents assigned to Ethyl Corporation addresses these wear problems associated with the use of alcohol fuels and solves them with the addition to the alcohol-ignition accelerator fuel of various compounds which significantly improve their wear characteristics. The fuel used, in all these patents, is a monohydroxy alcohol containing 1 to 5 carbon atoms. Methanol, ethanol, propanol, n-butanol, isobutanol, amyl alcohol and isoamyl alcohol may be used; ethanol is preferred.

In all cases, too, an ignition accelerator may be used in the alcohol fuel. This is preferably a substituted or unsubstituted alkyl or cycloalkyl nitrate having 2 to 10 carbon atoms. Most preferred are ethyl nitrate, propyl nitrate, amyl nitrate and hexyl nitrate. Preferred amounts of the ignition accelerator are from 0.5 to 3.0 wt % based on the total compression ignition fuel composition.

Containing a Fatty Acid Amine or Ester of Diethanolamine

R.E. Malec; U.S. Patent 4,204,481; May 27, 1980; assigned to Ethyl Corporation uses as an additive to the abovedescribed alkanol fuel, with or without an ignition accelerator, from 0.01 to 2.0 wt % (preferably 0.1 to 1.0 wt %) of a fatty acid amide or ester of diethanolamine.

These additives can be made by forming a mixture of a fatty acid and diethanolamine and heating the mixture to remove water. Optionally, a water-immiscible inert solvent such as toluene or xylene can be included to aid in the removal of water.

About 1 to 3 mols of fatty acid are used per mol of diethanolamine. The reaction proceeds to yield mainly amide according to the following formula

$$R{-}COOH + HN{-}(C_2H_4OH)_2 \rightarrow R\overset{\overset{O}{\|}}{C}{-}N{-}(C_2H_4OH)_2 + H_2O$$

wherein R is a hydrocarbon residue of the fatty acid.

Some of the diethanolamine can react to form an ester according to the formula shown on the following page.

$$RCOOH + HN(C_2H_4OH)_2 \rightarrow RC(=O){-}OC_2H_4NH{-}C_2H_4OH + H_2O$$

The components can be separated by distillation and used separately in diesel fuel compositions. Preferably, they are not separated, but are used as mixtures.

Preferred fatty acids used in making the wear-inhibiting additive are those containing about 8 to 20 carbon atoms. Most preferably the fatty acid is oleic acid. Thus, the preferred additives are N,N-bis(2-hydroxyethyl)oleamide, N-(2-hydroxyethyl)aminoethyl oleate and mixtures thereof.

Example 1: In a reaction vessel was placed 52.5 g (0.5 mol) of diethanolamine and 141 g (0.5 mol) of oleic acid (with caution since the reaction is exothermic). The mixture was stirred under nitrogen and heated to 188°C over a two-hour, 13 minute period while distilling out water. The resultant product was mainly N,N-(2-hydroxyethyl)oleamide containing about 35 wt % N-(2-hydroxyethyl)-aminoethyl oleate. These components can be separated by distillation.

Example 2: In a reaction vessel was placed 282 g of oleic acid, 105 g diethanolamine and a small amount of xylene. The mixture was stirred under nitrogen and heated from 165° to 185°C over a two-hour period while distilling out water and returning xylene. The xylene was then stripped from the mixture under vacuum leaving 363 g of a viscous liquid product consisting mainly of N,N-bis-(2-hydroxyethyl)oleamide and about 36 wt % of N-(2-hydroxyethyl)aminoethyl oleate.

The following examples illustrate the preparation of some typical fuel compositions of this type.

Example 3: To a blending vessel is added 1,000 parts of 190 proof ethanol, and 20 parts of a fatty acid amide or ester of diethanolamine. The mixture is stirred at room temperature until homogenous, forming a fuel composition useful for reducing and/or inhibiting the amount of engine wear in internal combustion reciprocating diesel engines operating on the fuel compositions.

Example 4: To a blending vessel is added 1,000 parts of 190 proof ethanol, and 1 part of a fatty acid amide or ester of diethanolamine. The mixture is stirred at room temperature until homogenous, forming a fuel composition useful for reducing and/or inhibiting the amount of engine wear in internal combustion reciprocating diesel engines operating on such a fuel composition.

The lubricity or wear properties of the fuel compositions were determined in the 4-Ball Wear test. This test is conducted in a device comprising four steel balls, three of which are in contact with each other in one plane in a fixed triangular position in a reservoir containing the test sample. The fourth ball is above and in contact with the other three. In conducting the test, the upper ball is rotated while it is pressed against the other three balls while pressure is applied by weight and lever arms. The diameter of the scar on the three lower balls are measured by means of a low power microscope, and the average diameter measured in two directions on each of the three lower balls is taken as a measure of the antiwear characteristics of the fuel. A larger scar diameter means more wear. The balls were immersed in base fuel containing the test additive. Applied load was 5 kg and rotation was at 1,800 rpm for 30 minutes at ambient temperature. Tests were conducted both with base fuel (190 proof ethanol) alone and base fuel containing the test additives. Results are as follows:

Additive* (wt %)	. . . Scar Diameter (mm) . . . Run 1	Run 2
None	0.89	0.90
1.0	0.45	0.45

*N,N-bis-(2-hydroxyethyl)oleamide

These wear-inhibiting agents are also effective in increasing the wear-inhibiting properties of fuel compositions comprising mixtures of monohydroxyalkanols having from 1 to 5 carbon atoms and fuel oil boiling above the gasoline boiling range, i.e., a mixture of hydrocarbons boiling in the range of about 300° to 700°F or diesel fuel oils devoid of any alcohol components. Such compositions may also contain ignition accelerators such as the organic nitrates referred to previously.

Dimerized Fatty Acid plus an Ester of a Phosphorus Acid

The additive used by *W.L. Perilstein; U.S. Patent 4,185,594; January 29, 1980; assigned to Ethyl Corporation* is a dimerized unsaturated fatty acid which is preferably a dimer of a comparatively long chain fatty acid, e.g., containing from 8 to 30 carbon atoms, and may be pure, or substantially pure, dimer. Alternatively, and preferably, the material known as dimer acid may be used. This latter material is prepared by dimerizing unsaturated fatty acid and consists of a mixture of monomer, dimer and trimer of the acid. A particularly preferred dimer acid is the dimer of linoleic acid.

B.T. Davis; U.S. Patent 4,177,768; December 11, 1979; also assigned to Ethyl Corporation uses a combination of the same dimer unsaturated fatty acid and the ester of an acid of phosphorus.

The phosphorus acid ester wear-inhibiting components found to be particularly effective for use in these antiwear compression ignition fuel compositions are the phenyl, benzyl, cresyl, or xylyl phosphates and phosphites. Similarly, higher homologous aryl, alkaryl, or aralkyl phosphates and phosphites may be used. Such esters can be obtained according to known methods which involve reacting an alcohol with phosphorus pentoxide or $POCl_3$ to form the phosphates and by reacting the appropriate alcohol with phosphorus trichloride to form the phosphites. A preferred phosphorus acid ester is diamyl phenyl hydrogen phosphate.

The amount of dimerized unsaturated fatty acid used in the compression ignition fuel composition is conveniently expressed in terms of weight percent of dimerized unsaturated fatty acid based on the total weight of the compression ignition fuel composition. A preferred range is from about 0.001 to 2.0 wt % dimerized unsaturated fatty acid. A most preferred range is from about 0.1 to 1.0 wt % dimerized unsaturated fatty acid.

The preferred amount of phosphorus acid ester wear-inhibiting agent used in the compression ignition fuel is between about 0.05 and 1.0 wt %.

Example 1: To a blending vessel is added 1,000 parts of 190 proof ethanol, 50 parts n-propyl nitrate and 5 parts of a blend of 45 wt % of the dimer acid derived from linoleic acid, 40 wt % kerosene, 10 wt % process oil and 5 wt % of diamyl phenyl hydrogen phosphate. The mixture is stirred at room temperature until homogeneous. This fuel resulted in 0.52 mm scar in the 4-Ball Wear Test.

Example 2: To a blending vessel is added 1,000 parts of 190 proof ethanol and 1 part (0.1 wt %) of a blend of 45 wt % of the dimer acid derived from linoleic acid, 40 wt % kerosene, 10 wt % process oil and 5 wt % of diamyl phenyl hydrogen phosphate. The mixture is stirred at room temperature until homogenous.

The additive of Example 2 was compared to the base fuel (190 proof ethanol) in the 4-Ball Wear test previously described. Results were as follows:

	. . . . Scar Diameter (mm)	
Additive	**Run 1**	**Run 2**
None (baseline)	0.89	0.90
Example 2	0.21	0.32

Dimerized Fatty Acid plus up to 25% Fuel Oil Fraction

W.L. Perilstein; U.S. Patent 4,227,889; October 14, 1980; assigned to Ethyl Corporation has found that from 0.05 to 1.0 wt % of the dimerized fatty acid (such as linoleic acid) described in the two previous patents is particularly effective as an antiwear additive in a fuel composition containing from 70 to 98.45 wt % of a monohydroxy alkanol (preferably ethanol) and from 1 to 25 wt % of a fuel oil boiling above the gasoline boiling range, preferably an oil which boils in the kerosene and gas oil range with an initial boiling point of 400°F and an end boiling point of about 700°F.

The wear properties of a base fuel and one containing an additive as described above was tested in the 4-Ball Wear test.

The base fuel was a blend of 85.5 wt % 200 proof ethanol, 9.5 wt % No. 2 diesel fuel, and 5.0 wt % n-propyl nitrate.

The test fuel without any additive gave a scar diameter of 0.51 mm. A mixture of 40 wt % dimer acid of linoleic acid and 60 wt % kerosene at a concentration of 1.0 wt % significantly reduced the wear index to 0.38 mm.

Straight Chain Aliphatic Primary Amine

R.E. Malec; U.S. Patent 4,208,190; June 17, 1980; assigned to Ethyl Corporation has also found that the addition of from 0.1 to about 1 wt % of a straight chain aliphatic primary amine provides wear protection to this type of fuel.

The long chain primary amines which are suitable for use are those having the formula RNH_2, in which R is an alkyl or alkenyl radical having 8 to 50 carbon atoms. The amine to be employed may be a single amine or may consist of mixtures of such amines. Examples of long chain primary amines which can be used in the process are 2-ethylhexyl amine, n-octyl amine, n-decyl amine, dodecyl amine, oleyl amine, linoleyl amine, stearyl amine, eicosyl amine, triacontyl amine, pentacontyl amine and the like. A particularly effective amine is oleyl amine (Armak Chemicals Division of Akzona, Inc.) known as Armeen O or Armeen OD.

Example: To a blending vessel is added 1,000 parts of 190 proof ethanol, and 5 parts of oleyl amine. The mixture is stirred at room temperature until homogeneous forming a fuel composition useful for reducing and/or inhibiting the amount of engine wear in internal combustion reciprocating diesel engines.

The lubricity or wear properties of the base fuel (190 proof ethanol) was tested alone and with the additive in the 4-Ball Wear test.

In two separate tests, the test fuel without any additive gave scar diameters of 0.89 and 0.90 mm, respectively. The addition to the base fuel of oleyl amine at a concentration of 1.0 wt % significantly reduced the wear index to 0.50 mm.

N-Hydroxy Oleamide

R.E. Malec; U.S. Patent 4,198,931; April 22, 1980; assigned to Ethyl Corporation suggests as a wear-inhibiting additive for this type of alcohol fuel a N-hydroxy-hydrocarbonamide having the formula

$$R\text{-}[CO\text{-}NH\text{-}OH]_n$$

wherein R is a hydrocarbon group having a valence of n and containing from 6 to about 51 carbon atoms and n is an integer from 1 to 3.

Representative N-hydroxy hydrocarbonamides are: N-hydroxy hexanamide, N-hydroxy decanamide, N-octadecanamide, N-hydroxy oleamide, N-hydroxy linoleamide, N-hydroxy linoleamide dimer and trimer, which usually occur in mixtures. The most preferred additive is N-hydroxy oleamide. The N-hydroxy hydrocarbonamides can be made by conventional methods.

A preferred concentration N-hydroxy hydrocarbonamide is from about 0.1 to about 1.0 wt %.

Example 1: To a blending vessel is added 1,000 parts of 190 proof ethanol and 20 parts of an N-hydroxy hydrocarbonamide. The mixture is stirred at room temperature until homogeneous forming a fuel composition useful for reducing and/or inhibiting the amount of engine wear in internal combustion reciprocating diesel engines operating on the fuel composition.

Example 2: To a blending vessel is added 1,000 parts of 190 proof ethanol, and 1 part of an N-hydroxy hydrocarbonamide. The mixture is stirred at room temperature until homogeneous forming a fuel composition useful for reducing and/or inhibiting the amount of engine wear in internal combustion reciprocating diesel engines operating on the fuel composition.

The amounts of each ingredient in the foregoing compositions can be varied within the limits aforediscussed to provide the optimum degree of each property.

The lubricity or wear properties of the fuel compositions were determined in the 4-Ball Wear test.

In two separate tests, the test fuel (190 proof ethanol) without any additive gave scar diameters of 0.89 and 0.90 mm, respectively. The addition to the base fuel of oleyl hydroxamic acid at a concentration of 1.0 wt % significantly reduced the wear index to 0.44 mm. Thus, the incorporation of N-hydroxy hydrocarbonamides into alcohol or alcohol containing fuels significantly increases the wear-inhibiting properties of these fuels.

C_{12-30} Hydrocarbyl Succinic Acid or Anhydride

In further work, *R.E. Malec; U.S. Patent 4,242,099; December 30, 1980; assigned to Ethyl Corporation* describes another satisfactory wear-inhibiting additive for these alkanol diesel fuels. It is a C_{12-30} hydrocarbyl succinic acid or anhydride, used preferably in an amount of from 0.1 to 1.0 wt % based on the total weight of the fuel composition.

A particularly preferred method for preparing the reaction product is the addition of the oligomer tetrapropylene to maleic acid anhydride or acid. The most preferred hydrocarbyl succinic acid for use as additive is tetrapropenyl succinic acid.

The lubricity or wear properties of the fuel compositions were determined in the 4-Ball Wear Test.

In two separate tests, the test fuel (190 proof ethanol) without any additive gave scar diameters of 0.89 and 0.90 mm, respectively. The addition to the base fuel of tetrapropenyl succinic acid at a concentration of 1.0 wt % significantly reduced the wear index to 0.67 mm. Thus, the incorporation of a hydrocarbyl succinic acid into alcohol or alcohol-containing fuels significantly increases the wear-inhibiting properties of these fuels. The hydrocarbyl succinic acids and anhydrides of the process are also effective antiwear agents when used in fuel compositions comprising mixtures of monohydroxy alkanols having from 1 to 5 carbon atoms and fuel oil boiling above the gasoline boiling range, i.e., a mixture of hydrocarbons boiling in the range of from about 300° to 700°F. Such compositions may also contain ignition accelerators such as the organic nitrates referred to previously.

Further, the hydrocarbyl succinic acids and anhydrides of the process are also effective antiwear agents when used in diesel fuel compositions comprising a mixture of hydrocarbons boiling in the range of from about 300° to 700°F devoid of any alcohol components. Such fuel oil compositions comprise both the heavy and light diesel fuel oils which are commonly used as fuels in diesel motor vehicles.

C_{8-20} Aliphatic Monocarboxylic Acids

Further research by *R.E. Malec; U.S. Patent 4,248,182; February 3, 1981; assigned to Ethyl Corporation* has shown that the addition of a C_{8-20} aliphatic monocarboxylic acid to a compression ignition fuel adapted for use in diesel engines comprising a monohydroxy alkanol having from 1 to 5 carbon atoms and optionally containing an ignition accelerator, such as an organic nitrate, can significantly improve the wear characteristics of the fuel.

The antiwear components of the fuel composition are aliphatic monocarboxylic acids having from 8 to 20 carbon atoms. Preferred acids are caprylic acid, pelargonic acid, capric acid, undecylic acid, lauric acid, tridecoic acid, myristic acid, stearic acid, linoleic acid and the like. More preferably the acid is an unsaturated fatty acid such as oleic, linoleic or erucic acid, most preferably the acid is oleic acid.

The amount of carboxylic acid used in the compression ignition fuel is conveniently expressed in terms of weight percent of acid based on the total weight of the compression ignition fuel composition. A preferred range is from about 0.1 to 1.0 wt %.

Example 1: 9.9 g of 190 proof ethanol and 0.1 g of oleic acid were deposited in a blending vessel. The mixture was stirred at room temperature until homogenous, forming a fuel composition useful for reducing and/or inhibiting the amount of engine wear in internal combustion reciprocating diesel engine operating on the fuel composition.

Example 2: 49.95 g of 190 proof ethanol, 0.05 g of oleic acid and 2.5 g of n-propyl nitrate were deposited in a blending vessel. The mixture was stirred at room temperature until it was homogeneous.

The lubricity or wear properties of the fuel compositions were determined in the 4-Ball Wear Test.

In two separate tests, the test fuel (190 proof ethanol) without any additive gave scar diameters of 0.89 and 0.90 mm, respectively. The addition to the base fuel of oleic acid at a concentration of 1.0 wt % significantly reduced the wear index to 0.32 mm. The addition of oleic acid at a concentration of 1.0 wt % to the base fuel containing 2.5 g of n-propyl nitrate reduced the wear index to 0.51 mm.

An additional 4-Ball Wear Test in which the applied load was increased to 10 kg showed a reduction in scar diameter from 0.92 mm for 190 proof ethanol to 0.45 mm for 190 proof ethanol containing 1.0 wt % oleic acid. Thus, the incorporation of oleic acid into alcohol or alcohol-containing fuels significantly increases the wear-inhibiting properties of these fuels.

GASOLINE-WATER COMBINATIONS

Nonionic Ethoxylated Alkyl Phenol Surfactant

A.I. Feuerman; U.S. Patent 4,158,551; June 19, 1979 has directed his efforts towards producing a gasoline-water emulsion which can be used as a replacement for conventional gasoline with only minor modifications of the carburetion system of the automobile engine.

This gasoline-water emulsion contains not more than 22.5% by volume of water, and a surfactant including 1 to 3.5% by volume of a nonionic ethoxylated alkyl phenol surfactant having from 1.5 to 30 mols of ethylene oxide for each mol of nonylphenol. Surfactants of this type have a hydrophobic-hydrophilic balance and have previously been proposed for use in preparing jet fuel-water emulsions wherein water is the major component.

This fuel preferably uses a secondary surfactant consisting of coconut diethanolamine Super Amide in quantities no greater than the alkyl phenol.

Within the range of 1.5 to 30 mols of ethylene oxide per mol of nonylphenol, the nonylphenol has a sufficient hydrophobic-hydrophilic balance to form the desired emulsion. An equal quantity by volume of the diethanolamine, combined with nonylphenol, will enhance the completion of the emulsion.

Other additives may also be used. One of the advantages of the emulsion is that it can be both clarified and deterred from separating from emulsion during temperature changes by the addition of a very small quantity of a polyoxyethylated dialkyl phenol; specifically, at least one-third of 1% by volume of this clarifier-depressant may be used with the emulsified fuel.

Increases in fuel economy in excess of 25% have been observed using these emulsions with a test engine.

Example 1: 1 ml of Igepal CO530 and 1 ml of Calamide C were poured into 78 ml of gasoline and then 20 ml of tap water was added. A slight shaking of the container formed a clear emulsion. The Igepal (GAF Corporation) is a nonionic ethoxylated alkylphenol containing 6 mols of ethylene oxide per mol of nonylphenol. The Calamide C (Pilot Chemical Company) is a coconut oil diethanolamine Super Amide.

Example 2: 3.5 of Varonic N30-7 and 3.5 ml of Varamide MA-1 were mixed with 70.5 ml of gasoline and 22.5 ml of water. The Varonic N30-7 (Ashland Chemical Company) contains 30 mols of ethylene oxide per mol of nonylphenol. The Varamide MA-1 (Ashland Chemical Company) consists of a coconut oil diethanolamine Super Amide.

Coal Slurries in Fuel Oil with a Mixture of Surfactants

J.B. Yount, III; U.S. Patent 4,162,143; July 24, 1979; assigned to ICI Americas Inc. describes water-in-oil type fuel oil emulsions which are particularly useful in firing boilers in ships, locomotives and industrial power plants. Furthermore, this process involves aqueous particulate slurries dispersed in fuel oil which improve the burning characteristics of fuel oil. Most specifically the process is directed to a blend of cationic, nonionic and anionic surfactants useful in forming these water-in-fuel oil emulsions.

The major ingredient of the surfactant blend, a known corrosion inhibitor, is a cationic surfactant which is made by reacting

(1) from about 1 to 300 mols of an alkyleneoxide; and

(2) one mol of polyamine condensation product characterized by the general formula:

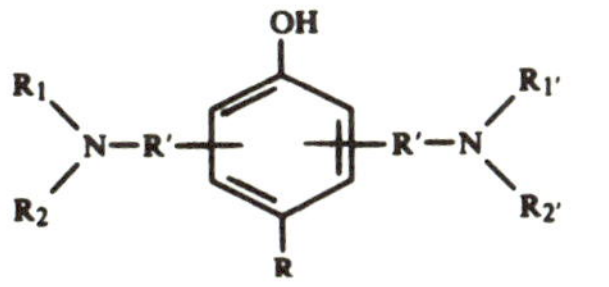

OH
R_1 $R_{1'}$
N—R′ R′—N
R_2 $R_{2'}$
R

wherein R is an organic radical selected from the group consisting of alkyl and cycloalkyl radicals having from 4 to 12 carbon atoms, wherein R' is an alkylene radical, wherein R_1 and R_1' are each selected from the group consisting of hydrogen and an acyclic hydrocarbon radical having from 1 to 18 carbon atoms and wherein R_2 and R_2' are each organic alkylene polyamine radicals containing from 1 to 3 amine groups selected from the group consisting of primary and secondary amine groups, each of the amine groups being separated from any other amine group in the composition by 2 to 6 carbon atoms, or an organic carboxylic acid salt of the reaction product.

Also included in the blend is at least one of the compounds selected from the group consisting of:

(1) a nonionic compound prepared by reacting from about 10 to about 300 mols of an alkyleneoxide with hydroxyl containing compounds having 1 to 8 carbon atoms and alkyl phenols, wherein the alkyl substituents have 2 to 15 carbon atoms; and

(2) an anionic compound selected from alkyl, aryl, and polyoxypropylene and polyoxyethylene ether esters of sulfuric, sulfonic and phosphoric acids and inorganic and organic neutral salts thereof.

Such a composition might advantageously contain, for example, 60% by wt polyoxyethylene (12) nonylphenol/formaldehyde/diethylenetriamine condensate monooleate; 10.4% calcium dodecylbenzene sulfonate; 15.3% of a polyoxyalkylene glycol ether consisting of a hydrocarbon moiety of an aliphatic monohydric alcohol having 1 to 8 carbon atoms, the hydrocarbon moiety having attached thereto through an etheric oxygen linkage a heteric mixed chain of oxyethylene and 1,2-oxypropylene groups, the average molecular weight of the hydrophobe being at least 1,000, and attached to the mixed chain a hydrophile consisting of a chain of oxyethylene groups, the weight ratio of hydrophile to the hydrophobe being from 0.8:1 to 1.2:1; and 4.0% polyoxyethylene (11) nonylphenol dissolved in 10.3% inert organic solvent.

These emulsifier blends, including the weight of the inert solvent, are useful for forming stabilized water in fuel oil emulsions in combinations containing up to 50% water. Normally, concentrations of 2 to 15% by wt of the emulsifier blend based on the amount of water provide a suitable emulsion. Preferably 2 to 10% by wt of the blend is used when a bituminous coal dust suspension is to be formed.

Fuel oils of grades 4, 5, and 6 are of particular interest for these blends since these are most often used in the stoking of commercial boilers. These materials have a flash point of about 55° to 60°C and a Saybolt viscosity measured at 38°C of 45 to 9,000.

The water employed in manufacturing the slurries employing the emulsifier blend is not considered critical since brackish, pure, and seawater have been employed satisfactorily. Fuel oil emulsions containing up to 30% water and more are obtained with the abovedescribed blend.

While bituminous coal has a lower fixed carbon content than other commonly available coals, such as anthracite, its density is sufficiently low due to its high porosity to permit its suspension more readily using an aqueous emulsifier blend of the process. Water is soaked up in the pores and surrounds the particle surface. Compositions having 0.5 to 10% water are most operable as fuel oil suspensions. Coal dust-containing oil slurries are best operated at 2 to 7% by wt water with optimum results at 4 to 6% water. Stable water-in-fuel oil emulsions are provided by the addition of 0.75 to 15% by wt of the water of the above-described emulsifier blend. The heat content of the oil is normally enhanced provided that the combustible materials dispersed therein are ground to a fine particle size. Best results are obtained with particles of less than 0.5 mil and preferably less than 0.074 mil, that is, less than 30 mesh and preferably less than 200 mesh U.S. sieve series.

Oleic Acid Derivatives as Emulsifiers

The process of *G.E. Fodor, W.D. Weatherford, Jr. and B.R. Wright; U.S. Patent 4,173,455; Nov. 6, 1979; assigned to the U.S. Secretary of the Army* relates to diesel fuel emulsions which are used as engine fuels and which are self-extinguishing even if the liquid temperature is above the flash point of the base diesel fuel.

In the storage of diesel fuel or the utilization of the fuel to drive a vehicle, a potential fire hazard is presented. While in storage for military use in war time, for example, the fuel may be subjected to an enemy incendiary projectile attack. When in actual combat, the fuel system of a military vehicle may be struck by a projectile and set afire. In addition, accidents such as rear-end collisions also result in the loss of lives and property through fire. The process formulation minimizes the potential for fire in the abovementioned instances. In addition, these diesel fuels are stable for an extended period of time.

The aqueous diesel fuel emulsion contains 2 to 6% by volume of an emulsifier consisting of at least 60% oleyl diethanolamide, 8 to 14% by wt diethanolamine soap of oleic acid, 16 to 24% by wt free diethanolamine, to which emulsifier has been added 0 to 7.5% by wt oleic acid under conditions causing it to react with an equivalent amount of free diethanolamine, thereby leaving essentially no free oleic acid in the final emulsifier blend. Up to 0.5% by wt of an antimist agent, based on the weight of the emulsion, may also be utilized.

The emulsifier is prepared utilizing a commercial grade oleyl diethanolamide, diethanolamine, and diethanolamine soap of oleic acid. It is typified by products known as Schercomid ODA (Scher Chemicals, Inc.) and LT-17-43-1 (Clintwood Chemicals). However, equivalent products are available from other manufacturers. The emulsifier incorporated into the diesel fuel may contain additional oleic acid derivatives formed from addition of up to about 7.5% by wt of additional oleic acid.

The emulsion is prepared by adding at room temperature the emulsifier to the diesel fuel in an amount of 2 to 6% by volume with mechanical agitation. Water in an amount of 1 to 17% is subsequently added with additional mechanical agitation to produce the desired emulsion. Intense homogenization is not required.

The oleic acid is incorporated into the emulsifier prior to the addition of the emulsifier to the diesel fuel. In a specific example, 7.5 g of commercially available oleic acid were added to 100 g of Schercomid, ODA. The mixture was heated to a temperature of about 55°C and held at that temperature for about 10 minutes. The emulsifier was permitted to cool to room temperature. The emulsifier was then ready to be added to the diesel fuel. Mechanical agitation was also used in the incorporation of the oleic acid in the emulsifier.

Antimist agents are long-chain, high molecular weight polymers, i.e., with average molecular weights in excess of 5 million, that were developed to improve the flow of oil through pipelines. The presence of an antimist agent in this emulsion prevents the fuel from atomizing on impact when a fuel container is ruptured. Instead of forming droplets which develop into an explosive fireball, the fuel is expelled in sheets and strings of beads and therefore does not provide sufficient surface area for explosive combustion. The addition of about 5% water is adequate to make ground fires self-extinguishing at temperatures above the flash point of the diesel fuel.

The antimist agent may be added at any time during the preparation of the emulsion, or when all of the components have been added, i.e., after the addition of the water. The antimist agent is added with low-shear agitation, i.e., less than 100 rpm with a propeller-type mixer, to the diesel fuel or the resultant emulsion. A commercial antimist agent known as CDR or AM-1 (Continental Oil Company) was utilized. Since the agent is proprietary, the composition is not known. Similar products are available from other manufacturers.

This process results in the production of a stable emulsion of extremely fine droplets of water in a fuel. The mean droplet size is less than 1,400 angstroms. Since the stability of an emulsion is related to the size of the water droplets, it would be expected that the fuel mixtures would have a shelf-life comparable to that of the unblended fuel. Although some diesel fuels have a shelf-life of up to 5 years, most fuels are utilized within 9 months. The microemulsions of the process remained unchanged at ambient temperatures for over 6 months.

Surface Active Agent to Control Nitrogen Oxide Emissions

N. Moriyama, Y. Aoki and Y. Fukuyama; U.S. Patent 4,182,614; January 8, 1980; assigned to Kao Soap Co., Ltd., Japan have developed a surface active agent for use in an emulsion fuel prepared by adding the agent with water to a fuel oil to reduce the concentration of nitrogen oxides in the exhaust gas. The surface active agent has the following general formula:

$$\begin{array}{l} CH_2O(C_2H_4O)_{n^1}X \\ | \\ CHO(C_2H_4O)_{n^2}X \\ | \\ CH_2O(C_2H_4O)_{n^3}X \end{array}$$

wherein at least 1.25 on the average of the three X's stand for an acyl group having 10 to 18 carbon atoms, the remainder being a hydrogen atom, and the sum of numbers n^1, n^2 and n^3 is in the range of from 2 to 50.

The lauric group and the oleic group are most preferably used as the acyl group.

When a water-in-oil-type emulsion fuel formed by incorporating this surface active agent in an amount of 0.05 to 0.5% by wt (all references to percent given here are by weight unless otherwise indicated), preferably 0.1 to 0.4%, based on the total system (fuel oil plus water) is used as a fuel for a boiler, generation of NO_x is prominently controlled as compared with the case of fuel oil alone (water is not incorporated) or an emulsion fuel prepared in the same manner as above by incorporating a commercially available surface active agent in an amount of 0.03 to 0.7%.

Example: At room temperature, 80 parts by weight (all of references to parts given below are by weight unless otherwise indicated) of kerosine (N content is 0.015%; product of Quignus Oil), 20 parts of water and a predetermined amount of a surface active agent were agitated by an appropriate agitator (for example, a line mixer) to form a water-in-oil-type emulsion (emulsion fuel), and this emulsion was used as a fuel for a boiler. The NO_x and O_2 contents in the exhaust gas from the boiler were determined by an apparatus for measuring the nitrogen oxide concentration in an exhaust gas (Yanagimoto Apparatus Model ECL-77) and an apparatus for measuring the oxygen concentration in an exhaust gas (Yanagimoto Apparatus Model EMG-77). The NO_x content was evaluated based on the conversion value to 4% O_2. The NO_x values obtained when the emulsion fuel and kerosine alone were used under the same boiler load at the same air/kerosine ratio were compared with each other, and the degree of reduction of the NO_x value by the emulsion fuel was calculated.

The table below shows the surface active agent used, the amount (% by wt based on the total system) of surfactant and the NO_x reduction ratio (ratio of reduction to value of combustion of the fuel oil alone).

Surfactant	Amount (wt %)	NO_x Reduction Ratio
Sorbitan monolaurate	0.3	13
Sorbitan monolaurate	0.5	20
Sorbitan monolaurate	1.0	41
Trilaurate of adduct of ethylene oxide (3 mols) to glycerol	0.1	25
Trilaurate of adduct of ethylene oxide (3 mols) to glycerol	0.2	39
Trilaurate of adduct of ethylene oxide (3 mols) to glycerol	0.4	48
Trilaurate of adduct of ethylene oxide (3 mols) to glycerol	0.6	51
Trilaurate of adduct of ethylene oxide (10 mols) to glycerol	0.1	22
Trilaurate of adduct of ethylene oxide (10 mols) to glycerol	0.2	25
Trilaurate of adduct of ethylene oxide (10 mols) to glycerol	0.4	38
Trilaurate of adduct of ethylene oxide (10 mols) to glycerol	0.6	45
Trioleate of adduct of ethylene oxide (10 mols) to glycerol	0.1	21
Trioleate of adduct of ethylene oxide (10 mols) to glycerol	0.2	34
Trioleate of adduct of ethylene oxide (10 mols) to glycerol	0.4	43
Trioleate of adduct of ethylene oxide (10 mols) to glycerol	0.6	48

Five-Ingredient Surfactant

P.S.T. Fung; U.S. Patent 4,199,326; April 22, 1980 has as objectives the provision of an emulsified fuel composition which (1) gives good heat per pound of oil; (2) burns clearly with reduced pollution; (3) uses economical amounts of surfactant; and (4) can be used in older boilers while still meeting current antipollution standards. This is accomplished by providing an emulsified fuel composition comprising fuel oil, water, and a surfactant. The surfactant comprises nitrobenzene, benzene, a fatty monocarboxylic acid, an alkylene oxide condensate, and a hydrocarbon oil.

The nitrobenzene aids in the dispersion of the surfactant and preferably comprises from 5 to 30 wt % of the surfactant.

The benzene improves the color of the resultant composition and tends to increase the Btu output per pound of oil. It is present in an amount preferably equal to 5 to 30 wt % based on the weight of the surfactant.

Another essential ingredient of the surfactant is the fatty monocarboxylic acid. The preferred acids are saturated acids containing from 8 to 30 carbon atoms. Examples of suitable unsaturated fatty monocarboxylic acids include among others oleic, linoleic, and linolenic acid. Examples of suitable saturated fatty monocarboxylic acids include among others palmitic acid and stearic acid which is the most preferred. The fatty monocarboxylic acid preferably comprises from 5 to 15 wt % based on the weight of the surfactant.

Yet another essential ingredient of the surfactant is the alkylene oxide condensate. The alkylene oxide can be propylene oxide or ethylene oxide but the latter is preferred. The alkylene oxide is condensed preferably with lower alcohols, fatty alcohols, fatty esters of lower alcohols, fatty acids and alkyl phenol. The alkylene oxide chain can be of any length but generally comprises from 10 to 40 alkylene oxide units. Nonionic surfactants are preferred. An especially preferred subclass is the ethylene oxide condensate of the formula $H(CH_2CH_2O)_nR$ wherein R is a member selected from the group consisting of residues of lower alcohols and residues of monocarboxylic acid esters of lower alcohols, and n is an integer from 10 to 40 inclusive. The alkylene oxide condensate is preferably equal to 5 to 15 wt % based on the weight of the surfactant.

According to another aspect of the process, the surfactant preferably comprises a hydrocarbon oil in an amount sufficient to dissolve and disperse the other ingredients of the surfactant. The hydrocarbon oil can be of identical composition to the fuel oil or can be of a different composition. Examples of suitable hydrocarbon oils include among others No. 4 fuel oil, No. 6 fuel oil, Bunker C fuel oil, kerosene, and diesel oil. The hydrocarbon oil is generally present in an amount sufficient to dissolve the other ingredients and generally comprises from 2 to 50 wt % of the surfactant.

The fuel oil useful in the composition is generally a hydrocarbon fuel oil having a boiling point of 200° to 400°C. Examples of suitable fuel oils include among others No. 4 fuel oil, No. 6 fuel oil, and Bunker C fuel oil.

The water and the fuel oil can be mixed in widely varying ratios but good results are achieved when the water comprises from 30 to 60 wt % of the composition. Optimum results are achieved when the water comprises from 40 to 50 wt % of the composition.

The emulsified fuel composition may be either a water-in-oil or an oil-in-water emulsion though preferably the fuel oil is the continuous phase and the water is the discontinuous phase.

The surfactant is generally present in the emulsified fuel composition in an amount equal to 10 to 1,000 cc and preferably equal to 50 to 500 cc per metric ton of composition.

Example: *Synthesis of the Surfactant* – The percentages of the following ingredients are combined as indicated in the table below.

Items E1 and E2 are mixed and heated to 30°C whereupon Item C is added followed by the addition of the other ingredients listed.

The PEG 400 is a polyethylene glycol condensate with a molecular weight of 400.

The Tween 80 is the ethylene oxide condensate of sorbitan monooleate.

Item	Ingredient	Percent
A	Nitrobenzene	20
B	Benzene	20
C	Stearic acid	8
D1	PEG 400	8
D2	Tween 80	4
E1	Bunk oil	20
E2	Diesel oil	20

OTHER COMPOSITIONS

Sulfonated Aromatic Compound Condensed with Formalin as Coal-Fuel Oil Stabilizer

Many stabilizers for coal-oil mixtures (COM) have been suggested, but none has proved completely satisfactory. *N. Moriyama and M. Tsuchihashi; U.S. Patent 4,171,957; October 23, 1979; assigned to Kao Soap Co., Ltd.; Japan* have provided a stabilizer for a mixed fuel of coal and fuel oil, which comprises a compound represented by the following general formula:

$$A{+}\underset{\substack{|\\ SO_3M}}{Z}{-}CH_2{+}_n{-}B$$

wherein n is a number of from 1.2 to 30, M stands for a cation, Z stands for an aromatic residue having a saturated or unsaturated hydrocarbon group as a substituent, or an aromatic residue to which a substituent containing an atom other than carbon is bonded, A is H or $-CH_2OH$; and B is OH or $-ZHSO_3M$.

The stabilizer represented by the general formula above consists of a product formed by sulfonating a compound having at least one aromatic ring and condensing the sulfonated compound with formalin and/or a salt of such product. The value (mean value) of n is preferably from 2.0 to 10. M stands for H or a cation such as NH_4, a lower amine, an alkali metal, e.g., Na or K, or an alkaline earth metal, e.g., Ca or Ba. It is especially preferred that M be H and that excess H^+ be present. Z stands for a residue of a compound having at least one aromatic ring. More specifically, Z stands for a residue of a compound having in the molecule as a substituent on the aromatic ring a hydrophobic group necessary for the stabilizer to have an affinity or compatibility with a fuel oil.

As such a hydrophobic group, there can be mentioned saturated and unsaturated hydrocarbon groups, with an alkyl group having 1 to 20 carbon atoms and a naphthalene ring group preferred as the hydrophobic group. As the aromatic residue, there can be mentioned residues of benzene, naphthalene, anthracene, phenanthrene, derivatives thereof and compounds having as a substituent $-OH$, $-NH_2$, $-COOH$, halogen or other functional groups on such an aromatic ring.

When the stabilizer is incorporated in a coal-oil mixture (COM) in an amount of

0.005 to 0.5% by wt, preferably 0.02 to 0.2% by wt, fine particles of coal can be stably dispersed in a fuel oil and the stabilizer exerts a function of converting coal particles precipitated in the lower layer to very soft precipitates that can easily be redispersed.

In general, the C/O (coal to a fuel oil) ratio in the COM is adjusted preferably to 40/60 to 55/45, though this ratio is changed to some extent depending on the kinds of coal and fuel oil to be combined.

From the viewpoints of stability, redispersibility, transportation efficiency and other factors, the water content in COM is adjusted preferably to 3 to 8% by wt, based on COM. Relating to the order of addition of powdered coal, water and fuel oil, water may be first added to fuel oil and powdered coal may be then added to the mixture, but the addition can be performed more effectively by adding water to powdered coal and then adding fuel oil to the mixture.

Example: A solution of a predetermined amount of a stabilizer in 16.85 g of water was added to 250.0 g of Vermont coal (produced in Australia) pulverized so that 80% of particles could pass through a 200-mesh sieve, and the mixture was sufficiently stirred by a spatula to form a Vermont coal-water mixture having an appearance resembling that of slightly wet Vermont coal, and the appearance of the powder was hardly different from the appearance of the powder before the addition of water. The resulting coal-water mixture was added to 250.0 g of Middle East heavy oil containing a predetermined amount of a stabilizer dissolved therein and heated at about 75°C and the mixture was allowed to stand in an oil bath maintained at 75°C for 30 minutes. Then, the mixture was stirred for 5 minutes at 300 to 400 rpm by means of a laboratory mixer to prepare COM. The so-formed COM was charged in a glass bottle and the bottle was sealed and stored in an oil bath maintained at 75°C. The hardness of the formed precipitate was measured at predetermined intervals, as was the redispersibility after 15 days' standing.

The following stabilizers were among those giving good results: dodecylbenzene-sulfonate-formalin condensate, dodecyl naphthalene-sulfonate-formalin condensate and its calcium and sodium salts, hexyl naphthalene-sulfonate-formalin condensate, etc.

Light Fuel from Cellulose

H. Rothlisberger; U.S. Patent 4,178,154; December 11, 1979 describes a process developed for producing a light synthetic fuel for internal combustion engines which is derived from cellulosic materials.

The cellulose source can be cereal straw, esparto, maize husks, wood waste, sawdust, paper, packing wood and the like. The raw material is crushed, composited and treated to solubilize the lignin component of the raw material. The cellulose component remains insoluble during the treatment and accordingly, the two components which occur together in plants, can be readily separated.

The cellulose constituent is acid hydrolyzed to produce a glucose as illustrated in the equation:

$$(C_6H_{10}O_5)_n + n(H_2O) \xrightarrow{\text{acid}} n(C_6H_{12}O_6)$$

The glucose is then placed in a basic medium such as sodium carbonate and is then amenable to fermentation, which is preferably carried out at a temperature in the range between 75° and 90°C at a pressure in the range from 30 to 50 bars. (It is noted that a bar is equal to 0.987 atm and within the range of the process the two units of measure are essentially equal and may be used interchangeably.) The fermentation produces 100 proof isopropyl alcohol which is converted to isopropyl ether, using conventional, well known methods.

Isopropyl ether, like the prior art alcohol-based fuels, produces unsatisfactory results, being too volatile to be a satisfactory fuel. Aside from methane, alkane stabilizers do not provide adequate results. The straight chain alkanes, in particular n-heptane, have proven to be inoperative in the present system. However, branched lower alkanes, preferably isooctane, in association with the isopropyl ether can be mixed with air and subjected to a high compression ratio without undergoing spontaneous ignition. Best results can be obtained with 2,2,4-trimethylpentane.

It has been found that to produce a commercially acceptable fuel, it is necessary to pass the fuel mixture through a catalyst. Cobalt aluminate and cobalt acetate have been found to provide the desired critical end result. The use of either acid resins or the ion exchange phenolsulfonic aldehyde resin known as Amberlite (Rohm & Haas Co.) is extremely advantageous.

It has been found that a remarkably porous body can be derived from scrap or waste from the treatment of bauxite. The surface area of the porous body is on the order of 250 to 600 m^2/100 kg. The material, $4(SiO_2)\cdot Al_2O_3\cdot H_2O$ serves as a substitute or carrier for the cobalt catalyst which is deposited on the porous carrier using known techniques.

The catalysis is carried out at a temperature on the order of 500°C, at a pressure on the order of 50 atm, in an inert atmosphere. The use of nitrogen to provide the inert atmosphere has been found to provide the desired results. The time period for the catalysis is on the order of from 10 to 20 seconds.

Addition of Hydrogen Carriers to Ammonia and Amine Fuels

It has already been recognized that ammonia and amines are fuels which upon combustion form combustion gases with considerably greater thermal energy than the initial combustion reactants. However, these fuels have failed to achieve an important position as an energy source because of their unsatisfactory combustion.

To enhance the combustion of a number of conventional fuels and fuels of poor combustibility, it has been previously suggested to add gaseous hydrogen to the combustion mixture. The addition of hydrogen to combustion mixtures can provide additional thermal energy release, lower ignition temperatures, advance flame speeds, reduce the undesirable emissions of nitrogen oxides and carbon monoxide and generally effect a more efficient combustion. However, the previously proposed methods of adding hydrogen to combustion mixtures have consisted of adding gaseous hydrogen to the volatilized fuel at the time of ignition or just prior thereto. The systems proposed heretofore for injecting gaseous hydrogen into a combustion mixture have been complex, costly and of questionable reliability.

H. Osborg; U.S. Patent 4,201,553; May 6, 1980 has devised a method by which hydrogen is made available to the combustion mixture by dissolving a hydrogen carrier in the base fuel. The carrier releases hydrogen for combustion at the time of ignition and thus obviates the need for a separate hydrogen gas injection system, dual fuel supply system, special carburetion devices, fuel mixing controls and hydrogen gas releasing or generating and storage equipment. The fuel compositions of this process are also advantageous in that the hydrogen carrier employed is a chemical compound which has chemically bound hydrogen. The release of hydrogen from the carrier occurs when the chemical bond is broken with a consequent release of energy. This energy release serves as an "energy kick" to assist ignition and boost combustion of the base fuel and the additive.

The process comprises a method of improving the combustion of a base fuel selected from the group consisting of ammonia and organic amines having a molecular weight of from about 17 to 110, which comprises mixing into the base fuel from 0.5 to 15% by wt of a hydrogen carrier and combusting the resulting mixture. The hydrogen carrier is a compound of hydrogen and at least one element selected from the group consisting of sodium, potassium, magnesium, boron, and nitrogen. This compound has a molecular weight of from 8 to 125, a heat of formation of from about 10 to about 100 kcal per mol and compatibility with the base fuel selected, at ambient temperatures, meaning herein a temperature within the range of from -20° to about 250°F.

These fuel compositions are particularly useful fuels for internal combustion engines, turbine engines, turbine jet engines and for combustion in conventional space heating apparatus.

The base fuels employed in the process are well known materials characterized in part as liquids which are largely volatile at ambient temperatures. More particularly, ammonia may be employed in its liquid form as a base fuel or in combination with other fuels such as alcohols, amines and/or hydrocarbons.

Organic amines which may be employed as base fuels in the process are represented by methylamine, dimethylamine, diethylamine, triethylamine, aniline, cyclohexylamine, and the like.

Preferred as the base fuel according to the process are ammonia and the lower aliphatic amines mentioned above.

Preferred hydrogen carriers for ammonia-based fuel compositions are monomer compounds wherein the element compounded with hydrogen is one or more of sodium, potassium, boron or nitrogen. Exemplary of such hydrogen carriers are:

(a) hydroxylamine;

(b) iminoalcohols, like N-(2-hydroxyethyl)ethyleneimine;

(c) hydrazines including alkyl substituted hydrazines as represented by methylhydrazine, phenylhydrazine, butylhydrazine, and the like;

(d) boranes such as diborane, pentaborane, borazine and the like;

(e) borohydrides such as magnesium borohydride and the like;

(f) certain borane-amine complexes such as borane-tert-butylamine;

(g) hydrazinoalkanols such as 2-hydrazinoethanol and the like;

(h) borohydride-ammonia adducts such as sodium borohydride monoammoniate (U.S. Patent 3,108,431) and the like; and

(i) ammonia-BH_3 adducts (coordination complexes) of the formula $(NH_3)_x \cdot (BH_3)_y$ wherein x and y are each integers of from 1 to 3.

A broad range of hydrogen carrier compounds may be admixed with organic amine base fuels to improve their combustion. In general, all of the aforementioned carrier compounds may be used. Preferred carriers for admixture with organic amine base fuels are compounds of hydrogen and at least one element selected from the group consisting of sodium, potassium, magnesium and boron; such as the borohydrides [group (e) above], and the borane amine complexes [group (f) above].

The base fuels described above may also be used in admixture with one another. For example, ammonia may be dissolved in the amine fuel and serve as a solvent for the hydrogen carrier which in this case advantageously may be represented by unsymmetrical dimethylhydrazine or one of the amine-borane adducts. As a further example, kerosene and an amine fuel such as triethyl amine may be admixed with an amine-BH_3 adduct as the hydrogen carrier such as, for example, an amineborohydride adduct (amineborane) as a hydrogen carrier. A preferred hydrogen carrier for use in such a mixture of base fuels is the adduct of equimolar proportions of ammonia and one-half B_2H_6, $(NH_3 \cdot BH_3)$ or a methylamine or ethylamine-BH_3 adduct, i.e., $CH_3NH_2 \cdot BH_3$; $(C_2H_5)NH_2 \cdot BH_3$; $(C_2H_5)_2NH \cdot BH_3$; $(C_2H_5)_3N \cdot BH_3$ or corresponding hydrazine adducts like $(CH_3)_2NNH_2 \cdot BH_3$.

In a preferred embodiment, the hydrogen carrier is a hydrate such as, for example, hydrazine hydrate (N_2H_5OH) or a hydrazino alcohol. In this manner, oxygen as well as steam are contributed to the ignition and combustion process.

In another preferred embodiment of the process, the hydrogen carrier is an aminoalcohol, like N-hydroxyethyl ethyleneimine. Such compositions have the further advantage of improving combustion efficiency and lowering the emission of noxious by-products.

Example 1: A suitable pressure reaction vessel is charged with 100 lb of liquid ammonia. To this charge there is added with stirring 1 lb of sodium borohydride. The resulting mixture is stirred for about 15 minutes and then transferred to a pressure container where the liquid mixture is maintained. The resulting fuel is useful to power turbine engines.

Example 2: A suitable vessel is charged with 100 lb of dimethylamine. To the charge there is added with mixing 4 lb of unsymmetrical dimethylhydrazine (UDMH). The resulting fuel may be used to power heating plants.

Example 3: The pressure container prepared in Example 1 above and containing 99% by wt ammonia with 1% by wt of sodium borohydride is attached to a pressure reducing valve previously connected to an air mixing burner. The valve is opened to permit the fuel composition to enter the burner head and to be admixed with air in the ratio of about 75 to 25 parts of air to the fuel composition. The air fuel composition is ignited and found to burn evenly with a bright yellow flame.

H. Osborg; U.S. Patent 4,197,081; April 8, 1980 enlarges the concept of adding a hydrogen carrier to improve the combustion of fuels based on ammonia or organic amines to include improving the combustion of fuels such as petroleum distillates and alcohols having 1 to 16 carbon atoms and molecular weights of 17 to 275.

The hydrogen carriers preferred for petroleum distillate fuels are the hydrazines, especially alkyl hydrazines such as unsymmetrical dimethylhydrazine, boron-hydrogen adducts like amine-BH_3 or higher homologues. One of the significant improvements resulting from the presence of the additives described herein relates to significant reduction of undesirable emission components. Also combustion efficiency is increased.

Representative of preferred hydrogen carriers employed as components of the compositions employing alcohols as a base fuel are the hydrazines described in group (c) in the previous patent and the borane-amine complexes described in (f), provided they are nonreactive with the alcohol and with water.

Example 1: A suitable reaction vessel is charged with 100 lb of methanol. To this charge there is added with mixing 3 lb of unsymmetrical dimethylhydrazine. The resulting mixture is a fuel which may be used in internal combustion engines.

Example 2: A suitable vessel is charged with 100 lb of kerosene. To this charge there is added with stirring 5 lb of hydrazine base. The resulting fuel may be used in a kerosene burning heating plant.

Alcohol as Petroleum Fuel plus Carbon Black

B.L. Denker and F.D. Hoffert; U.S. Patent 4,249,911; February 10, 1981; assigned to Hydrocarbon Research, Inc. have developed a method for increasing the heating value of heating liquids by adding to them up to about 20% by wt of carbon black particles. The combustible liquid may be a lower alkyl (1 to 4 carbons) alcohol such as methanol or a hydrocarbon liquid such as kerosene or naphtha. Up to 20 wt % carbon black particles may be added without causing a substantial increase in the volume of the mixture (5% or less).

The fuel composition comprising carbon black mixed with a combustible liquid provides a storable clean-burning liquid fuel for those applications where use of such type fuel is essential, such as for gas turbine combustion applications in densely populated areas, and contains acceptably low levels of sulfur. Fuel compositions produced in this way are clean-burning and contain insignificant concentrations of nitrogen and sulfur.

Carbon black particles which may be used have a particle size of from about 5 to 600 millimicrons, more preferably from about 8 to about 100 millimicrons, and a surface area of from about 5 to 1,000 m^2/g, more preferably from about 25 to 250 m^2/g. Among the various carbon blacks which may be used, furnace process oil blacks are preferred, although any carbon black is operable.

In a preferred embodiment, methanol is used as the combustible liquid and the carbon black is present in the composition in an amount of from about 0.05 to 18% by wt.

In another preferred embodiment, the hydrocarbon liquid is kerosene having a gravity rate of about 30 to 50°API and the carbon black is present in the composition in an amount of from about 0.5 to 10% by wt, more preferably approaching the condition whereby the liquid is saturated with the carbon black as reliably as possible to avoid exceeding such a saturation condition, such as 18 to 20% by wt carbon black in liquid.

The fuel composition may be prepared by adding the carbon black particles to the combustible liquid and mixing to form the product. It is of importance to note that by adding the carbon black to the liquid, the volume of the mixture is not substantially increased until the mixture contains about 20% by wt of carbon black particles. Visual observations of the carbon black/liquid mixture show that the carbon black appears to dissolve or disappear into the liquid, involving no significant volume increase in the mixture. When coal particles are added to a combustible liquid such as methanol or kerosene, the particles are not dissolved therein, thus causing an increase in the total volume of the mixture. In addition, a carbon black-methanol mixture appears to be pumpable whereas a mixture of coal and methanol has the appearance of a sludge, and can be pumped only with difficulty, if at all.

The heating value of combustible hydrocarbon liquids on a volumetric basis can be increased substantially by adding thereto carbon black particles. For example, for a carbon black/methanol mixture, wherein the carbon black is present in an amount of about 15% by wt, the heating value of the mixture has been found to be greater than that of methanol alone by about 33%. The heating value for methanol alone is 63,900 Btu/gal. The calculated heating value of such a carbon black/methanol mixture is about 85,000 Btu/gal. This increase in heating value without increasing the fuel mixture volume is significant, in that it makes the use of methanol as a fuel more economically feasible.

In the instance where carbon black particles are added to kerosene without causing an increase in the total volume of mixture, i.e., the kerosene is saturated with the carbon black particles, the heating value of the mixture exceeds that of kerosene by about 10%.

COMPANY INDEX

The company names listed below are given exactly as they appear in the patents, despite name changes, mergers and acquisitions which have, at times, resulted in the revision of a company name.

Akzona Incorporated - 194
Ashland Oil - 107
Aviex Corp. - 266
BASF AG - 10, 33, 269
BASF Wyandotte Corp. - 34
Betz Laboratories, Inc. - 215
Calgon Corporation - 260
California-Texas Oil Company - 248
Calumet Industries, Inc. - 85
Chevron Research Co. - 40, 42, 44, 45, 48, 50, 218
Edwin Cooper, Inc. - 239, 241, 242
Dow Chemical Company - 233
E.I. Du Pont de Nemours & Company - 20, 22, 158, 160, 161, 213
Elf Union - 150
Ethyl Corporation - 64, 223, 236, 237, 282, 284, 285, 286, 287
Exxon Research & Engineering Co. - 95, 102, 104, 113, 118, 119, 122, 125, 129, 132, 135, 136, 138, 140, 180, 194, 226, 228, 230, 231, 232, 234, 268
Gulf Canada Limited - 173
Gulf Research & Development Co. - 53, 116, 154, 163, 166
Hydrocarbon Research, Inc. - 300
ICI Americas Inc. - 289
Institut Francais du Petrole - 100, 150
Kao Soap Co., Ltd. - 292, 295
King Industries, Inc. - 188
Lubrizol Corporation - 56, 66, 68, 73, 87, 108, 196, 200, 205, 208, 243
Milliken Research Corporation - 98
Mobil Oil Corporation - 31, 81, 83, 130, 179, 217, 219, 221, 222
Mooney Chemicals, Inc. - 157
Morton-Norwich Products, Inc. - 264
Nalco Chemical Co. - 211
National Research Development Corp. - 270
National Resources Guardianship International, Inc. - 245
Nippon Oil Company, Ltd. - 171
Orogil - 63
Petrolite Corporation - 62, 181, 184, 186, 253, 257
Phillips Petroleum Co. - 29, 69, 70, 72, 92, 164, 255
Rigs Corporation - 169
Rohm and Haas Company - 58, 60, 76, 80, 96
Shell Oil Co. - 37
Sherex Chemical Company, Inc. - 246
Standard Oil Company - 24
Standard Oil Company (Indiana) - 28, 89, 91, 149, 255, 266
Tenneco Chemicals, Inc. - 75, 190
SA Texaco Belgium NV - 145
Texaco Inc. - 2, 5, 7, 8, 12, 14, 16, 18, 19, 25, 52, 105, 142, 143, 146, 147, 148, 176, 192, 202, 204, 207, 212, 262, 274, 276, 280
Texaco Development Corporation - 4
Tri-Pak, Inc. - 279
Union Oil Company of California - 263
U.S. Secretary of the Army - 291
United Technologies Corporation - 199
University Patents, Inc. - 155
UOP, Inc. - 111, 112

INVENTOR INDEX

Abdul-Malek, A.B. - 149
Ali, A. - 75, 190
Alink, B.A.O. - 62
Alkaitis, A. - 157
Allen, E.A. - 270
Amick, J.W. - 266
Andress, Jr., H.J. - 31, 179
Aoki, Y. - 292
Arnold, J.D. - 85
Audeh, C.A. - 130
Barrer, D.E. - 200
Bello, C. - 53
Berry, J.D. - 85
Bialy, J.J. - 212
Biasotti, J.B. - 8, 18, 19
Bollinger, J.M. - 76, 80
Bonazza, B.R. - 69, 70, 72
Bove, F.S. - 176, 274, 276
Braid, M. - 219, 221, 222
Brewster, P.W. - 230, 231
Broeckx, W.P. - 145
Brois, S.J. - 102, 119, 180, 231, 234
Bryant, C.P. - 68, 208
Cahill, P.J. - 89
Campbell, C.B. - 44
Caruso, R. - 190
Cells, P.L. - 157
Chase, J.D. - 173
Cheng, W.J. - 181, 184, 185, 186, 253
Chibnik, S. - 81, 83, 217
Chikul, O.S. - 259
Chikul, V.I. - 259
Chmovzh, V.E. - 259
Clason, D.L. - 87
Cohen, C. - 100
Cohen, J.M. - 56, 87, 108, 208
Coupland, K. - 226, 228, 232
Cullen, W.P. - 52, 212
Cummings, W.M. - 2, 5, 7, 25, 192
Cusano, C.M. - 12
Daumiller, G. - 10
Davis, B.T. - 236, 284
Davis, K.E. - 73
Davis, M.E. - 204
Dawans, F. - 150
Deffner, J.F. - 154
Deinet, A.J. - 190
Denker, B.L. - 300
DesMarais, Jr., R.C. - 169
deWaal, W. - 118
Diehl, R.C. - 260
Dille, K.L. - 8, 204
Dooley, M.F. - 132
Dorn, P. - 8, 52, 202, 212
Dulog, L.G. - 145
Durand, J.-P. - 150
Elliott, R.L. - 95
Fair, H.J. - 85
Fair, L.V. - 85
Faure, A. - 150
Feldman, N. - 118, 119, 125, 132, 135
Feuerman, A.I. - 288
Fischer, A. - 190
Fodor, G.E. - 291
Fodor, L.M. - 164
Forsberg, J.W. - 205
Frost, Jr., K.A. - 40
Fujiwara, Y. - 171
Fukuyama, Y. - 292

Fung, P.S.T. - 293
Gainer, A.B. - 211
Gallacher, L.V. - 188
Galluccio, R.A. - 58, 60, 96
Gardiner, J.B. - 95
Garth, B.H. - 20, 213
Gatti, A.R. - 37
Graefje, H. - 10
Griffin, Jr., T.J. - 233
Guthrie, D.B. - 181, 184, 186, 253
Gutierre, A. - 231
Gutierrez, A. - 119, 194
Hansen, G. - 269
Hanson, J.B. - 24, 91
Harle, O.L. - 218
Hartle, R.J. - 53, 163, 166
Heilman, W.J. - 116
Herbstman, S. - 8, 142, 176, 274, 276, 280
Hoffert, F.D. - 300
Hoffman, H. - 10
Hokanson, J.S. - 37
Holtz, H.D. - 70, 72
Honnen, L.R. - 42, 44, 50
Hunter, S.A. - 113
Illnyckyj, S. - 122
Ippolito, A.L. - 16
Ito, Y. - 261
Jones, R.E. - 12
Kaiser, Jr., F.E. - 266
Kanao, M. - 274
Karn, J.L. - 243
Kaspaul, A.F. - 248
Keller, J.L. - 263
Kennedy, B.W. - 173
King, R.G. - 188
King, S.B. - 276, 277
Kluger, E.W. - 98
Knepper, J.I. - 257
Kwong, G.W.Y. - 111, 112
Langdon, W.K. - 34
Levy, J. - 112
Lewis, R.A. - 42, 44, 48
Lewtas, K. - 138
Lilburn, J.E. - 45
Lynch, T.J. - 116
Machleder, W.H. - 76, 80
Maldonado, P. - 150
Malec, R.E. - 64, 223, 282, 285, 286, 287
McCormack, W.B. - 158
Merger, F. - 269
Mih, L.C. - 142
Miley, J.W. - 98
Miller, H.N. - 102, 136, 180
Miller, K.D. - 143
Milligan, J.D. - 279
Minezaki, T. - 251
Minieri, P.P. - 190
Moriyama, N. - 292, 295
Moss, P.H. - 4
Mueller, H. - 10
Naiman, M.I. - 257
Nebzydoski, J.W. - 212
Nestler, G. - 269
Newman, T.C. - 194
Nichols, Jr., W.H. - 272
Nottes, G. - 10
Nozaki, K. - 37
O'Brien, J.P. - 239, 241, 242
O'Halloran, R. - 104
Oppenlaender, K. - 33
Orelup, R.B. - 264
Osborg, H. - 298, 300
Papay, A.G. - 239, 241, 242
Pappas, P.G. - 149
Parker, D.G. - 279
Parker, L.C. - 143
Parlman, R.M. - 164
Peck, R.A. - 142
Perilstein, W.L. - 284, 285
Petrosian, R.A. - 259
Peyla, R.J. - 44
Phillips, P.C. - 270
Pike, R.A. - 199
Pindar, J.F. - 56, 87, 208
Polyakovskaya, V.I. - 259
Powell, J.C. - 2
Retzloff, J.B. - 236
Rhodes, R.P. - 136
Ripple, D.E. - 66
Rossi, A. - 140
Rothlisberger, H. - 296
Rubin, I.D. - 12
Ryer, J. - 102, 132, 180, 194, 231, 234, 268
Salva, J.M. - 228, 232
Sandven, O.A. - 169
Sandy, C.A. - 158, 160, 161
Scheule, H.J. - 22
Schiff, S. - 69
Schlicht, R.C. - 105, 192
Schmidt, F.H. - 213
Scholtz, M.T. - 250
Shaub, H. - 194
Shields, T.C. - 107, 246

Shubkin, R.L. - 237
Sidi, H. - 190
Sievers, R.E. - 155
Sillion, B. - 100
Singerman, G.M. - 53, 166
Smirnov, A.S. - 259
Smith, C.R. - 226, 228, 232
Smithson, T. - 270
Sonnenfeld, R.J. - 29
Soula, G. - 63
Spence, J.R. - 255
Sprague, H.G. - 261
Stambaugh, R.L. - 58, 60, 96
Starke, K. - 33
Starn, Jr., R.E. - 22
Steckel, T.F. - 196
Stover, W.H. - 113, 118
Su, T.K. - 98
Sujdak, R.J. - 214
Sung, R.L. - 52, 202, 207, 212
Sweeney, W.M. - 146, 147, 148, 176, 261, 276, 280
Tack, R.D. - 138
Takezono, T. - 171
Tamarkin, L.Z. - 259
Thomas, S.P. - 92
Thompson, N.E.S. - 62, 257
Trepka, W.J. - 29
Trufanov, P.A. - 259
Tsuchihashi, M. - 295
Tyagunov, B.I. - 259
Udelhofen, J.H. - 28, 89
Ulanen, Y.S. - 259
Vartanian, P.F. - 12, 14, 16, 18, 19
Vogel, H.-H. - 33
Walker, J.L. - 260
Watson, R.W. - 28
Weatherford, Jr., W.D. - 291
Webb, H.M. - 245
Wenzel, T.J. - 155
Willcox, K.W. - 255
Wilson, R.F. - 142
Winans, E.D. - 104, 194, 231, 234
Wisotsky, M.J. - 125, 129, 136
Woods, H.J. - 173
Wright, B.R. - 291
Yeakey, E.L. - 4
Yount, III, J.B. - 289
Zeger, K.E. - 259
Zeidler, G. - 269
Zielinski, J. - 102, 180
Zimmerman, A.A. - 268

U.S. PATENT NUMBER INDEX

Copies of U.S. patents are easily obtained from the U.S. Patent Office at 50¢ a copy.

4,132,531 - 2
4,132,663 - 116
4,133,648 - 154
4,134,846 - 80
4,135,887 - 140
4,138,227 - 142
4,140,492 - 118
4,141,692 - 263
4,142,865 - 145
4,142,866 - 119
4,144,034 - 5
4,144,035 - 4
4,144,036 - 7
4,144,181 - 95
4,145,190 - 245
4,145,297 - 37
4,146,489 - 96
4,147,520 - 122
4,147,641 - 80
4,147,643 - 56
4,148,605 - 179
4,149,853 - 169
4,153,066 - 233
4,153,422 - 129
4,153,423 - 125
4,153,424 - 125
4,153,425 - 10
4,153,564 - 217
4,153,567 - 98
4,155,719 - 146
4,156,434 - 143
4,157,243 - 180
4,158,551 - 288
4,159,898 - 100
4,160,459 - 146
4,160,648 - 42
4,160,739 - 58
4,161,392 - 12
4,161,452 - 60
4,162,143 - 289
4,162,986 - 157
4,163,645 - 253
4,163,646 - 62
4,163,728 - 181
4,164,472 - 186
4,164,473 - 226
4,164,474 - 188
4,166,724 - 274
4,166,725 - 266
4,166,726 - 218
4,168,242 - 63
4,171,957 - 295
4,171,958 - 255
4,171,959 - 14
4,172,707 - 20
4,173,094 - 272
4,173,455 - 291
4,173,456 - 22
4,175,926 - 129
4,176,073 - 231
4,176,074 - 232
4,177,040 - 266
4,177,768 - 284
4,178,154 - 296
4,178,950 - 147
4,178,951 - 148
4,179,271 - 16
4,179,383 - 181
4,179,385 - 190
4,180,385 - 259
4,180,386 - 158
4,182,613 - 113
4,182,614 - 292
4,182,913 - 171
4,183,732 - 18
4,184,851 - 125
4,185,594 - 284
4,185,965 - 192
4,186,102 - 64
4,189,306 - 161
4,192,757 - 230
4,193,769 - 186
4,193,770 - 173
4,194,885 - 130
4,194,886 - 66
4,195,976 - 102
4,195,977 - 194
4,197,081 - 300
4,197,091 - 211
4,197,409 - 45
4,198,303 - 219
4,198,306 - 48
4,198,931 - 286
4,199,326 - 293
4,199,463 - 234
4,200,545 - 87
4,201,553 - 298
4,201,554 - 19
4,201,683 - 231
4,202,671 - 260
4,202,784 - 89
4,203,730 - 24
4,204,481 - 282
4,204,841 - 8
4,205,959 - 261

4,205,960 - 68
4,206,172 - 214
4,207,076 - 274
4,207,077 - 276
4,207,079 - 8
4,208,190 - 285
4,209,302 - 264
4,209,411 - 194
4,209,422 - 268
4,210,414 - 269
4,210,424 - 132
4,210,425 - 25
4,211,534 - 135
4,211,535 - 163
4,211,663 - 221
4,212,754 - 81
4,214,876 - 213
4,215,997 - 160
4,216,099 - 196
4,217,111 - 40
4,218,385 - 199
4,219,431 - 83
4,222,746 - 262
4,222,884 - 223
4,224,180 - 214
4,225,319 - 92
4,225,446 - 85
4,225,448 - 222
4,225,449 - 240
4,226,739 - 184
4,227,889 - 285
4,229,309 - 181
4,230,588 - 69
4,231,756 - 276
4,231,757 - 73
4,231,759 - 28
4,233,035 - 270
4,235,730 - 105
4,236,020 - 44
4,236,898 - 236
4,237,020 - 242
4,237,022 - 200
4,238,202 - 29
4,239,497 - 111
4,240,802 - 272
4,240,803 - 31
4,240,804 - 246
4,242,099 - 287
4,242,100 - 279
4,242,101 - 33
4,242,212 - 91
4,243,538 - 237
4,244,703 - 248
4,244,704 - 176
4,246,125 - 239
4,247,300 - 72
4,247,301 - 50
4,247,404 - 107
4,248,182 - 287
4,248,720 - 228
4,249,911 - 300
4,249,912 - 70
4,251,232 - 119
4,251,233 - 155
4,251,670 - 25
4,252,541 - 176
4,252,542 - 255
4,252,659 - 75
4,252,745 - 112
4,252,746 - 111
4,253,976 - 205
4,255,158 - 277
4,255,159 - 136
4,255,160 - 149
4,256,465 - 171
4,256,596 - 108
4,257,779 - 52
4,257,780 - 202
4,259,086 - 76
4,259,087 - 257
4,260,500 - 205
4,261,702 - 280
4,261,703 - 138
4,261,704 - 34
4,263,014 - 204
4,263,015 - 212
4,263,153 - 104
4,264,334 - 150
4,264,335 - 53
4,264,336 - 164
4,265,639 - 250
4,266,944 - 207
4,266,945 - 243
4,266,947 - 166
4,270,930 - 44
4,272,254 - 251
4,273,891 - 208
4,274,837 - 45

NOTICE